Also in the Variorum Collected Studies Series

RICHARD W. UNGER
Ships and Shipping in the North Sea and Atlantic, 1400–1800

JOHN KOMLOS
The Biological Standard of Living in Europe and America, 1700–1900

JOHN H. MUNRO
Textiles, Towns and Trade: Essays in the Economic History of Late-Medieval England and the Low Countries

JOHN HARRIS
Essays in Industry and Technology in the 18th Century: England and France

PHILIP LAWSON
A Taste for Empire and Glory: Studies in British Overseas Expansion

MICHAEL NORTH
From the North Sea to the Baltic: Essays in Commercial, Monetary and Agrarian History, 1500–1800

G. V. SCAMMELL
Ships, Oceans and Empire: Studies in European Maritime and Colonial History, 1400–1750

RALPH SHLOMOWITZ
Morality and Migration in the Modern World

ANTHONY PAGDEN
The Uncertainties of Empire: Essays on Iberian and Ibero-American Intellectual History

DENNIS O. FLYNN
World Silver and Monetary History in the 16th and 17th Centuries

FRANÇOIS CROUZET
Britain, France and International Commerce: From Louis XIV to Victoria

A. RUPERT HALL
Science and Society: Historical Essays on the Relation of Science, Technolgy and Medicine

PIETER EMMER
The Dutch in the Atlantic Economy, 1580–1880: Trade, Slavery and Emancipation

VARIORUM COLLECTED STUDIES SERIES

Artisans, Entrepreneurs and Machines

In memory of
Theresa Alison née Fromow
(1943–1991)
my wife
whose love and support enabled me to write these essays

David J. Jeremy

Artisans, Entrepreneurs and Machines

Essays on the Early Anglo-American Textile Industries, 1770–1840s

Ashgate
VARIORUM

Aldershot · Brookfield USA · Singapore · Sydney

Published in the Variorum Collected Studies Series by

Ashgate Publishing Limited
Gower House, Croft Road,
Aldershot, Hampshire GU11 3HR
Great Britain

Ashgate Publishing Company
Old Post Road,
Brookfield, Vermont 05036–9704
USA

ISBN 0–86078–663–3

British Library CIP Data
Jeremy, David J.
Artisans, Entrepreneurs, and Machines: Essays on Technology Transfer in the Early Anglo-American Textile Industries, 1770–1840s. – (Variorum Collected Studies Series; CS 608)
1. Textile Industry – Great Britain –Technology Transfer – History – 18th Century.
2. Textile Industry – United States – Technology Transfer – History – 18th Century.
3. Textile Industry – Great Britain – Technology Transfer – History – 19th Century.
4. Textile Industry – United States – Technology Transfer – History – 19th Century.
I. Title
338.4'7'746'0941'0903

US Library of Congress CIP Data
Jeremy, David J.
Artisans, Entrepreneurs, and Machines: Essays on Technology Transfer in the Early Anglo-American Textile Industries, 1770–1840s / David J. Jeremy.
p. cm – (Variorum Collected Studies Series: CS 608)
Includes bibliographical references and index.
1. Textile Industry – Technological Innovations – United States – History.
2. Textile Industry – Technological Innovations – Great Britain – History.
3. Technology Transfer – United States – History. 4. Technology Transfer – Great Britain – History. 5. Industrial Revolution – United States. 6. Industrial Revolution – Great Britain. I. Jeremy, David J. II. Series: Variorum Collected Studies: CS 608.
HD9855.A78 1998 98–2786
338.4'7677'00973–dc21 CIP

The paper used in this publication meets the minimum requirements of the American National Standard for Information Sciences – Permanence of Paper for Printed Library Materials, ANSI Z39.48–1984. ∞ ™

Printed by Galliard (Printers) Ltd, Great Yarmouth, Norfolk, Great Britain

VARIORUM COLLECTED STUDIES SERIES CS608

CONTENTS

Part 3: Modifications

This volume contains x + 352 pages

PUBLISHER'S NOTE: *The articles in this volume, as in all others in the Collected Studies Series, have not been given a new, continuous pagination. In order to avoid confusion, and to facilitate their use where these same studies have been referred to elsewhere, the original pagination has been maintained wherever possible.*

Each article has been given a Roman numeral in order of appearance, as listed in the Contents. This number is repeated on each page and quoted in the index entries.

PREFACE

We all stand on others' shoulders. I wish to acknowledge those to whom I am intellectually indebted for leading me into the areas covered by these essays. It all began in 1961 when, on leaving the University College of North Staffordshire (now Keele University), my tutor Hugh R. Leech encouraged me to undertake master's degree research part-time whilst teaching (first at Park Grammar School, Swindon, later at Herefordshire Technical College, Hereford). That thesis[1] was written under the supervision of Peter D. Marshall and John A. Cannon at Bristol University and centred on a Wiltshire clothier who travelled to America in 1794. Before following his example in 1967, I learned much about the textile industry from correspondence with Julia de Lacy Mann and Chris Aspin.

Most of the essays in this volume were written in the 1970s and early-1980s and resulted from an intellectual itinerary which led me to the USA where I was privileged to meet a number of giants in their chosen fields of historical research. I learned much of the craft of the scholarly editorial historian by working with Julian P. Boyd, Editor of *The Papers of Thomas Jefferson* at Princeton University and a pioneer of the series of edited American presidential papers. Influential in shaping my understanding of the history of technology were a number of other historians. The first was Philip W. Bishop, in whose department at the National Museum of American History (as it is now named) at the Smithsonian Institution, Washington DC, I had the freedom to pursue my own research interests for a year. Eugene S. Ferguson and George Rogers Taylor at the Hagley Museum, near Wilmington, Delaware, and Brooke Hindle when he was visiting MIT, led research seminars which advanced my insight into American economic and technological history. Through those scholars and seminars, and a year teaching in the History Department at the University of Delaware, I encountered the fledgling Society for the History of Technology and its dynamic mainspring, Mel Kranzberg, who offered inspiration and encouragement. During my three years as Curator of the Merrimack Textile Museum at North Andover, Massachusetts (now the Museum of American Textile History and

---

[1] Published as David John Jeremy (ed.), *Henry Wansey and His American Journal, 1794* (Philadelphia: American Philosophical Society, *Memoirs* vol. 82, 1970).

located in Lowell, Massachusetts) I was further privileged to learn about early American textile technology first-hand from the splendid integrated collections founded by Caroline Stevens Rogers and her husband Horatio Rogers and then under the directorship of Thomas W. Leavitt. Weekly trips from North Andover into Boston and Harvard Business School's Baker Library brought me into the scholarly world of early American business records and books where leads and comment were provided by Robert W. Lovett, Kenneth E. Carpenter, and Glenn Porter (then Editor of the *Business History Review*). After six years in the USA I and my family returned to England in 1973. A further dimension to my learning about the diffusion of technology opened when I worked part-time (while teaching at the Cecil Jones High School, Southend-on-Sea) on a PhD thesis at the London School of Economics under the supervision of Charlotte Erickson. Her unrivalled knowledge of nineteenth century transatlantic migration history guided me into a pursuit of the mass of artisans who took their skills from Britain to America. To all these scholars, and others who are acknowledged in footnotes, I remain grateful for numerous privileged opportunities of learning. That I was enabled to accept them was due to my late wife to whose memory this volume is dedicated.

Eight of the eleven essays in this volume were written either during the preparation or since the completion of my doctoral thesis. The thesis and its published version[2] drew on some of these articles but used none of them in full, and most of them are now out of print. Hence I have the temerity to think that there might be an interest in seeing them collected and reprinted here.

DAVID J. JEREMY

Whaley Bridge, 9 October 1997

[2] David J. Jeremy, *Transatlantic Industrial Revolution: The Diffusion of Textile Technologies between Britain and America, 1790–1830s* (Cambridge MA: MIT Press, 1981).

ACKNOWLEDGEMENTS

Grateful acknowledgement is made to the following persons, editors, publishers, institutions and journals for their kind permission to reprint the articles included in this volume: *Business History Review*, Boston (II, III, V); Manchester University Press (IV); *Textile History*, University of Nottingham (VI, VII); Hagley Museum and Library, Wilmington (VIII); The University of Chicago Press (IX); Association for Industrial Archaeology, Leicester (X); *Éditions de l'École des Hautes Études en Sciences Sociales*, Paris (XI).

I

Introduction

1. Technology transfer: past and present

Without the transfer of technology, the theme linking all the essays that follow, technical change in the modern world would be massively retarded and the prospects of seeing technology check world poverty massively reduced. Granted those propositions, the reader may well ask what the transfer of technology in the early industrial period, the age of steam-driven factories and primitive railways, has to do with the spread of technology in the closing years of the twentieth century, the age of information and space technologies. One contention in this essay is that they have much in common. If technology transfer is regarded as a process then there are persisting and shared, as well as novel and different, elements between the two eras.

Differences are quite apparent. The prevailing technologies are far apart: almost exclusively mechanical engineering in the industrial revolution; today, an accumulation of different technologies based on new materials, energy sources, the computer, genetics, and space. The human skills are also very different: craft skills in the eighteenth century; mathematical, scientific, and philosophical knowledge today. The commercial contexts of technology transfer have likewise changed drastically: the partnership and the family firm before the 1850s; the limited company and the multinational corporation today. So too have communications networks: the painfully slow horse-drawn conveyance and sailing ship before 1800; the globe-girdling jet airliner and the Internet today.

However, there are a number of important facets of the transfer process which have been constant over time. Technology remains organised in systems. Technological systems (following Hughes) 'contain messy, complex, problem-solving components' that are 'both socially constructed and society shaping'.[1] Those components include physical artefacts (tools, machines),

---

[1] Thomas P. Hughes, 'The Evolution of Large Technological Systems' in Wiebe E. Bijker,

organizations, scientific knowledge, laws, and people (inventors, scientists, engineers, financiers, managers, workers). The components have inputs and outputs and interact in various ways, including problem-solving, evolution, invention, development, innovation, and transfer. Technical change comes through salients and reverse salients – where some components develop beyond others or else fall behind others (e.g. in an aeroplane system where the power of the engine has been developed beyond maximum forces tolerated by the airframe and wings, and the sophistication of the controls).

Secondly, the key component in these technological systems remains the human input. This means not only inventors, scientists, and engineers but also managers, capitalists, politicians, and government officials. The transfer of technology is affected not only by technicians, entrepreneurs, and the state. It is modified by cultural differences between countries, regions, and localities. Employees and consumers, as instruments and targets of technical change, have impinged heavily on technology transfer. Differences in language, customs, values, religious beliefs have been and remain of the greatest importance in limiting or liberating possibilities for the adoption and modification of imported, new technologies.

Thirdly, the relationship of leader and follower remains to this day. In the early-nineteenth century world the industrial leader was Britain, quickly followed by nations in Europe, the USA, and later Japan. By the late-1980s the world's economies outside the Soviet bloc could be grouped into four sets according to per capita GNP (gross national product). At the bottom were 35 countries with a median per capita income of $290. Middle income countries split between lower and upper groups, of 34 and 14 countries respectively, with per capita incomes ranging between $1,000 and $5,000. High income countries (most industrialised, apart from the oil economies) numbered 25 and had a median per capita income of just under $15,000.[2] Clearly there are still leaders (perhaps around two dozen) and followers (at least 70). Whether the followers tread the same path of economic growth as leaders is debatable.[3] To grow they will certainly have to import and modify to their own circumstances some of the technologies that have benefited the world's high income economies.

Lastly, the shape of the technology transfer process in the late-twentieth

---

Thomas P. Hughes and Trevor J. Pinch (eds), *The Social Construction of Technological Systems: New Directions in the Sociology and History of Technology* (Cambridge MA: MIT Press, 1987), p. 51.

[2] Peter Dicken, *Global Shift: The Internationalization of Economic Activity* (2nd edn, London: Paul Chapman Publishing, 1992), pp. 441–42.

[3] No longer is it fashionable to assume that the growth path is a single track, or flight, as perceived by Marx or, in the opposite corner, by Walt W. Rostow, *The Stages of Economic Growth. A Non-Communist Manifesto* (Cambridge University Press, 1960).

century continues to exhibit many of the features evident in the early industrial period. Inappropriate technologies, bottlenecks, secretiveness, dependence on key managers and technical staff, choices between organizational structures, clashes of values, state-imposed inhibitions, legal contests, struggles with climate and weather: all these and others were associated with technology transfer in the late-eighteenth and early-nineteenth centuries. Many remain familiar to those trying to transplant Western technology to China, Africa, and Latin America, or those managing the transfer of Japanese technology to the West, today.

2. Models of technology transfer

This raises the question of patterns and explanations (models) for the technology transfer process. What is the dominant pattern if there is one? The borrower is one metaphor. Another analogy is the epidemic, with 'infection' carried along major trade and communication routes. Neither allows much originality to the adopter or 'victim'. In real life, however, many adopters have made major modifications to imported technologies. The borrower and epidemic models therefore fail. In his illuminating synthesis of recent work on the development of Western technology, Robert Adams concluded, 'Contingent elements, path dependency, and variants of adoptive behaviour all converge virtually to refute the possibility of order and regularity in technological diffusion.'[4] In short he implies that formulating a model of technology transfer is impossible.

He is probably right when considering the transfer of technology over a range of entities as disparate as two (or more) economies, two multinational firms, two domestic regions, or two engineers or craftsmen. However, if the process is considered as a movement across international and cultural boundaries, of at least two societies or economies, then the elements involved may be arranged in relationships which constitute a model, analogous to the real-world experience.

Without claiming universality for it, the model I favour is derived from the communications literature of the sociologists and from the notion of stages which has an even longer pedigree.[5] Rogers and Shoemaker,[6] in an exhaustive survey of the pre-1970 literature, offered a communications-based model of the diffusion of innovations. The elements are detailed in Table 1.

---

[4] Robert McC. Adams, *Paths of Fire: An Anthropologist's Inquiry into Western Technology* (Princeton NJ: Princeton University Press, 1996), p. 29.

[5] Going back at least to Marx in the mid-nineteenth century.

[6] Everett M. Rogers and F. Floyd Shoemaker, *Communication of Innovations: A Cross-Cultural Approach* (New York: Free Press, 1971), p. 20.

Table 1
SMCRE Communication Model

Elements in SMCRE communications model	Meaning/example in communications model	Corresponding elements in the diffusion of innovations
Source	Inventor; change agent; opinion leader responsible for the innovation	Inventor; scientist; engineer; manager; change agent; opinion leader, in the originating country, society or economy
Message	Innovation (idea, practice, device) with perceived advantages	Technological system or its elements (knowledge; skill; artefact; organization, &c)
Channel	Carrier of innovation	Migrant or itinerant inventor; scientist; skilled worker; engineer; manager; change agent; or opinion leader: crossing boundaries between countries/societies/economies Also, machine parts; machine model; drawings or plans; formulae; verbal or written description of innovation, published or unpublished
Receiver	Social system receiving the innovation	Members of the recipient country, society, economy
Effects	Consequences over time in terms of knowledge, changed attitudes, and changed behaviour	Modifications to the innovation to meet the requirements of the receiving country, society, or economy, (e.g. to meet different prevailing social values; or different relative costs of land, capital, and labour)

To these elements of communication may be added four stages which I found when examining the movement of textile technologies across the Atlantic in the early industrial period.[7] Table 2 details the stages.

To capture something of the pattern of technology transfer simple matrices may be constructed as follow in Tables 3 and 4. They attempt to show how the cross-cultural technology transfer model may be applied to both an early and a later stage in economic development. In the former, the agents of transfer are craftsmen; in the latter they are multinational corporations.

Table 2
Stages in the cross-border diffusion of technology

Stage	Name	Characteristics
1	Demonstration of potential	The arrival of skilled technicians, managers, plans, machines and so forth from the originating society/economy
2	Pilot plant	The establishment of pilot production units in the host society/economy
3	Internal diffusion	The replication and spread of these production units in the host society/economy
4	Modification	The modification of the imported technology to suit the factor endowment (relative prices of land, labour, and capital) and social structure (including values) of the host society/economy

These matrices are no more than ways of categorising a mass of interrelated elements. There is some danger that they would impose a straitjacket on processes that are by no means as neat and tidy as the matrices imply. Another way of trying to identify a framework for understanding technology transfer is to isolate the elements that comprise its major components. These I

[7] David J. Jeremy, *Transatlantic Industrial Revolution: The Diffusion of Textile Technologies between Britain and America, 1790-1830s* (Cambridge MA: MIT Press, 1981). pp. 4, 138–40, 254–55, 262. Hereafter *TIR*

I

Table 3
Diffusion from a more developed to a less developed socity/economy

	Source	Message	Channel	Receiver	Effects
Build-up of potential	Developed economy	New technology	Arrival of immigrant technicians, manager-stools, machines, plans, parts in less developed economy	Less developed economy	Awareness of commerical opportunity in less developed economy
Pilot plant	As above	As above	Immigrant skilled workers and managers,	Entrepreneur or capitalist in less developed economy	Investment in pilot plant in less developed economy
Internal diffusion	Pilot plant in less developed economy	As above	Immigrant workers plus indigenous workers in less developed economy	Other entrepreneurs and capitalist in less developed economy	Pilot plants became typical best practice plants in less developed
Modifi-cation	Problems in factor supply for arising from cultural mix in the less devel-oped economy	Adjustment of imported new technology in the less developed economy	Technician or manager, immigrant or indigenous, in one of the new typical plants in the less developed economy	A pilot plant among the typical plants in the less developed economy	Modification to the imported new technology in the less developed economy

Table 4
Diffusion from one multinational facility to another across country borders

	Source	Message	Channel	Receiver	Effects
Build-up of potential	Leading MNE facility in country A	New technology	Migrant technicians, managers, tools, machines, plans, parts	MNE facility in country B	Awareness of commercial opportunity in country B
Pilot plant	As above	As above	As above	Addition to MNE facility in country B	Demon-stration of effectiveness of pilot plant in country B
Internal diffusion	Pilot plant in country B	As above	Training of local indigenous workers	Investment in additional facilities or sale of technology to other capitalists in country B	Replication of pilot plant produces new typical plant in country B
Modifi-cation	Problems in factor supply or arising from cultural mix in country B	Changes to new technology imported from country A into country B	Technician or manager, immigrant or indigenous, in one of the new typical plants in country B	A pilot plant among the typical plants in country B	Modification to the imported new technology in country B

explored in 1981[8] and then tried to persuade others to develop in 1991[9] as follows:

i. Source economy/society

- What kinds of **inhibiting factors in the originating economy** slowed down (or accelerated) transfer? Were they **internal to the technology**: for example, were they regional and distortive (like technical vocabularies or measuring systems), or were they due to technical sophistication (like a fast pace of technical change as signalled, for example, by the rate of patent registration).
- If **inhibiting influences were non-technical**, were they related to economic conditions (such as the high price of skilled labour or capital equipment in the originating economy)? Or did they spring from the social environment of the technology (for example, attitudes of secretiveness among factory owners, or policies of state protection)?

ii. Movement

- What does a case study of transfer reveal about some or all of the following:
- **Vehicles or channels of transfer**: was the technology carried by verbal descriptions, drawings and plans, patents, machine parts, or much more by skilled workers, technicians, managers?
- **Networks of access into the originating economy**: how did followers get hold of technical information in a leader economy? Did they rely on stray visitors, consular officials, students serving apprenticeships, or who else? What problems or assistance did these agents encounter: foreign language barriers, cultural misunderstandings? Or did they share the social and political stances of the prevailing ruling or professional groups in the leader economy?
- **Information goals of acquirers**: what kinds of target were set for agents seeking new technology: were they largely technical, or were they economic (such as costs of production by a new method), or did they include social considerations (for example, an estimate of the human cost of shifting from one production or distribution method to another)?

---

[8] *TIR*.

[9] David J. Jeremy (ed.), *International Technology Transfer: Europe, Japan, and the USA, 1700–1914* (Aldershot: Edward Elgar, 1991), pp. 3–5.

- **Methods of information collection**: were these transparent or surreptitious, literary or mathematical, pictorial or scientific? Did they involve single agents or groups of agents acting together?
- **Speed of transfer of a given technology**: can this be measured in length of time and can it be explained?

iii. The adoption of imported technology in a receptor economy/society

- **Choice of technology**. Much of the technology exported from a leader to a follower economy may not be appropriate for the follower to adopt. Selection has to take place. In making the selection, entrepreneurs and capitalists in the follower economy have to consider the conditions which appropriate technologies have to meet in their society. Those conditions would relate to natural resources, to the supply and price of factors of production (land, labour, capital), to social structures and values.
- **Rate of adoption**: is it possible to plot the rate of adoption? Does it conform to the classic S-shaped curve,[10] with adoption comencing slowly, accelerating, and then tailing slowly over a period of time?
- **Networks of distribution into the receptor/follower economy**: what part did government (national or local) play in promoting imported technology. Or did private enterprise predominate? If the latter, what ownership forms predominated, single proprietor, partners, incorporated business? What contractual arrangements did any multinational activity take: share ownership, licensing agreement, joint venture, direct investment, or something else?
- **Obstacles in the receptor economy**: were there technical obstacles (such as an admixture of technical traditions from the originating economy; or a clash between science and intuitive technical tradition), or were there cultural hindrances (like a fear, or a disrespect, of nature which motivated irrational responses to weather or tectonic activity)?

iv. Modifications of the imported technology

What does case evidence reveal about the conditions which shaped incoming technology?

- Were **economic conditions** such as differentials between factor costs or the size and structure of available markets of prime importance?

---

[10] Rogers and Shoemaker, *Communication of Innovations*, pp. 176–79.

- Or, were **social factors** more important, for example, the dictates of fashion or the taboos of religion or the rigidities of a social system (which might prohibit the employment of women and so induce their replacement by more automatic machinery)?
- Were **conditions in the physical environment** important in modifying an imported technology?

v. Reverse flows

To what extent were the technological modifications made in a receptor economy applicable to conditions in the originating economy? Were they taken back and were they adopted?

A number of case studies have been published illustrating many of these aspects of technology transfer.[11] Policy makers, sponsors of the badly-needed movement of technology from the North to the South of the globe,[12] and managers in multinationals, frequently find that orchestrating technology diffusion is fraught with a variety of problems. If anything can be learned from their predecessors in the early industrial period it is that the process is a multi-faceted one. Managing it requires attention to many aspects at various stages of transfer. The greatest mistake is to assume that it is either smooth or solely contingent upon the physical movement of information from country A to country B.

3. Elements in technology transfer

The essays reprinted here consider a variety of aspccts of the transfer of early industrial textile technology between Britain and the USA. They have been grouped under the three headings of obstacles and inducements, channels, and modifications.

---

[11] For example, eighteen essays in Jeremy (ed.), *International Technology Transfer* and in David J. Jeremy (ed.), *The Transfer of International Technology: Europe, Japan and the USA in the Twentieth Century* (Aldershot: Edward Elgar, 1992). More recently the sort of model outlined here has been applied to the transfer of accounting techniques from Britain to Australia: see Garry D. Carnegie and Robert H. Parker, 'The Transfer of Accounting Technology to the Southern Hemisphere: The Case of William Butler Yaldwyn' *Accounting, Business and Financial History* 6 (1996).

[12] See Willy Brandt, *North-South: A Programme for Survival. The Report of the Independent Commission on International Development Issues* (London: Pan, 1980), pp. 187–200.

Obstacles and inducements

a. The state

The first of the obstacles preventing technology transfer in the early industrial period was the range of attempts by the British government to halt the emigration of skilled artisans and the export of machinery (Chapter 2). The laws themselves, as John Harris has shown,[13] were imposed earlier in the eighteenth century in response to political pressures exerted by manufacturers and trade organizations in Britain. They and the government were horrified by instances of the organised emigrations of dozens of skilled workers to France and Russia. To prevent foreign rivals from securing British skill and technology, the emigration of artisans and manufacturers was checked by heavy financial penalties and even imprisonment. In peacetime the laws appear to have been applied with something less than vigour.

However, after the outbreak of war against France in 1793 the continued outflow of workers increased government worries and a situation not dissimilar to the Cold War years of the twentieth century developed. Customs men were empowered to search emigrant ships. Consular officials were urged to persuade expatriate workers to return home. Masters of Post Office packets were instructed to give free passage to returning manufacturers and artisans. All efforts were in vain. Neither sticks nor carrots could check the movement of skilled workers eastwards to Continental Europe or westwards to the USA. Nearly a decade after the end of the Napoleonic Wars, the legal restraints on the emigration of manufacturers and skilled workers from Britain were repealed (1824).

The state also tried to stop the export of machinery and machine parts. High technology was distinguished from low technology in textile engineering and the latter was allowed to be exported while the former was not. Low technology and the grey area between low and high were covered by a licensing system. Rational as such a filtering procedure sounded, it broke down for two reasons. One was due to difficulties in defining technology in the 'grey' area. It was identified in various ways: by its destined ownership and location, by obsolescence, and by its processing position (anything at a preliminary production stage being allowable; as was some equipment at the final production stage). These criteria proved hard to apply. Even worse, because the criteria were shaped in response to the pressures of the textile manufacturers (i.e. users of the machinery), the licensing system overlooked many classes of machine tool. Since the machine tool makers were eager to export

---

[13] John R. Harris, *Industrial Espionage and Technology Transfer: Britain and France in the 18th Century* (Aldershot: Ashgate, 1998), chapter 1.

I

their products abroad, the machines that were needed to make embargoed machines fell into the hands of foreign rivals and thus undermined the great objective of the prohibitory laws: to prevent foreign competitors from catching up with British textile manufacturers.

The second reason why the licensing system had to be given up in less than two decades (in 1843) was its impracticability. The break-up of machines into unidentifiable parts, the cunning packaging techniques of exporters, the 7,000-mile coastline of the British Isles, the slender resources of the Customs officers: all militated against a successful sealing of the boundaries of Britain, the workshop of the world, against foreign predators.

On the other hand, the state in the USA offered inducements, rather than obstacles, to European mechanics bringing their knowledge and skills to the infant republic. As Chapter 5 relates, the Pennsylvania Assembly in 1788 awarded £100 to John Hague, apparently a former weaver of Hayfield, north Derbyshire, England, for building a successful carding machine in Philadelphia (a newspaper report of which stiffened Samuel Slater's resolve to leave Derbyshire for the USA the following year). Nor were members of the federal government shy about encouraging American agents and officials abroad in the business of aiding mechanics on a westward crossing of the Atlantic. Both Thomas Jefferson, when American Minister to France in the 1780s, and Alexander Hamilton, when Secretary of the Treasury in the 1790s, were implicated in various schemes.[14]

b. Technical traditions

The third chapter examines yarn numbering. Although this seems a rather obscure topic, it is one of many examples of the specialised mental knowledge, the software, which inevitably accompany any technology, and which may cause problems in the process of international technology transfer. Yarn numbering was a technical measurement system, with many localised variations, used in the textile trades in order to estimate the fineness of yarn. Because such technical software is much less visible than implements or machines, it is easily overlooked in the standard accounts of technical change, pre-industrial change especially. As a mental knowledge system, yarn measuring continued when manual technology gave way to machine technology, so it is also an interesting case of the impact of industrialisation on pre-industrial technical tradition.

---

[14] See, in addition to Chapter 4, Anthony F. C. Wallace and David J. Jeremy, 'William Pollard and the Arkwright Patents' *William & Mary Quarterly* 3rd ser. 34 (1977); Jacob Cooke, *Tench Coxe and the Early Republic* (Chapel Hill, NC; University of North Carolina Press, 1979).

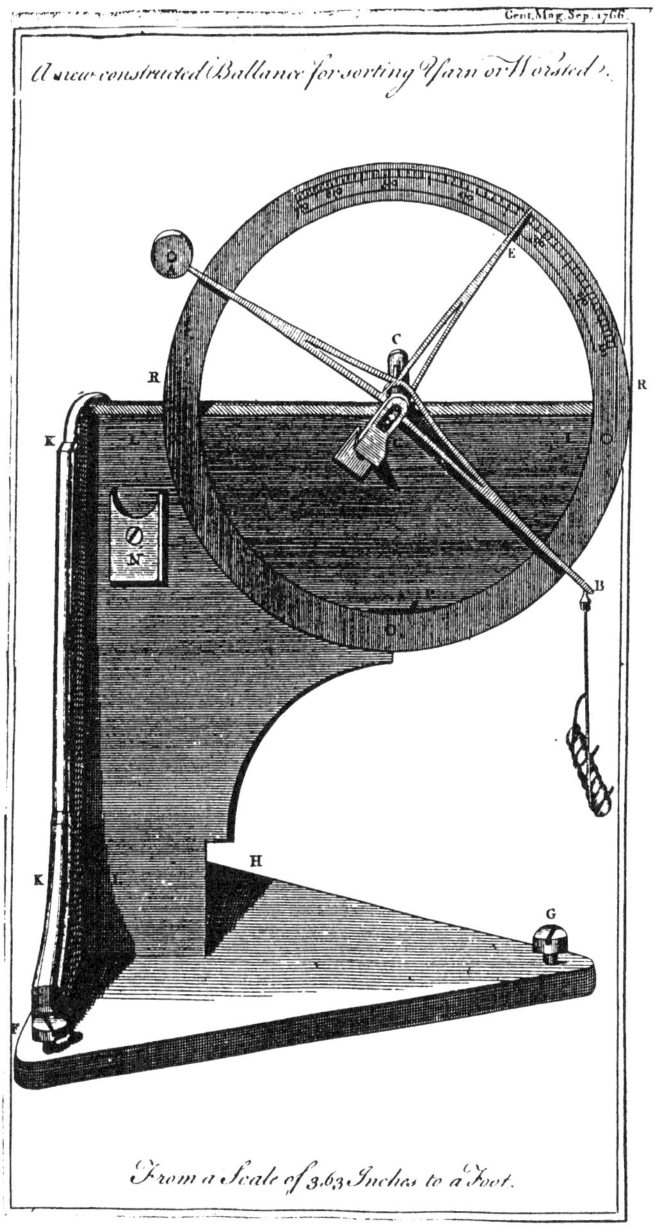

Fig 1: W. Ludlam's yarn scales

Source: *The Gentleman's Magazine* 36, (Sept. 1766), ff. 408–10

Knowledge of the fineness of yarn was (and still is) essential to spinners, who would be paid acccording to the quality (as well as the quantity) they spun. Weavers also needed to know the quality of fineness in order to choose a yarn appropriate to the structure of the cloth to be made; to gauge how much yarn was needed for that cloth; and also to guide them in selecting the appropriate reed (separator) and heddles (controlling the warp yarns) in the loom. Of course the yarn would need to be reasonably uniform. This would not be a problem in a machine technology but could have been of some significance in hand-spun yarn. The only method of checking the uniformity of hand-spun yarn was by visual inspection of samples. However, long-experienced handspinners were able to produce sufficiently regular yarns.

Yarn fineness was usually expressed by relating length units to a fixed weight unit (in Britain and the USA often one pound avoirdupois). The higher the number of length units, the finer the yarn. So instruments for recording length units were the key to understanding yarn numbering. In the pre-industrial world there were two of these, the spinning wheel (on which yarn was made) and the reel (on which it could be packaged).

My efforts to understand how the fineness of yarn was estimated started (when I was at the Smithsonian Insitiution in 1969–70) with a postal survey of reels in 200 British and American museums with textile collections. The aim was to discover the length units of the yarn measuring systems prevailing in various textile districts. The results were inconclusive, as explained in Chapter 3, note 62. An examination of the 60 spinning wheels in the Merrimack Textile Museum's collections was slightly more illuminating. The problem was that the artefacts, the hardware, were no more than approximations of the measuring codes, the software, of the technological system. The answer therefore lay in tracing these codes. As Chapter 3 shows, they emerged to meet the needs of technical processing, managerial control, and the changing interests of the pre-industrial state. Faced with the problems of fibre and yarn embezzlement, merchants and clothiers pressured the government to lay down standards and so curb dishonest practices. Perhaps it was no coincidence that the design for a nicely-calibrated yarn scale was published (twice, in 1765 and 1766 and seen in Fig 1) in a period when relations between spinners and weavers and their masters deteriorated (following the 1757 repeal of a rate-regulating act of 1756) and embezzlement intensified in the West of England.[15]

Four findings were illuminating. First, in the interests of checking embez-

---

[15] See Chapter 2, note 18 and also *The Gentleman's Magazine* 36 (1766), pp. 408–10 for the yarn scale design. For the background see Julia de L. Mann, *The Cloth Industry in the West of England from 1640 to 1880* (Oxford: Clarendon Press, 1971), p. 114.

zlement, the state intervened and contributed to the shaping of the technological software. Second, the greaty variety of measuring systems, totalling 38 in Britain,[16] was largely a product of the regionalisation of the preindustrial economy. Third, although many of these systems died out with the coming of industrial technology, regional specialisation, and national markets, a surprising number of yarn measuring systems persisted into the twentieth century. Last, and of great relevance to technology transfer, evidence emerged to show that when this regional software was taken from Britain and intermixed by the mingling of immigrant workers in the USA, the result was a measure of chaos. Apparently it most affected the woollen industry and lasted until one or other of the systems gained acceptance as a regional and then a national (and eventually an international) standard.

c. *Cultural differences*

Divergent entrepreneurial cultures represented quite the reverse of an obstacle looking westwards across the Atlantic. In Chapter 4 the argument, rather than exhaustive empirical evidence, is that a widening gap between American and British values provided a growing inducement to British migrants with entrepreneurial inclinations to seek their fortunes in the USA rather than in Britain. By the 1830s, the evidence suggests, religion and law offered fewer restraints to capitalists in the USA than in Britain.

On Britain's religious life, where the deepest values were formed, a strong, though not unified vision was projected in the sophisticated and traumatised world of post-Revolutionary and post-Napoleonic Europe, by British Evangelicals. The moderates among them, as Boyd Hilton[17] has shown,

---

[16] To the 36 covering all fibres and found at different times in the UK which are listed in the Appendix to Chapter 2, must be added two more that have come to my notice: a fixed weight system for silk (reeled or organzine) using a weight of one ounce and a one yard length unit; and a fixed length system for wool used at Selkirk in Scotland where the length unit was a cut of 300 yards and the weight unit was the dram. The latter was in use in the George Roberts & Co. Ltd., Forest Mill at Selkirk until 1968 when the Tex System was adopted. See Thomas Oliver, *Weaving and Designing* (Galashiels, 1911), p. 10 and information from Mr William P. H. Sinclair.

On the American side, further evidence of a replication of numbering systems emerged in the federal census of manufacturers in 1820. The returns showed that Peter Gough's factory in Saint Mary's County, Maryland, used a one-cut reel in its cotton department. Since the cut was a length of 300 yards (used in both the Scottish wool and linen industries) it seems that here was another divergence from the standard Lancashire cotton count based on the fixed one-pound weight and 840-yard hank. See USA, National Archives, Record Group 29, Bureau of the Census: Returns of the 1820 Census of Manufactures, on USNA Microfilm M.279, Maryland, document 314.

[17] Boyd Hilton, *The Age of Atonement: The Influence of Evangelicalism on Social and*

defended the traditional class structure. Moreover, they saw God at work through inexorable moral and economic laws. The Malthusian nightmare of over-population was paralleled in business by the trough of the trade cycle. The intemperance of the speculator brought bankruptcy as certainly as sexual vice incurred the punishment of sexual disease. To economic aspirants among the lower classes this world view taught that life was a moral struggle to be faced with docility, conformity, caution, satisfaction with the status quo. By contrast, in the USA abundance rather than poverty beckoned.[18] Life was a material, not a moral, trial. Church and state were separate and ministers of religion were more beholden to the propertied orders. Upwards social mobility into the entrepreneurial ranks in the community was both possible and approved by the voices of religion. The USA was the Promised Land and Old England the land of bondage.

The economic values embedded in the laws of the American Republic also made the USA appear, to observers, more attractive than early nineteenth century Britain. Mercantile and entrepreneurial influence helped to reshape the Common Law in the interests of economic development. The principle of eminent domain was one notable example. By this concept land could be swiftly redistributed for socially beneficial economic uses such as roads, canals, and railways, in sharp contrast to Britain where private property rights could delay and impose substantial additional charges on investment in new infrastructure.[19] The usury law cap on the rate of interest was lifted slightly earlier in the USA than in Britain. As important as the limits to agrarian property rights was the American courts' determination to curb monopoly and thereby promote competition. Ready access to joint-stock, limited liability company formation, and the enlarged possibility of capital accumulation to

---

Economic Thought, 1785–1865 (Oxford University Press: Clarendon Press, 1989), pp. 36–162.

[18] Political freedom and material opportunity were two of the great appeals of the infant USA and occur time and again in the contmporary literature. One early statement was the essay 'What Is an American?' in Hector St John de Crevecoeur, *Letters from an American Farmer* (London: Dent, 1912) first published in London in 1782. That agrarian view was soon transmuted into a vision of manufacturing opportunity when Americans (led by Alexander Hamilton, Secretary of the Treasury in the early 1790s) realised that economic independence was a necessary complement to political independence. On American abundance see David M. Potter, *People of Plenty: Economic Abundance and the American Character* (Chicago: University of Chicago Press, 1954).

[19] With their private acts of parliament UK turnpike, canal, and railway promoters could make compulsory purchases of land on their planned routes. However, the private landowner was protected by legal rights and processes which gave him a full hearing and enabled him to exact ample compensation. Kellett estimated that 16.25 percent of the cost of 26 railways built 1825–50 was spent on land. See John R. Kellett, *Railways and Victorian Cities* (London: Routledge & Kegan Paul, 1979), pp. 25–43, 427–31, passim.

which it led, came sooner in the USA (in the 1780s) than in Britain (the 1850s). Likewise, invention in the USA was protected by a patent law to which, as Chapter 8 argues, a much wider section of the population in America than in Britain had access.

To pro-entrepreneurial biases in their religious and legal frameworks the American business community in the early nineteenth century added a much more positive self-assessment than their British counterpart. The difference is evident in the very different public posthumous profiles given to the respective pioneers of the British and American cotton industries, Sir Richard Arkwright (1732–1792) and Samuel Slater (1768–1835). While Arkwright received a very mixed press, particularly in his native Lancashire, Slater was hailed as 'the Father of American Manufactures' with an unstained reputation. In the latter half of the nineteenth century Slater's image also suffered an eclipse from which it has only recently been rescued in a definitive reappraisal.[20]

Diverging cultures were also evident in differing entrepreneurial attitudes to labour, to capital, to their future prospects, and their past achievements. In a nutshell, Americans were conscious of the moral evils of the factory system and struggled to find ways of reconciling it to their new-found democracy. The solutions that their political ideology and their relative shortage of labour drove them to adopt before the 1840s aimed to give greater independence and dignity to workers. As is clear in Chapters 8 and 9, this meant substituting capital for labour. At the same time American entrepreneurs avoided the

---

[20] James L. Conrad, Jr., '"Drive That Branch": Samuel Slater, the Power Loom, and the Writing of America's Textile History' *Technology and Culture* 36 (1995). Conrad demonstrates that Slater was neither heroic nor resistant towards powerloom weaving technology: quite the reverse, he did not single-handedly introduce Arkwright technology to the USA but he was closely associated with other new technologies as well as machine spinning. He invested in powerlooms and stationary steam engines.

The most influential commentator on the American cotton industry in the 1830s and 1840s was James Montgomery, the Glasgow cotton mill manager and textbook author who emigrated to New England in the mid-1830s and in 1840 published *A Practical Detail of the Cotton Manufacture of the United States*. That he echoed White's appraisals of Slater as pioneer of cotton spinning in the USA is not surprising. Nor is his inclusion of the ambiguous passage from White which reported that Slater declined to adopt powerlooms when they were first introduced to Rhode Island. For the intended second edition of his work Montgomery clearly wanted to revise his remarks about Slater and the powerloom. Perhaps because Slater was a proponent of steam power, Montgomery, an ardent advocate of water power, gave him relatively little attention in his newspaper controversy about the relative costs of steam and water power. See David J. Jeremy, *Technology and Power in the Early American Cotton Industry: James Montgomey, the Second Edition of His 'Cotton Manufacture' (1840) and the Justitita Controversy about Relative Power Costs* (Philadelphia: American Philosophical Society, *Memoirs* vol. 189, 1990), pp. 154–60, 226–27, 245.

pitfall of gentrification into which their counterparts in Britain, including the sons and grandsons of Sir Richard Arkwright, fell. An education completed at Harvard College was a preparation for a life in business and industry, in contrast to Oxford or Cambridge, a stepping stone for sons of the landed gentry and aristocracy on their way into politics or the higher professions, rather than a career in industry.

Channels

A second group of articles treats channels or vehicles of transfer. The need for channels of course arose from differences in the reservoirs of knowledge and skill between the leading and the following society. The channels or agents by which knowledge and skill flowed from leader to follower were numerous and varied. However, some were much more important than others in the early industrial period.

a. Machines and models

Unaccompanied by an artisan with a knowledge of how they worked, machines or models of machines had a limited utility in the early years of the American Republic. A disassembled spinning mule was exported to Philadelphia in 1783 (as explained in Chapter 5) but, without its knowledgeable owner, the parts defeated all attempts to rebuild a working machine. Four years later, the purchasers sent the parts back to England. Nor were New England mechanics much more successful with machines, supposedly of Arkwright's design, brought over by Robert and Alexander Barr from Scotland. When Samuel Slater saw replicas of these machines four years later in 1790 he pronounced them 'good for nothing in their present condition, nor can they be made to answer'.[21] However, machines were an efficient vehicle for technology transfer when they were accompanied by skilled mechanics or craftsmen, as importations of calico printing machinery in the 1820s illustrated.[22]

b. Printed drawings and verbal information

Of even less value as a stand-alone vehicle of technology diffusion were

---

[21] George S. White, *Memoir of Samuel Slater, the Father of American Manufactures* (Philadelphia, 1836), p. 74; William R. Bagnall, *The Textile Industries of the United States* vol. 1 (Cambridge MA: Riverside Press, 1893), pp. 150-52.

[22] Caroline Sloat, 'The Dover Manufacturing Company and the Integration of English and American Calico Printing Techniques' *Winterthur Portfolio* 10 (1975); *TIR*, pp. 104-17.

printed drawings and verbal accounts of new technology. Published descriptions of Britain's new textile technology emerged over the course of forty years, from the 1790s. A proportion, varying from a fifth to a third, of patents (specifications and drawings) was published from 1794.[23] Static descriptions appeared in cyclopaedias from 1812 (as detailed in Chapter 5) but the first set of operating rules for mechanised cotton spinning was not issued until 1832 when James Montgomery published his *Carding and Spinning Master's Assistant*.[24] When operating rules, proven by experience and accepted by the community of mechanics in the originating society, were published it was possible for mechanics in the follower society to begin to understand the further incremental improvements (as registered in patents) conveyed across international boundaries. At this stage there already had to be a reservoir of mechanics and skilled operatives (who had learned from skilled immigrants and then 'by doing') in the follower society.

c. Skilled immigrants

By far the most important vehicle of technology transfer in the early industrial period were skilled workers – operatives, managers, machine makers. Together, they knew how to build machines, how to run and maintain machines, and how to manage technological systems to meet market demand and make profits. They were vital in the early stages of technology transfer because they had mental and manual knowledge/skill and because they were able to interact with the technological system on the one hand, with the natural and cultural environment of the receptor society on the other hand, and so make the multitude of adjustments that were necessary in accommodating a new technological system to fresh, cross-cultural surroundings.

Human carriers inevitably have their limitations. The immigrant textile workers in New England presented problems in the 1820s because of their conservative attitudes (Chapter 4).

In the early industrial transatlantic context these key personnel comprised a relatively small number of people. Three-quarters or more of the immigrant textile workers arriving in the USA in the late-eighteenth and early nineteenth centuries were handloom weavers, carriers of obsolete skills. By the 1820s the picture was changing. For that decade some empirical evidence permits a comparison between immigration and workforce expansion in the textile factories of the United States. While the American cotton and woollen industries' workforces grew by 56,811 between 1820 and 1831–32, immigration in those years brought 4,999 textile workers, a difference of over 51,000

---

[23] *TIR*, p. 47.

[24] Jeremy, *Technology and Power*, pp. 6–13.

workers. Likewise, while the number of American cotton and woollen mills rose in those years by 628 establishments, the number of textile managers arriving in the USA was 354. As for machine makers, the 1820s saw a rise of 2,449 positions in the USA, compared to the arrival of 1,001 textile machine makers from the UK.[25] Imperfections in both kinds of data (passenger lists and manufacturing censuses) mean that these numbers are indicative rather than precise. It is certain, however, that by the 1820s American labour was supplying most of the country's textile industry needs for skilled operatives, managers, and machine makers. The United State's own reservoir of technical knowledge and skills in the textile industry had both spread beyond the confines of the immigrant community and risen in level. By the 1820s immigrants were needed only at the 'demonstration of potential' and 'pilot plant' stages of technology transfer for any new devices within the existing (basically mechanical) technological system.

The roles of immigrant workers are further explored, in the case of machine makers, in Chapter 6. In brief, this shows that their introduction of new technology was not without its problems. Immigrant machine makers tended, naturally, to stick to the designs and technical solutions with which they had grown up in England. This was not always appropriate for American circumstances. More positively, they played an important role not only in introducing new technical knowledge but also in lowering machinery prices, in training young Americans, and in supplying American machine makers with key components for the new machines.

d. Industrial spies

Throughout the period under review, the first four or five decades of the American Republic, American agents visited Britain's industrial districts, both to recruit skilled workers and to engage in industrial espionage. Chapter 7 provides a series of examples of these activities. In many respects the situation was similar to the Cold War world of the 1950s–1980s. Each side tried to counter the other at every level, from the individual factory to the Customs clearing shed. Webs stretching across commercial and diplomatic relationships facilitated the movement of American agents. Equipment was illegally exported by devious routes. Problems arose when machines needed to be accompanied by skilled workers, and skilled immigrants proved as secretive as the agents who recruited them. As already seen, however, it proved impossible to check the outlow of machinery. Prohibition and policing succeeded only in pushing up the price of smuggled equipment, rather than in sealing the UK's reservoir of technological capabilities. Many of these points are

[25] *TIR*, pp. 144–175, 256, 269–73.

more fully illustrated in the Anglo-French context by the late Professor Harris's last volume.[26]

Modifications

The theme linking the last four essays is that of adjustment of imported techniques to American factor conditions (supplies of land, labour, and capital) and cultural conditions (prevailing social values). American inventive effort would, expectedly, produce modifications to the imported technological system. Hence Chapter 8 is an attempt to plot American invention in textile technology from the patent evidence. A comprehensive study is impossible because the federal Patent Office burned down in 1836 and afterwards only those patents still in force were 'reconstructed' and preserved. No complete publication of patent drawings and specifications (even in summary form) had been made before 1836 so many of the patents were lost for ever. Only their dates, authors' names and addresses, and the subject of the patent were earlier published, and hence survive. Published lists, in subject and chronological sequences, contain some discrepancies and these are noted in Appendices IIA and IIB. Of more significance is the list of patents taken out by John Thorp, inventor of the ring spindle, the device on which twentieth century spinning technology in nearly every country in the world has rested. Some of these were reconstructed after 1836, so some glimpses of the alternatives and steps he took in the months before and after the award of his ring spindle patent (20 November 1828) may be glimpsed in the patent drawings.

The patents recorded descriptions of the textile technology that was offered but not entirely selected by users. In Chapter 9 an attempt is made to analyse the main trends behind the textile technology that *was* selected by users. This essay was written before I had seen the patent litigation papers in the Circuit Court Case Files for Massachusetts and Rhode Island and before I had read the 303 patents awarded in Britain between 1790 and 1830 for mechanical devices in textile technology — evidence used in *Transatlantic Industrial Revolution*.[27] Several comments must therefore be added.

First, many of the technical solutions chosen by the northern New England engineers at Waltham and then Lowell had earlier been registered in British patents. Sometimes these were known to the Americans, sometimes they were not (as in the case of several of the stop motions). One of the most explicit statements recorded to promote the use of stop motions was made by John Leigh Bradbury, a 'gentleman' of Gloucester (and later of Manchester)

---

[26] Harris, *Industrial Espionage*.

[27] *TIR*, pp. 54–66, 184–202, 304–6, 324–27.

in his UK patent of 1816. In view of the labour-saving value of stop motions to early American textile manufacturers, it is worth citing in full: "When a thread breaks, the roving or material to be spun continues to be delivered out and to run to waste till seen by the person who attends the frame (who is for the most part a child, and of course subject to all the causes of inattention, which act more forcibly in early years) which waste amounts, in even the best regulated mills, to a very large quantity annually; but by my improvement for preventing waste this evil is completely removed, the machinery becomes self-regulated and totally independent of the care of the children or others employed to piece the threads. In this particular and on this account the persons employed in a spinning mill for said purpose may made to attend each a great number of spindles, and thereby much fewer of them be required, and the profits be increased on this account, as well as by the saving of waste, and prevention of the damage which the waste causes to machinery."[28]

The principle Bradbury enunciated was relatively neglected by British cotton manufacturers but was pursued extensively by American textile manufacturers. It is clear, therefore, that Americans were deliberately shaping their textile technology to suit their own circumstances. All this has been long-known, not least from the work of Habbakkuk[29] who drew attention to the differences in factor proportions as an explanation for American endeavours to substitute capital for labour in their technological choices.

Second, the influence of the British immigrant mechanic played a part in reshaping technology developed in northern New England, where it was designed to meet minimal skill levels, as well as as in Rhode Island, where the technology more closely resembled English models. For example, Thomas Charlton Faulkner, who was recruited from Manchester in 1816, helped Paul Moody the Waltham engineer to build a double speeder (a device to produce cotton roving (a thick rope, preparatory to spinning into a yarn).[30]

Third, labour-saving stop motions were not confined to the drawing and twisting frames and the warper (as claimed in Chapter 9, p. 64). In addition, stop motions were used by northern New England manufacturers in the filling frame, the dresser, and the powerloom.[31] Later it was recorded in lap doublers.[32]

Fourth, several attributions need to be amended if not corrected. The warping frame stop motion, ascribed to the engineer Jacob Perkins (Chapter

---

[28] GB Patent 3990 (9 March 1816).

[29] H. J. Habakkuk, *American and British Technology in the Nineteenth Century: The Search for Labour-saving Inventions* (Cambridge University Press, 1962).

[30] *TIR*, p. 191.

[31] *TIR*, p. 198.

[32] Jeremy, *Technology and Power*, p. 85.

9, p. 66), was attributed back to Paul Moody when James Montgomery prepared the second edition of his *Cotton Manufacture*.[33] The latch-type stop motion for the drawing frame credited to Samuel Batchelder (Chapter 8, p. 66) was probably only partly his. The spring bolt section of the motion was patented in the USA by Lewis Cutting of Lowell on 15 May 1834.[34] Batchelder, through the agency of Daniel Treadwell, a Harvard science professor who visited Britain in 1835, patented his drawing frame stop motion (without Cutting's bolt-and-spring action) under the name of John Houldsworth, cotton spinner of the Glasgow.[35] Furthermore, the latch type of stop motion was patented earlier in Britain, by John Leigh Bradbury, on 9 March 1816 (for the bobbin-and-fly frame) and again by him on 3 July 1824 (for the spinning frame).[36] Who was first in the invention of the various stop motions is hard to tell. At this time both the UK and the USA had patent systems in which novelty was proved by the courts, rather than by professional examiners. At the least, it may be concluded that multiplying transatlantic connections took ideas relatively swiftly between the English-speaking centres of innovation. In those centres competing hives of mechanicians selected the technology most appropriate to the costs and qualities of labour, power, land, materials, capital in their respective textile districts.

That simultaneous invention or re-invention was possible is evident from the case of the differential gear, the subject of Chapter 10. Remarkably, this arrangement of gears was invented half-a-dozen times or more in the space of two thousand years, from the ancient Greeks and Chinese to eighteenth-century clockmakers and millwrights and on into the nineteenth century when textile machines, steam carriages, and motorcars incorporated the differential. In fact some of these later 'inventions' were no more than new applications of the differential gear train.

The final essay, Chapter 11, is an attempt to compare the water-powered textile manufacturing systems of New England with the steam-powered systems of Lancashire's cotton industry. By 1841 the USA, with a population of 17.1 million, for the first time overtook England and Wales, at 15.9 million. However, Massachusetts, the centre of northern New England's textile industry, with a population of 738,000 had less than half the population of Lancashire, which had 1.7 million. Since Massachusetts had an area 4.5 times that of Lancashire, the population density of Lancashire was about ten times that of Massachusetts. In other words, New England entrepreneurs in the textile manufacturing trades faced a sharper choice between rural and urban sites that did Lancashire manufacturers.

---

[33] Jeremy, *Technology and Power*, p. 113.

[34] Jeremy, *Technology and Power*, p. 233, n. 16.

[35] GB Patent 6951 (9 December 1835); Jeremy, *Technology and Power*, pp. 17–20

[36] GB Patent nos. 3990 and 4984.

The essay argues that American manufacturers chose rural sites because although the capital cost of waterpower was four times that of steam power, its overhead costs were only 60 per cent of the overhead costs of steam. Most textile mills in early nineteenth century New England used waterpower. The question then arises, how did they survive against the competition of the giant mills set up at Lowell? They apparently did so by several means, including the holding of smaller stocks of raw materials and finished products, flexible family labour, room-and-power sharing, and avoidance of the psychic costs to families moving into the city. However, as the piece concludes, much more needs to be done to understand the strategies by which rural mills survived even when competing in national markets. In particular, we do not know the extent to which they used the more labour-saving technology of northern New England or the less labour-saving technology of southern New England.

Conclusion

The transfer of technology examined in this introduction and most of the essays is concerned with the transfer of whole technological systems (in Hughes's sense) across international cultural boundaries. Some essays (e.g. on yarn numbering or the differential gear) have taken elements in the technological system and examined their diffusion as closely as possible, but the assumption is that these aspects only make sense when seen as part of a greater whole, the technological system, in this case mechanised textile manufacture.

Several conclusions arise from the essays. One is that whole technological systems, being large and complex, will need a variety of complementary skilled workers and a variety of media to convey them from one society to another. A second conclusion is that in the early industrial period (and probably much later) the skilled migrant with new industrial capabilities was a key carrier of new technology. He may have moved as an individual but it was the combined effect of numbers of such individuals which achieved a successful transfer. Just how important such migrants were depended on the stage at which the transfer process had reached. Migrants were more necessary at the first and second (demonstration and pilot plant) stages, rather than the third and fourth (internal diffusion and modification) stages. A third conclusion follows: publishing information is not sufficient in itself to accomplish transfer.

A fourth conclusion is that people, as individuals and groups, govern what technology can be imported into a new social environment. Immigrant and indigenous technicians and entrepreneurs may overcome supply-side difficulties, and demonstrate what new goods and services and methods for their production are available on the international market. However, employees and consumers and politicians will have a crucial role on the demand side. Unless the technology and its products are accommodated to their preferences, habits, values, and social arrangements, transferors will have to engage in a costly demonstration and marketing exercise which may not in every case succeed. Last, technology transfer will have to take heed of the natural, physical endowment of the receptor society. As in New England, for a time water-power was much more appropriate than the steam power used in Lancashire. This too had implications for the ways in which it was possible and desirable to modify the imported technology.

I

Wharf shed of the Trafalgar Dock, Liverpool, before 1843:
One place where Customs men impounded smuggled machines and machine parts

Source: *The Land We Live in: A Pictorial and Literary Sketch Book of the British Empire* (4 vols., London: Charles Knight, c. 1847–50)

II

Damming the Flood: British Government Efforts to Check the Outflow of Technicians and Machinery, 1780–1843*

In the waning decades of the age of mercantilism, Great Britain intensified her efforts to keep at home the new industrial technology—technicians and machines—that the inventiveness of her people had produced. From his researches in the records of the Board of Trade, which played a major role in coordinating these efforts, Mr. Jeremy shows that as time wore on the policy of prohibiting emigration and exports became more and more internally contradictory and incapable of enforcement despite great ingenuity on the part of those responsible.

Once British entrepreneurs had demonstrated the superiority of machinery in the manufacture of textiles, in the decades following the inventions of Hargreaves, Arkwright, Crompton, and Cartwright, traditional efforts to contain British technology within the kingdom were intensified. Checks against the outflow of Britain's early industrial technology were applied both by private businessmen and the government. The attitudes of suspicion and secrecy displayed by numerous early-nineteenth-century manufacturing firms are well known. Even more familiar are numerous aspects of the protectionist system

Business History Review, Vol. LI, No. 1 (Spring, 1977).

* I am grateful to the Pasold Research Fund for a grant to work on Board of Trade records at the Public Record Office in summer 1974. I am also indebted to Dr. Charlotte Erickson, Reader in Economic History at the London School of Economics, to Miss Julia de L. Mann, and to my wife Theresa for their keen criticisms of earlier and longer drafts of this article. This research is part of a doctoral thesis being written under Dr. Erickson's supervision, on the transmission of cotton and woolen manufacturing technologies from Britain to the United States, 1790–1840.

The cover illustration, entitled "Wharf Shed of the Trafalgar Dock, Liverpool," shows one of the scenes where Customs officers attempted to prevent the illegal export of machinery before 1843 – Trafalgar Dock being opened in 1836. This print, drawn by Hair and engraved by Jackson, was published in Charles Knight (publisher), *The Land We Live In: a Pictorial and Literary Sketch Book of the British Empire* (4 vols., London, *ca.* 1847–1850). The author is indebted to H.M. Customs & Excise, in the person of Mr. P.J.T. Machin, Deputy Librarian, for providing this print for reproduction, and to Mr. J. Smith of the Liverpool Record Office for identifying its source.

constructed by seventeenth– and eighteenth–century governments. But relatively little has been written on the working of British legislation aimed at the retention of technology.[1] With what consistency and effectiveness were the prohibitive laws applied? When modification came in 1824–1825, what sort of logic freed artisan emigration but continued governmental control over machinery exports? What policies did the Board of Trade follow in licensing machinery for export in the period 1825–1843? How effective were the continuing prohibitions? These are some of the questions that surround the importance of governmental checks on the international movement of technology, especially textile technology, in this critical period.

I. The Period of Maximum Legislative Prohibition, 1780s–1824

In the early 1780s no skilled artisan or manufacturer was legally free to leave Britain or Ireland and enter any foreign country outside the Crown's dominions for the purpose of carrying on his trade. Textile printing workers were even forbidden to leave the British Isles, the implication being that other workers could at least travel within British possessions. It was an offence, moreover, to entice artificers or manufacturers to emigrate to foreign parts. It became illegal to export or to prepare to export to any place outside Britain and Ireland any pre-industrial or industrial textile, metal-working, clock-making, leather-working, paper-making or glass manufacturing equipment. But whereas the laws relating to textile implements and machinery were thoroughly comprehensive, encompassing existing equipment and any invented in the future, those dealing with metal-working tools and utensils were not exhaustive. As we would expect, the penalties for removing the living instruments of trade, the artisans, were much higher than for exporting the machines they were capable of reproducing with their own hands. Emigrants lost their nationality and property and recruiting agents were fined £500 per migrant enticed and received twelve months' imprisonment. In contrast, a £200 fine, forfeiture of equipment and twelve months' imprisonment (or a £500 fine and forfeiture in the case of textile printing tools and machinery) was laid down for the export or attempted export of the hardware. These massive penalties threatened a middle class manufacturer more than an unpropertied artisan.[2]

[1] The major exception is A. E. Musson, "The 'Manchester School' and Exportation of Machinery," *Business History*, XIV (January, 1972), 17-50, which looks at the question from a significant local vantage point.

[2] The textile-related prohibitory statutes were 7 & 8 Wm. 3, c. 20 (1695); 5 Geo. 1, c. 27 (1718); 23 Geo. 2, c. 13 (1749); 14 Geo. 3, c. 71 (1774); 21 Geo. 3, c. 37 (1781); 22 Geo. 3, c. 60 (1782); Ireland 25 Geo. 3, c. 17 (1785); 26 Geo. 3, c. 76 (1786), which repealed clauses in the 1774 and 1781 statutes regarding cheap wood cards.

At least six government departments were involved in enforcing the prohibitive laws. The Foreign Office collected consular reports of artisans domiciled in foreign countries or of machinery illegally exported. The actual policing of the ports and coasts was the province of the Customs Commissioners. The Home Office, to whom the local magistrates reported, intermittently received information about foreign recruiters operating in the British manufacturing districts or artisans who were making arrangements to emigrate. Permissions to emigrate or licenses to export machinery were issued by both the Privy Council and the Treasury, while the former was also capable of issuing orders that could clarify the statutes. The key administrative department was the Board of Trade, as reconstructed in 1786 by William Pitt. It advised the Treasury on Customs policy and on commercial relations with other nations,[3] and it is clear from the Board's Minutes between 1780 and 1843 that it was usually informed or consulted by all the other departments except the Privy Council about infractions, interpretations, or modifications of the prohibitive laws. Thus, records of the Privy Council and the Board of Trade provide a good picture of central government policies, the Privy Council being concerned with these at the highest level, particularly in wartime, and the Board of Trade, with their commercial implications.[4]

After the 1780s, the loss of the American colonies, the prolonged European war with France (1793–1815), and expanding industrialization at home made the problem of emigrant workers critical. Foreign demand for the advanced technology created by Britain's pioneering industrialization grew rapidly, while a wide variety of incentives made hundreds of individual British industrial workers willing to risk the penalties of the prohibitory laws to meet, east and west, the rising demand for their skills.[5] From 1783 to 1812, some

---

The metal trades were covered by 25 Geo. 3, c. 67 (1785) and 26 Geo. 3, c. 89 (1786) as well as the 1718 Act. The act 26 Geo. 3, c. 89 included the prohibitions on the export of paper-making, glass blowing and cutting, pottery and leather-working equipment. The act 39 Geo. 3, c. 56 (1799) placed the seducing of colliers to emigrate on the same footing as the seducing of manufacturers.

3 Lucy Brown, *The Board of Trade and the Free Trade Movement, 1830–1842* (Oxford, 1958), 20–21.

4 This article is based on the following collections in the Public Record Office: P.C.1 (Privy Council papers, unbound); P.C.2 (Privy Council Minutes), vols. 128–226 (1782–1844); B.T.1 (Board of Trade In-Letters), letters from the period 1791–1830s; B.T.5 (Board of Trade Minutes), vols. 1–45 (1784–1838); B.T.6 (Board of Trade Miscellanea), vols. 151 and 152 (Machinery Books, 1825–1843). I should like to thank the staff of the North Room and the Rolls Room at the P.R.O. for their cooperation in locating and producing documents.

5 On the emigration of industrial workers to Europe, see W. O. Henderson, *Britain and Industrial Europe, 1750–1870* (3rd ed., Leicester, 1972). On the emigration of industrial workers to the United States, see Mildred Campbell, "English Emigration on the Eve of the American Revolution," *American Historical Review*, LXI (October, 1955), 1–20; Herbert Heaton, "The Industrial Immigrant in the United States, 1783–1812," *Proceedings of the American Philosophical Society*, CV (1951), 519–527; David J. Jeremy, "British Textile Technology Transmission to the United States: the Philadelphia Region Experience, 1770–

100,000 persons left Ulster for the United States, an average of 3,000 a year. Between 1785 and 1800, however, immigration from Ulster was at a rate of 5,000 a year.[6] Incomplete Returns of Enemy Aliens made during the War of 1812 show about 7,500 male Britons resident in the United States in 1812–1814; of this number some 1,300 were workers in the textile or textile-related trades.[7] How aware was the central government of this widespread defiance of its prohibitory laws and what actions did it take to reduce it?

Relatively few consular reports about British artisans working abroad were received by the Board of Trade during the period. This is somewhat surprising because we know that at least one consul, Phineas Bond at Philadelphia, bombarded the Foreign Office with missives on the subject between 1787 and 1805.[8] The matter of emigration was first brought to the Board's attention by a letter dated October 5, 1791 from Sir John Temple, Consul General at New York. It complained of an influx from Ireland. A few months later another communication to the Governor of Lower Canada, suggesting that the Irish migration to Philadelphia, estimated at 1,500 persons a year, might be deflected to Canada, brought no response from the Board of Trade, which, according to the minutes of its meetings, feared that action might "have the appearance of encouraging still further Emigration." Such passivity disappeared when war against France broke out in 1793.[9]

The period of the negotiations leading up to the Peace of Amiens in March 1802 found the subject again under consular consideration. This time there were reports of English artificers in Germany, detained on the Continent by fear of criminal prosecution should they return home. Apart from information on English artificers in Austrian factories in 1805 and a general comment on West of England workers in the United States in 1812, the consular service provided the Board of Trade with very limited indications of the leakage of skill.[10] Even so, the Board was hardly ignorant of the scale of the leakage, since a variety of informal means of communication, such

---

1820," *Business History Review,* XLVII (Spring, 1973), 24–52; Charlotte Erickson, *Invisible Immigrants. The Adaptation of English and Scottish Immigrants in Nineteenth Century America* (Coral Gables, Fla., 1972), 229–389. Anthony F. C. Wallace and David J. Jeremy, "William Pollard and the Arkwright Patents,' *William and Mary Quarterly,* forthcoming.

6 Maldwyn A. Jones, "Ulster Emigration, 1783–1815," in E. R. R. Green, ed., *Essays in Scotch-Irish History* (London, 1969), 46–68.

7 My analysis of U.S., Marshals' Returns of Enemy Aliens, War of 1812 Papers, Record Groups 45 and 59, National Archives, Washington, D.C.

8 Joanne L. Neel, *Phineas Bond. A Study in Anglo-American Relations. 1786–1812* (Philadelphia, 1968), 75–76, 129, 149, 151.

9 B.T. 5/7, p. 310; 8, pp. 102–104.

10 B.T. 5/13, pp. 62–65; 15, p. 292; 21, p. 117.

as manufacturers, customs officials, and informers, could bring news from foreign sources.

Internal reports of specific enticements or migrations were first recorded in 1807 with advance news of the visit of an industrial spy from Russia, Charles Baird (an agent of Sir Charles Gascoyne), who was on his way to obtain information and machinery from the Scottish and Manchester manufacturing districts. Notification came from Joseph Seddon, Chairman of the Society for the General Protection of Trade, a Manchester organization with correspondents in Glasgow. In the years just before and after the depression of 1819, moreover, manufactures in the Midlands metal trades made complaints of workers emigrating.[11] Customs reported goldsmiths leaving for the United States in 1809, and in 1811 they relayed an anonymous report about a Bristol mechanic taking machine-making tools to Falmouth with the intention of embarking for the United States. The same year Customs submitted reports on the emigration of Bristol glass workers and Lancashire cotton workers, and the following year they relayed an anonymous letter about workers from the Birmingham arms manufacturers taking their tools with them to America.[12]

There appear to have been no financial inducements to inform against skilled workers, as there were in the case of emigrant seamen after 1793, and informers sometimes risked intense odium. One tradesman found that he had been named as a prosecution witness after he patriotically informed in 1809 against a mechanic who openly bragged of plans to emigrate. Fearing for his reputation and his business, the tradesman had to make a desperate appeal to the Board of Trade for his discharge from appearing at the Stafford Assizes.[13]

Efforts to Improve Enforcement

Cognizant of this extensive emigration during a period of rapid industrial change and European war, the government was obliged to consider improving the enforcement of the laws. Logically, questions of definitions and interpretation came first. Could any exceptions to the laws governing the emigration of artisans be allowed and, if so, who should legally be permitted to go abroad? A number of limited exceptions were made, although the arguments behind the permissions are rarely recorded. Presumably they were thrashed out

11 B.T. 5/17, p. 277; 26, p. 419; 27, p. 337; 30, p. 406; 31, p. 436. For this Manchester society, precursor of the Chamber of Commerce, see Arthur Redford, ed., *Manchester Merchants and Foreign Trade, 1794–1858* (Manchester, 1934), 67–68.

12 B.T. 5/19, p. 64; 21, pp. 70, 364. B.T. 1/76, ff. 29–38.

13 P.C. 2/137, pp. 468–470. B.T. 1/44, f. 176; 45, f. 292.

orally. For at least one category of artisan – steam engine erectors – the government's hand was forced when an anomalous situation arose. Whereas the prohibition against the emigration of artisans was technically universal, that against the export of machinery was selective. For reasons that are not clear, neither Savery's, Newcomen's nor Watt's steam engine ever came into the machinery clause of the prohibitory statutes. Obviously, if the government permitted a foreign customer to buy a British steam engine, it could hardly deny him the services of British millwrights and mechanics to set it up. The Privy Council, which apparently controlled the licensing of emigrant artisans during the French Wars, allowed two erectors to set up three steam engines at St. Petersburg in 1806. Later that summer Samuel Davies was licensed to set up a steam engine at New Orleans, while three other men were permitted to sail to Sweden for the same purpose. Between 1815 and 1824 the Board of Trade approved only three applications for steam engine erectors to emigrate. Presumably most of those who went abroad did so without government licenses.[14]

After it was decided in 1797 to allow the export of mint machinery, and especially after 1799 when Matthew Boulton obtained an act of Parliament to send both coining machinery and millwrights to Russia, millwrights were often released to complete similar foreign orders. Assuming that the same principle could cover the printing of bank notes, the Swedish government obtained Privy Council approval in 1806 for a person to go to Sweden and set up a paper mill for bank-note-paper making.[15] During the French Wars, the Privy Council received applications from the Swedish government for the release of four sinkers to help open up some new coal mines, and two manufacturers of farming implements. All were allowed abroad. It seems likely that political considerations played a part, but evidence is lacking.[16]

No artisans at all were licensed by the Privy Council to emigrate between 1814 and 1824. And in the period 1780–1824, apart from the three permits to steam engine erectors, the Board of Trade allowed only one skilled worker, Richard Smith, to go abroad. Halted at the Liverpool Customs in 1817, he admitted that he had once been a master spinner, was presently assistant to a Staffordshire land surveyor, and was going to Philadelphia to recover unspecified property taken thence by James Slater, late of Cheadle.[17] Only one application (this to the Board of Trade) for permission for artisans to emi-

14 P.C. 2/169, p. 629; 170, pp. 429, 674. B.T. 5/25, p. 337; 26, p. 140.
15 B.T. 5/10, pp. 435–437. P.C. 2/160, p. 384; 170, pp. 543–544.
16 P.C. 2/164, pp. 128, 177; 175, p. 67.
17 P.C. 2/205 pp. 420–421. B.T. 1/119, ff. 8–10. B.T. 5/26, p. 140.

grate was rejected before 1824. The Turkish government was emphatically refused (in 1802) permission to hire British artisans. But these were cannon founders, and Britain was still at war against France.[18]

One other important question arose on the interpretation of the laws regarding artisans. Did the definition of artificer and manufacturer cover the non-manufacturing trades? In wartime the old prohibitive laws might be interpreted more rigorously than in peacetime, and only the positively unskilled be let through Customs. Perhaps for this reason the issue did not reach the central government until 1817. That year the Liverpool Customs officers squabbled with the Liverpool magistrates over the interpretation of the law on this point. The Liverpool Collector wrote to the Customs Commissioners asking, "whether common workmen, such as shipwrights, coopers, ropers, joiners or house carpenters, brick makers &c. [are covered by the laws]." The Board of Trade wrote to the Customs Secretary asking what the established practice had been, and to the Attorney General and Solicitor General for their opinion of the law, but Board of Trade minutes record no settlement.[19]

The greatest loophole in the prohibitory laws related to the problem of detection. The Chairman of the Customs readily acknowledged in 1824 that his officers had no way of identifying a skilled worker.[20] Even if the embarked passengers were mustered, as was possible under the 1795 proclamation or the 1803 Passenger Act, determined artisans would willingly perjure themselves to escape abroad.[21] Some joined ships after they had been cleared from English ports. In 1807, it was claimed, "emigrant artists & manufacturers" took small boats out to ocean-going vessels cleared from Liverpool "on their passage along the Cheshire shore." Revenue cutters or Royal Navy vessels might occasionally intercept these emigrants, but the problem of identification remained.[22]

Despite the odds against success, the Privy Council determined to

[18] B.T. 5/13, pp. 114–115.

[19] B.T. 5/26, p. 261. B.T. 1/120, ff. 90–95.

[20] G.B. *Parl. Papers (Commons)*, 1824 (51), V, "Six Reports from the Select Committee on Artizans and Machinery," 50.

[21] Few cases of artisans who perjured themselves are documented. One of them, Samuel Mearbeck, later recalled his questioning by Boston Customs officers who wondered how he had managed to escape from England with his cutlery tools: "They asked how did I to get away from England if I did not take a false oath. I told them what Butler says: 'He that imposes an oath makes it, not he who for convenience takes it.' They laughed at me and said that would do and desired me to tell it them again and they wrote it down."

Samuel Mearbeck to his mother, from Boston, Mass., January 1, 1817, Mearbeck Letters, typescript copies made from those in Sheffield Central Reference Library and kindly brought to my notice by Dr. Charlotte Erickson. The quotation is from Samuel Butler's seventeenth–century satirical poem, *Hudibras*.

[22] B.T. 1/35, ff. 192–193.

apply the laws more effectively in the early years of the wars against France. An Order in Council of April 8, 1795 forbade foreign vessels to sail without their captains first submitting to the port officer a list of passengers, specifying names, ages, occupations, and nationalities. Any British artificers, manufacturers, or seamen discovered were to be arrested. This new measure against the emigration of artisans apparently owed much to the indefatigable reporting of Phineas Bond, Consul General at Philadelphia. Doubts arose over its legality, but Grenville and the President of the Board of Trade were satisfied and made the order.[23]

Although the Customs officers implemented this Order so well that at first few foreign vessels were cleared at all and the Privy Council began receiving complaints from masters, the problem of accurate identification defeated them. The obvious badge of the artisan was his bag of tools, but this he generally concealed carefully, or sent by a separate ship, or never took at all. Otherwise the indiscreet possession of some kind of documentation might announce the artisan. It was often only by accident that true identities were revealed, as with David Wooding, a Manchester spinner who, in March 1811, boarded the ship *Union* at Liverpool, bound for New York. During the Customs mustering of passengers, his marriage certificate was found in his sea chest and letters of introduction to a Philadelphia inhabitant in his pocket. Wooding, haled before the magistrates, betrayed his father-in-law, who was also on the *Union,* and the pair were committed for trial at the next Lancaster Assizes.

James Kipping, the Surveyor on Special Service, suggested, after his detection of Wooding, another way of solving the identification dilemma. Wooding's marriage certificate and the testimony of Manchester magistrates had given him away. Why not require all aspiring emigrants to produce a certificate, attested by their local magistrate or parish officer, "stating that in [his or their] opinion the Parties were not prohibited emigrants." Essentially this meant extending and reinvigorating the old Settlement and Removal Laws to cover the skilled as well as the poor, emulating the Prussian (1731) and French (1803) workers' passport laws. The Board of Trade received this proposal in May 1813, some two years after the Liverpool officers wrote to their superiors in London, and silently passed the papers to the Privy Council, which was expected to advise the Treasury. No more was heard of the plan. The bureaucratic procrastination, countenanced perhaps because of the War of 1812 (which tem-

[23] P.C. 2/137, pp. 468–470; 142, pp. 361–362. P.C. 1/25, A48 and A49. Neel, *Phineas Bond*, 75, 149.

porarily closed emigration to the United States), suggests an unwillingness to pursue protectionist policies to their logical conclusion.[24]

While improving the means of detection, the authorities also tried the tactics of deterrence. Prosecutions followed arrests so consistently for English artificers on the Continent, as late as the 1820s, that they feared returning home to meet fines and imprisonment. Fear of loss of citizenship was the major legal deterrent. If the country was at war, moreover, some emigrants would be open to the charge of treason.[25] But another type of legislation supplemented the prohibitory laws: the Passenger Act of 1803, patterned on Phineas Bond's proposals. The purpose of the first Act is disputed, but recent evidence seems to favor a Machiavellian interpretation. In limiting the number of passengers that British vessels could carry to one for every two tons of a ship's unladen capacity, the Act not only relieved the conditions of transatlantic travel but also struck a blow at emigrant traffic in general, and thereby also at emigrant artisans. According to Castlereagh (President of the Board of Control in 1803) in an admission to John Quincy Adams, this was the real purpose of the Act.[26]

In attempting to hold back the artisan from emigration, the central government adopted no wide-sweeping measures. Owing perhaps to contemporary ideas on the free circulation of labor, it opposed increasing legislative restrictions (like the Settlement laws) on labor migration. Instead, it limited itself to sniping actions against individuals. When informed, the Board of Trade could alert Customs of impending depredations of foreign recruiters, like those of Baird in 1807 and Edward Thomas in 1819 (the latter was returning to collect artificers to set up a steam engine factory near Elberfeldt). If recruiters were already active in manufacturing districts, or artisans were known to be preparing to emigrate, the Home Office could send out Bow Street runners (then the country's only police force) to bring miscreants before magistrates, as in the case of James Patterson in 1809. But the scale on which those procedures took place was insignificant, considering the size of the problem.[27]

24 For passport systems, see Arthur Redford, *Labour Migration in England, 1800–1850* (2nd ed. Manchester, 1964), 87–88. B.T. 1/76, ff. 33–40. B.T. 5/22, p. 276.

25 See P.C. 2/159, p. 254; 180, pp. 539–540 and B.T. 5/19, pp. 55–56 for prosecutions. G.B., *Parl. Papers (Commons)*, 1824 (51), V, 569.

26 Oliver MacDonagh, *A Pattern of Government Growth. The Passenger Acts and Their Enforcement, 1800–1860* (London, 1961), 57–58. This may be compared with Jones, "Ulster Emigration," 57–58 and Neel, *Phineas Bond*, 87, 149–150.

27 Redford, *Labour Migration* pp. 81–84. B.T. 5/17, p. 277; 26, p. 419; 27, p. 329. B.T. 1/44. f. 176.

Efforts to Reverse the Flow of Emigrant Artisans

Since the root of the matter was a complex of changing economic and social conditions, it was hardly possible for an enlightened and constitutional government to take adequate preventive steps against emigration. Indicative of the very limited possibilities for collective measures was the Board of Trade's reaction to the unemployment and emigration of the Birmingham arms manufacturers who, in 1812, were reported to be taking their tools with them to America. It suggested that the Board of Ordnance be asked to find work contracts for the 200 unemployed manufacturers. This was the solitary occasion in the records examined, when the central government tackled the emigration of artisans with a sound preventive economic measure, impelled by wartime conditions when arms manufacturers would plainly claim much greater government attention than handloom weavers.[28]

Apart from improved detection, deterrence, or prevention, the only other avenue in implementing the laws was to attempt a reversal of the migrational flow. Several times this tactic was tried or considered during the period of the French Wars. But if the law penalized the emigrant artisan, how could he be expected to return willingly to Britain? Either some clauses of the prohibitory laws must be set aside or else even more drastic and effective penalties must be involved to dislodge the fugitive artisan from his foreign abode. The former possibility was clearly more practicable. In 1797 Phineas Bond recommended a compensatory payment to returning artisan James Douglas (sometimes named as John or William), a native of Dumfriesshire, for his invention of three labor-saving machines (a cloth shearing frame, a sailcloth powerloom, and a brickmaking machine). Douglas had emigrated to the United States in 1792 or 1793. Ironically, after his return Douglas was recruited by Chaptal, Napoleon's Minister of the Interior and eminent chemist, and introduced the latest British carding and spinning machinery to the French woolen industry. The relevant point is that the Board of Trade acted on Douglas's memorial and Bond's covering letter with a surprising recommendation to the Treasury: that Douglas be rewarded not for his inventions, but "for his expences in returning to this country, by the recommendation of His Majesty's Minister and Consul at Philadelphia." [29]

[28] B.T. 5/21, pp. 221, 275, 364.

[29] Henderson, *Britain and Industrial Europe*, 30–31; G.B., *Parl. Papers (Commons)*, 1841

Twice again this kind of waiving of the law was considered by the Board of Trade. A fortnight before the Treaty of Amiens was signed on March 27, 1802, a number of letters from Continental sources indicating the activities of English mechanics in Germany and the Low Countries prompted the Board to recommend that a bill be proposed to Parliament offering amnesty to returning workers. Artisans availing themselves of the amnesty would escape the penalties of the prohibitory law 5 Geo. 1, c. 27 and the Traitorous Correspondence Act, 33 Geo. 3, c. 27, provided they came back within a fixed period and had not emigrated for treasonable or criminal purposes. At the same time the Board advised the Foreign Secretary to initiate the process of artisan re-emigration. British ministers and consuls, having located British manufacturers and artificers in their districts, were to be instructed "to warn such Artificers to return into this Kingdom within six months according to the provisions of the said Act of 5 Geo. 1 cap. 27." To facilitate the exiles' return, the Postmaster General was to instruct masters of Post Office packets to grant any artificers holding consular certificates (of evidence of identity) a free passage to Britain. Presumably this amnesty scheme derived from the impending reopening of the Continent under the Amiens Treaty. The proposed bill never reached the statute book. And though the instructions went to the Postmaster General, it is difficult to see how he could carry them through without the complementary act of Parliament. Still, the Board did not abandon the possibilities of attracting artisans back to Britain. In September 1805 it considered John Rae's report on the activities of British artificers in developing the cotton manufacturers of Austria and his plan to offer financial inducements to these machine builders and manufacturers to return home. But again, the proposals produced no response because two months later, after the battles of Ulm and Austerlitz, Napoleon for the third time in ten years overran Austria and removed her from the Third Coalition. Nevertheless in 1824 Sir Charles Stuart was reported to have sent many artisans back to England from the Continent, paying their fares for them.[30]

If re-emigration could not be induced by this method, the only alternative was to strengthen the sanctions against remaining in a foreign country. Loss of citizenship could be a serious matter, especially for those who migrated to Europe. As the 1793 proclamation

(201), VII, "First Report from the Select Committee Appointed to Inquire into the Operation of the Existing Laws Affecting the Export of Machinery," 64. B.T. 5/10, pp. 427-429. B.T. 1/15, ff. 120–126.

30 B.T. 5/13, pp. 62–65; 15, pp. 292–293, 322. G.B., *Parl. Papers (Commons)* 1824 (51), V, 581.

reminded seamen, if they lost their British nationality and fell into the hands of Algerine corsairs or Turks, they would not be "reclaimed" as British subjects. But only if the migrant happened to be captured while crossing the Mediterranean or while in the Eastern Atlantic when the Straits of Gibraltar were open to Algerian cruisers, was this a remotely likely, and very unpleasant, fate. The fortunes of war (as the foundation of a royal woolen factory at Seville in 1781 by English prisoners of war illustrates) increased the possibility of this sort of eventuality. Rae believed that if refusal to return home was elevated to a felonious offence some emigrants would come back from Europe.[31]

Migration to the United States was not susceptible to this kind of penalty. British immigrants could become naturalized Americans even though this required five years' residence, after Jefferson secured repeal of the Federalists' Naturalization Act of 1798. About this the British government could do little, until war broke out in 1812 and erstwhile British citizens served on the American side. With its forces invading America, the British government was in a position to flush out Britons who had become naturalized Americans, especially since British law at this time generally rejected foreign naturalizations and recognized only "natural Allegiance" or citizenship endowed by birth. Thus the royal proclamation of July 23, 1814 gave British citizens serving in the American forces four months to withdraw or else be deemed guilty of high treason, which no doubt tested the motivations and adaptation of some immigrant workers.[32]

Prohibiting the Export of Machines

The ease with which Continental and, despite the greater distance involved, American manufacturers could obtain British machinery is well known. Until the 1820s, during this early non-verbal period before the new skills and designs were reduced to writing and drawings, the role of the artisan in the diffusion of technology seems to have been more important than that of machines, models, or specifications. Even so there was a sizeable traffic of undetermined volume in the export of tools, machinery, parts, and plans. In 1794 the Wiltshire clothier, Henry Wansey, found all the most up-to-date British cotton and woolen equipment in operation in the United

[31] P.C. 2/137, pp. 468–470. H. G. Barnby, *The Prisoners of Algiers. An Account of the Forgotten American-Algerian War, 1785–1797* (London, 1966), 103–106. J. Clayburn La Force, "Technological Diffusion in the 18th Century: the Spanish Textile Industry," *Technology and Culture,* V (Summer, 1964), 330.

[32] P.C. 2/196, pp. 36–38.

States. Much was surely constructed in America, but some vital components, drawings, and specifications were clearly smuggled from Britain. Likewise the Continental industries obtained Britain's new technologies. At first, equipment was mostly accompanied by artisans. When overseas reservoirs of skill had developed, the foreign demand for British machinery could be sustained by three factors: (1) its cost competitiveness, for even with the charges involved in smuggling, British equipment could be cheaper than foreign-built machinery; (2) its superior construction, though the durable construction of British-built machines could retard innovation; and (3) in many cases, its improved or new design. The continuing movement of British machinery abroad was readily acknowledged when the Parliamentary Committee investigated the problem in 1824.[33]

The Board of Trade minutes show that the first report of unlawful machinery exports was made in 1799. Then the Home Secretary relayed to the Board a memorial from "The Committee of Merchants and Manufacturers of Manchester and the neighbourhood for the general protection of trade" complaining that its unrestricted export to Ireland allowed British machinery to be conveyed to foreign rivals. The problem was solved administratively, if not in reality, by the Act of Union the next year. In 1807 the same body of Manchester merchants and manufacturers notified the Board of the impending arrival of Charles Baird on a machine-collecting assignment for the Russian government. That year Customs reported its first seizure of illegally exported machinery, though significant quantities were escaping undetected through the Customs net. The central government seems to have been apprised of only a fraction of the leakage.[34]

Whereas before 1807 there was only one report of unlawful machinery exports, between 1807 and 1811 a handful of complaints came to the Board about workers emigrating and taking their tools with them to the United States – London goldbeaters, Birmingham arms makers, and a Bristol textile machinery maker. This was the period of the Embargo that hurt America's foreign trade but diverted capital and labor into the manufacturing sector of the economy. New industrial opportunities in America naturally offered fresh migrational "pulls." It was also the period of Napoleon's Continental system, which theoretically closed Europe to British trade and traffic.

33 David J. Jeremy, ed., *Henry Wansey and His American Journal, 1794* (Philadelphia, 1970), *passim; idem,* "British Textile Technology Transmission to the United States," 24–52. Henderson, *Britain and Industrial Europe, passim.* G.B., *Parl. Papers (Commons),* 1824 (51), V, 1, 2, 3, 6, 9, 10, 13, 36, for example.

34 B.T. 5/11 pp. 367–368; 17, pp. 277, 302–303.

After Waterloo the two or three complaints about unlawful export of machinery concerned Italy and France: a consignment of paper-making and cotton-stocking machinery reached Naples in 1816, and in 1819 English manufacturers were sending models of machinery to France.[35]

As with artisan emigration, reports on the export of machinery reached the Board of Trade through the Consular system, and domestically through informers, Customs, and manufacturers or masters. Anonymous tip-offs were an important source of "counter-intelligence." One informer, identifying himself as "Bill," in 1822 submitted a memorial that detailed a smuggling route to the Continent: "the means practiced for conveying off machinery is by concealing it in different Packages and sending them to Rye and other places from whence they are taken to the Continent by Fishermen." When, in 1824, leading Custom officers had the opportunity to summarize their experience of the problem, they observed that they had insufficient staff for searching; that wharf facilities for unpacking and inspecting items for export were usually lacking; that large cargoes could not be closely examined; and that, when parts of different machines were mixed up and packed together, it was impossible for Customs searchers to identify the original equipment. Manufacturers, on the other hand, when complaining to the Board of Trade, assumed that the leakage of hardware was capable of being stopped, and that an arrest could be made simply by notifying a magistrate of a suspected intention to export, but this the law did not allow. Trade associations at Manchester in 1783 and Birmingham in 1799 expressed strong feelings for the more effective enforcement of the prohibitions on machinery export. The 1806 Report of the Parliamentary Committee on the Woolen Manufacture, reflecting for the most part the opinions of the Yorkshire manufacturers, likewise expressed a protectionist verdict on the laws, as being "founded in sound policy."[36] But, in spite of manufacturers' and Customs' complaints, the Board shied away from tougher statutory measures. Instead it simply urged Customs officers to greater vigilance, and the Board and the Privy Council allowed a series of relaxations of enforcement, some serving the manufacturing interest.

35 B.T. 5/18, pp. 138, 159, 160, 205; 21, pp. 70–71, 364; 25, p. 153; 28, p. 136.

36 B.T. 1/169, bundle 26. G.B. *Parl. Papers (Commons)*, 1824 (51), V, 49–59. Redford, ed., *Manchester Merchants*, 7, 67–68. G.B., *Parl. Papers (Commons)*, 1806 (268), III, "Minutes of Evidence Taken before the Committee Appointed to Consider the State of the Woollen Manufacture," 4–5; Julia de L. Mann, *The Cloth Industry in the West of England from 1640 to 1880* (Oxford, 1971), 147.

Introductions of Exemptions

Simple pre-industrial "tools and implements of a plain and simple construction" were virtually removed from the terms of the prohibitory laws in 1785. Under this principle the Privy Council permitted the export of implements for the construction of canals to Sweden (1809) and, in 1811, allowed Messrs. Burley & Co. to despatch "a spinning wheel made of mahogany, ebony &c" to Malta, "the said wheel being intended to be sent to Egypt as a present to a female at Cairo." Similarly the Board of Trade allowed the export of farming utensils (including hemp implements and a threshing machine) to Quebec (1805) and permitted a wife in England to send cutlery tools to her husband in Canada (1819). Perhaps it was under the same principle that the export of hand calico printing blocks to Calcutta was permitted in 1824.[37]

The second protectionist exemption, established in 1797, was in the case of equipment used to process raw materials needed for the British cotton manufactures. The Privy Council pronounced that the prohibitory laws "[were] not intended to prevent the exportation of machines [in this case cotton packing presses], the object of which is to facilitate & increase the importation of the raw material into this Kingdom." Between 1797 and 1806 the Privy Council passed five applications to export cotton packing or bailing presses, to the West Indies, the United States, the East Indies, and India. The Board of Trade passed four more between 1814 and 1824, to Surinam and Egypt, among other places. Because of this economically prudent policy, mechanized cotton packaging opened the distribution bottleneck between Whitney's recently-patented American cotton gin (1794) at the harvesting level and the new manufacturing technology at the processing stages.

Under the 1797 decision the Privy Council made other exceptions, releasing a skin-packing press to Philadelphia in 1801, a cotton gin to New Orleans in 1806, and a wool press to Lisbon in 1810. The Board, most active after the French Wars, passed a wool press to Bilbao, a cotton gin to Valparaiso or Lima, and four almond presses to Bengal. Primary processing equipment could also support traditional handicraft manufactures in the colonies, and possibly this explains why cotton printing machinery went to India under license.[38]

[37] P.C. 2/130, p. 313; 184, p. 31; 218 (for Oct. 5, 1811). B.T. 5/15, p. 282; 28, p. 143; 32, pp. 152, 342.

[38] The sources for the preceding two paragraphs are P.C. 2/147, pp. 572–573; 148, pp.

Another exceptional case was created when the Board of Trade in 1797 advised the Treasury on the export of steel rollers for the U.S. Mint at Philadelphia. Since mint machinery would not compete with British manufactures, the Board urged the Treasury to allow its export. This led to further exports of mint machinery, mostly Matthew Boulton's, to Russia (confirmed by act of Parliament in 1799), Denmark (1804), and America again (1816).[39]

The last excepted type of equipment was the steam engine. In 1786 the Privy Council refused to prevent three "Fire Engines" from going to Spain, although allegedly "extremely injurious to the copper mines of this Kingdom," on the grounds that Parliament alone could prohibit machinery exports not covered by the law. Other permissions followed. Under Privy Council license Hugh and Robert Baird sent two to St. Petersburg in 1802; Robert Francis shipped one to New York in 1805; Baron Henry de Bode despatched three more to St. Petersburg in 1806; Messrs. Mann & Barnard exported a 14 h.p. one to New Orleans to drive a saw mill in 1806; and two more went to Lima in 1814 and 1816. Numbers were regularly exported by Boulton & Watt during the French Wars. Thereafter the Board of Trade licensed three for export to Germany and Austria.[40]

By the early 1820s imperial preference was the other ground on which central government allowed exemptions from the prohibitions. Whereas Customs were ordered to seize iron–working rollers destined for France, Alexander Galloway was licensed to export copper–working rollers to Egypt, and Peter Fary was even permitted "to carry machinery for spinning stocking worsted to Van Dieman's Land" (Tasmania).[41]

In the most significant area of machine designs, government found it quite impossible to apply the letter of the law. First laid under the prohibitions in 1781, machine plans became increasingly available to foreigners by three routes. Of private communications, we have only traces of foreign visitors and little idea of the extent to which technology transfer occurred through correspondence. Published drawings and sketchy specifications came with the appearance of Rees' *Cyclopaedia* (45 vols., London, 1802–1820) and operating rules with Montgomery's *Carding and Spinning Master's Assistant* (Glasgow, 1832).[42] Censorship could check the publication

432–433; 153, pp. 529, 539; 158, p. 112; 170, pp. 70–71; 189, p. 205. B.T. 5/11 pp. 210–212; 22, p. 449; 25, p. 378; 26, pp. 10, 260; 29, p. 80; 32, p. 15; 33, p. 43.

39 B.T. 5/10, pp. 435–438; 11, pp. 3–7; 14, p. 165; 25, p. 25. 39 Geo. 3, c. 96 (1799).

40 P.C. 2/131, p. 528; 161, p. 112; 167, p. 178; 169, p. 629; 170, pp. 70–71; 173, p. 233; 174, p. 252; 205, pp. 420–421. B.T. 5/24, p. 155; 25, pp. 24–25, 337.

41 B.T. 5/29, p. 304; 32, pp. 2, 15, 345–346; 33, pp. 24–25, 30.

42 Jeremy, "British Textile Technology Transmission to the United States," 37–42.

of technical information, but this the government left to the discretion of individual authors. Patents of invention, the third source of machine plans for foreigners, offered government its greatest opportunity to enforce the law on machine designs. But secret patents, not uncommon in contemporary France, were scarcely allowed in Britain. Of the 5,000 patents enrolled before 1824, only two were secret, i.e. placed under covers, sealed and lodged with the Lord Chancellor "in order to prevent the invention getting to foreign countries," a procedure involving statutory approval.[43] Ironically, both patents, for textile processes, turned out to be technological failures. Instead, British patents were loosely guarded from prying foreigners by a chaotic search system evolved over centuries of accretion in patent procedures. This was supplemented by the inadequacy of published patents, which were often abridged and haphazardly chosen for inclusion in trade journals, where only a quarter of the enrolled patents could be found. But all this would not prove insuperable to the persistent investigator, who, at the cost of time and money, could legally obtain accurate copies of any British patent, copies which might then go abroad.

Only twice, between 1785 and 1824, did the Board of Trade recommend any statutory alteration of the prohibitions on machinery exports. In 1785–1786 it heeded the pleas of wool card manufacturers from Bocking and Colchester in Essex and Southwark, London, for permission to export their products to the United States. They argued that Dutch cards rivalled English ones, that hogsheads of Irish cards were going to America, and that Americans had John Kay's card–clothing engines at Boston, leaving only middle and southern American markets to Britain. They further maintained that the inferior cards that the British exported, with teeth loosely set in thin leather, were suitable merely for coarse wool, and that the prohibition on exports was causing unemployment among poor families in England's textile districts. Following the Board's recommendation, Parliament passed an amending act allowing the export of coarse wool cards to the United States.[44] The other occasion when the Board expressed willingness to change the law came in 1798 over the export of primary processing machinery, as noted above.

Though the Board made limited efforts to enforce the prohibitory laws down to about 1807, after that date it simply recorded its in-

43 For the operation of the early–nineteenth–century patent system, see G.B., *Parl. Papers (Commons)*, 1829 (332), III, "Reports from the Select Committee on the Law Relative to Patents for Invention." Censorship of technical publications and the role of the British patent system in the process of technology transfer will be considered in more detail in my forthcoming doctoral thesis.

44 B.T. 5/2, pp. 290, 296, 300–305. P.C. 2/130, pp. 278, 309–314.

ability to halt the emigration of artisans and the export of machinery. Finally, in 1820 the Privy Council suspended its 1795 wartime Order-in-Council, which was cited by the Board as late as 1818 and used by Customs until early in 1821.[45] It seems to have been this administrative impracticability in applying the laws that spurred the changes of the 1820s.

II. Liberalizing Amendments of 1824 and 1825

All restrictions on artisan emigration were lifted in 1824, and the following year the prohibitions on machinery exports were consolidated and placed under a licensing system that allowed some relaxations. These changes came in the period of economic liberalism associated with Lord Liverpool's enlightened Tory government of 1822–1827 and in the years of expanding trade, 1823–1825.

Initiatives for the political assault on the prohibitory laws originated with two prominent free traders, the Ricardian economist J. R. McCulloch, and the Radical M.P., Joseph Hume. In the *Edinburgh Review* early in 1824, McCulloch ended a long essay against the Combination Laws (which in the 1823 session of Parliament the Radicals were planning to repeal) with condemnation of the restraints on emigration: all the great jurists pronounced them unjust; in a prospering economy they were unnecessary; in declining trades they prevented the relief of unemployment; they were unenforceable; and they deterred only the "poor and ill-educated class of artificers" from flocking to the United States, rather than the most skilled artisans. Regarding machinery export restrictions, he contended that these too were unenforceable; that they undermined the principles behind Britain's industrial growth; and that they spurred the construction of foreign machine shops.[46]

Less than a month after McCulloch's article appeared, Hume moved the appointment of a House of Commons Select Committee to investigate artisan emigration, machinery exports, and the combination laws. William Huskisson, President of the Board of Trade, spoke in his support. Both M.P.s unreservedly urged the removal of all prohibitions on the emigration of workers. Additionally, Hume pointed out that Parliament itself had, in attempting to reduce unemployment and spiralling poor–law rates, promoted emigration

[45] B.T. 5/17, p. 277; 19, p. 64; 26, p. 419; 27, p. 337. P.C. 2/202, p. 3. English Customs Board, *General Orders or Letters*, vol. 2 (1814–1823), 991–992.

[46] *Edinburgh Review* 39 (1823–1824), no. 78 (January, 1824), 315–345. For McCulloch's authorship, see Graham Wallas, *The Life of Francis Place, 1771–1854* (4th ed., London, 1925), 208.

schemes to the Cape and Canada. On machinery export, both Hume and Huskisson were more cautious. The law was inoperable, however, and if smuggling was enabling foreign rivals to construct their own manufacturing equipment, the British machine-making industry ought to be impeded no longer. Even so, Hume felt that the industry needed the encouragement of free export because it was hobbled by a patent system that allowed the piracies of foreign rivals.[47]

These views may be compared with those of the machine makers, manufacturers, and Customs officials who gave evidence before Hume's Select Committee between February 17 and May 20, 1824. Over the question of artisan emigration, witnesses from the textile industry, in contrast to the solidly protectionist iron masters, were divided. Those opposed feared the loss of technical secrets, a shortage of skilled workers, or the development of foreign competition. In the opinion of Thomas Ashton (a spinner and manufacturer in the Manchester area), only "rambling mechanics; no steady men, but loose sort of characters," and inferior workmen would resort to emigration; while William Fairbairn, the Manchester millwright, thought that repeal might even bring some artisans back to Britain. These diverging attitudes enabled Hume's Select Committee to recommend repeal of the prohibition on artisan emigration, on the grounds that the law was both unjust and impracticable. Repeal came on June 21, 1824.[48]

A very different response emerged from the textile industry over the proposed repeal of the prohibitions on the export of machinery. Apart from the London machinists, the textile manufacturers and machine makers before the Select Committee all registered their strong opposition to free export, fearing loss of overseas markets in the long term and, more immediately, curbs on the current expansion of business activity. Such resistance prompted Hume's Committee to advise a further enquiry. It came in the following session of Parliament, when business activity was just beginning to slow down. The poor response to Hume's appeals for witnesses possibly reflected the deepening depression. Only the Birmingham Chamber of Commerce and the master cotton spinners of Renfrewshire sent in evidence. Consequently, Hume's Committee was forced to rely on a reappraisal of the evidence given in 1824 and on the testimony

[47] *Hansard* 2nd ser. 10, cols. 141–151 (February 12, 1824); Adam Smith, *An Inquiry into the Nature and Causes of the Wealth of Nations*, Edwin Cannan, ed. (New York, 1937), 121.

[48] G.B., *Parl. Papers (Commons)*, 1824 (51), V. Unless otherwise stated, this section is based on these six reports on the export of machinery, emigration of artisans and on trades unions; and on the 1825 report on the export of machinery, G.B., *Parl. Papers (Commons)*, 1825 (504). Musson, "'Manchester School'," p. 24. 5 Geo. 4, c. 97 (1824).

of four London engineers (John Martineau, Timothy Bramah, Henry Maudslay, and Alexander Galloway), together with an assessment of the pertinent statutes. When the Report on the Export of Machinery finally appeared on June 30, 1825, the Privy Council's continued discretionary control over the export of tools and machinery was recommended. Hume fought for his bill another eighteen months, but eventually gave way to the manufacturers' protectionist sentiments.[49]

As Joseph Hume argued in 1825, free emigration made nonsense of the restraints on machinery exports. How can we explain the incongruity arising from repealing the prohibitions on the live instruments of trade while preserving those covering the dead instruments? Obviously many more people were immediately affected by emigration restraints than by the machinery export prohibitions, and the former were more difficult to enforce. Public opinion had long regarded freedom to travel as one of the free born Englishman's rights, a feeling reinforced by Adam Smith, Thomas Paine, and the French Revolution.[50] All of these points help to account for the freeing of emigration in the liberal climate of 1824.

A change in the economic and political climate strengthened the protectionists' hand in 1825–1826. Depression led the Manchester machine makers to secede from their alliance with the manufacturers who used their machinery, since free machinery exports would allow them to accept £20,000 worth of foreign orders. But jobs for machine makers now meant unemployment for manufacturers, and politically the manufacturers held the whip hand: the government dared not risk alienating wealthy new industrialists or aggravating the threat to law and order posed by rising unemployment in their segment of the labor force.[51] Economic depression may account for the postponement of repeal of machinery export prohibitions, but it does not explain the manufacturers' long-term faith in the practicability of the machinery prohibitions despite, and concurrent with, free artisan emigration.

The incompatibility of the policies of free emigration on one hand and restriction of machinery exports on the other reflects differing assumptions about the locus of technology. Protectionists based their views on Manchester machine-making practices. Here the division of labor was reaching levels unknown before, particularly in the

49 *Hansard* 2nd ser. 15, cols. 908–911 (May 5, 1826), 1118–1122 (May 11, 1826); 16, cols. 291-298 (December 6, 1826).

50 *Hansard* 2nd ser. 12, col. 652 (February 24, 1825); Smith, *Wealth of Nations*, 625; E. P. Thompson, *The Making of the English Working Class* (London, 1963), Chap. 4.

51 See sources in note 49.

handicraft arts of turning and filing. Consequently, specialization was diffusing traditional skills over a multitude of sub-trades, at least six of which, for example, were required to make a spinning frame. Now the international movement of machine–building initially required the emigration of teams of skilled artisans, in place of individual craftsmen. Since workers emigrated singly or in family groups, free artisan emigration seemed unlikely to represent a leakage of technology.

Hume and the free traders, drawing upon London's machine–building experience, arrived at the opposite conclusion. In the first two decades of the nineteenth century Henry Maudslay and his pupils had wrought a machine-making revolution in the London area. Their metal-cutting tools – screw cutters, planes, and lathes – permitted a precision in metal working that could hardly be matched by the most skilled hand craftsman. The new machines dispensed with the need for highly-developed skills and also facilitated accurate copying of machinery from models, specifications, or drawings. As Alexander Galloway testified in 1824, metal-framed cotton machines were as well built in Paris as in Manchester; the only difference was their price.[52]

Some Manchester machine makers, like Sharp, Roberts & Co., were operating the new machine tools in the 1820s, but northern manufacturers seem to have been unaware of this, assuming that the locus of technology still resided in teams of specialized craftsmen. In fact it had shifted from men to machines, and the manufacturers' imposition of regulated protection between 1825 and 1843, based upon the Board of Trade's licensing system, became a vain effort to shut the stable door after the horse had bolted.

III. Period of Discretionary Limitation of Machinery Exports, 1825–1843

In February 1825, four months before the Select Committee published its Report on the Export of Machinery, Board President William Huskisson admitted to the Commons that he had assumed discretionary powers and was allowing some prohibited machinery, such as hydraulic presses, to be exported. After a Commons debate on the subject in mid–June, it was agreed that the Board of Trade

[52] Besides the evidence in the 1824 (p. 17 for Galloway's comment) and 1825 reports, see L. T. C. Rolt, *Tools for the Job. A Short History of Machine Tools* (London, 1965), 83–121, for a general survey of the machine tool revolution. A. E. Musson and Eric Robinson, *Science and Technology in the Industrial Revolution* (Manchester, 1969), 459–509, deal with Manchester developments. Also Thompson, *Making of the English Working Class*, 245–246.

should exercise this kind of control. No statutory power was ever conferred on the Board; instead, the Customs Regulation Act of 1825 consolidated and continued all the earlier prohibitions on the export of machinery. They were renewed in 1833 and finally repealed in 1843. During this period the Board advised the Treasury on applications for machinery export licenses "under the general power of the Lord High Treasurer," but Huskisson apparently never received an increased staff to cope with the flood of applications that he expected.[53]

Under the new system an exporter petitioned the Board of Trade, disclosing the nature and value of his machinery, its port of shipment, and destination. If the Board believed that the machinery came under the prohibitions but might be exported, it wrote to the Treasury recommending a license. The Treasury then informed Customs at the port of exit, and the exporter learned of the license by contacting Customs there. The license, which was valid for twelve months, cost two guineas. If the Board did not decide to permit export, it simply notified the exporter and the case was closed.

Short of staff, the Board developed several tactics in doubtful cases to transfer the burdens of investigation and even of decision to other parties. It placed the burden of proof upon the applicant that his machinery came under the 1825 Act, and the onus of detecting inaccurate machinery descriptions upon Customs officials, who also advised the Board on the mechanical aspects of identification, relying on technical textbooks or the opinions of professional engineers. On rare occasions the Board noted in its Machinery Books, "left to Treasury to decide."

On the wharf, Customs men inspected machinery to verify that it corresponded with its license description. They made a record of the shipment in the Customs House books and cleared the equipment or, in doubtful cases, detained it. In clearly fraudulent cases, officials seized the shipment and placed it in the King's warehouse. If the shipper was convicted in the Court of the Exchequer (Common Pleas), his impounded machinery was auctioned off and the proceeds paid to the Crown and the Custom officers involved. No other penalty was exacted.

Occasionally the Treasury issued export licenses without apparent

[53] *Hansard* 2nd ser. 12, col. 652 (February 24, 1825); 13, cols. 1136–1138 (June 14, 1825). G.B., *Parl. Papers (Commons),* 1841 (201), VII, "First Report from Select Committee Appointed to Inquire into the Operation of the Existing Laws Affecting the Exportation of Machinery," 8. Unless otherwise stated, the material in the remainder of this paper derives from this report and from its companion, the Second Report on the Exportation of Machinery, *Parl. Papers (Commons),* 1841 (400), VII. Precise citations from these major sources are provided only for verbatim quotations.

reference to the Board of Trade. Between 1825 and 1839, the Board in every year except 1826 dealt with more applications than the Treasury, but in 1826 the Treasury handled sixteen applications untouched by the Board and in 1836, two. Nevertheless, the evidence suggests that the Board of Trade was the key member of the administrative triumvirate (the Treasury, the Board of Trade, and the Customs Board) responsible for running the licensing system.

The Board of Trade's Licensing Policy

Under the licensing system there were three broad categories of machinery: machines not specified in the 1825 Customs Act, and therefore freely exportable; machines listed in the Act but exportable under a Treasury license; and machines of which export was prohibited by the Act and by the Board's licensing principles. In a system that evolved slowly, these categories were not always clear-cut.

At the beginning of the licensing system, Huskisson tried to avoid charges of prejudice by laying down a principle to guide the Board in its licensing decisions. Addressing himself to a petition from the Manchester Chamber of Commerce, Huskisson explained during a debate at the end of 1826 on one of Hume's motions to repeal the prohibitory laws that "machinery of great bulk and much raw material" was exportable, but machinery of modern construction, depending on "the ingenuity and excellence of the mechanism, and where the raw material was trifling," was prohibited.[54] John Kennedy, a leading Manchester cotton spinner and machine maker, had first formulated this principle in his pamphlet *On the Exportation of Machinery* (dated April 6, 1824). It was hardly meaningful. According to A. E. Musson, Kennedy referred to steam engines and millwork when he wrote about heavy machinery.[55] But these had been exported for decades, as shown above. If, on the other hand, the principle applied to processing–machinery, it could only be confusing and naive. Spinning frames, for example, were both heavy and ingenious in design.

A different principle was formulated and relayed to Customs collectors by a General Order dated November 14, 1838. As Deacon Hume explained in 1841, licenses were withheld for all textile machinery used in the first, and all subsequent, processing stages in which the mass of raw material was divided up ready for spinning.

54 *Hansard* 2nd ser. 16, col. 293 (December 6, 1826).
55 Musson, "'Manchester School'," 25–26.

The only exceptions were made for foreign heads of state.[56] But this principle was not borne out by the Board of Trade's licensing decisions, for textile machinery was not the only type denied export licenses in years 1825–1835. Of twelve non-textile applications denied licenses in this decade, six were for iron rolling equipment, three were for machine tools (turning lathes and teeth cutting engines), and one for wire weaving reeds. However the late 1820s saw iron rolling and metal-working equipment and machine tools passed more and more often.[57]

As for textile machinery, the first ten years of the licensing system were marked by numerous violations of the 1838 rule. Spinning wheels went to Greece in 1827; more spinning apparatus to missionaries in New Zealand in 1829; cotton spinning machinery to missionaries in Madagascar in 1825; flax spinning machinery to Guernsey in 1825 and to Hamburg in 1831. The American ambassador, Albert Gallatin, obtained permission in 1827 to send a silk tramming machine to the United States. Mehemet Ali continued to gain as Egypt received over 500 power looms from Alexander Galloway in 1826 and 1828, and four drawing frames, 320 slubbing spindles, 580 fly frame spindles and 4800 mule spindles in 1835. Some 200 power looms went to Calcutta in 1828, and handlooms the next year, to be followed by power loom equipment in 1833. Three jacquard card cutters, though no jacquard looms, were exported, as were weavers' mails in 1838 and, earlier, shuttles and pickers in 1830. Marc Isambard Brunel was allowed to export his knitting machine (patented in 1816) to France in 1828, and the following year two stocking frames went to Upper Canada. Brunel's "tricoteur" was released on the grounds that it was a French invention that had not proved popular in England.[58] All of this suggests that the Board spent over ten years working towards the seemingly clear-cut criterion enunciated in 1838. A close study of the Board's Minutes for the first decade of the licensing system reveals that a rather more flexible and larger principle guided decision-making about machinery export licensing.

Hume and Huskisson recognized this principle in the Commons in 1825, and Huskisson took the opportunity to reiterate it early in 1826. He explained to a M. Andelle, who pressed for permission to export fifty cases of cotton spinning machinery to France, that the Board of Trade's "discretion was to be limited to such articles of machinery as might *in the opinion of the manufacturers* of this country, be per-

[56] Customs Board, *General Orders,* 1838 vol., for October 24, 1838. G.B., *Parl. Papers (Commons),* 1841 (201), VII, 4.

[57] B.T. 6/151, *passim.*

[58] *Ibid.* B.T. 5/34, pp. 48, 417. B.T. 1/254, no. 33.

mitted *to be exported without injury to their interests.*"[59] On other occasions, too, refusals were made on the grounds of "the sentiments of the manufacturers."[60]

Manufacturing opinion came mostly from petitions sent spontaneously to the Board. In one case the Board did solicit opinion, asking the Customs, through their officers, to "ascertain among the woollen trade" whether the export of shearing machines would be "injurious to that branch of trade." Within a month the Leeds merchants and woolen manufacturers sent a memorial to the Board objecting to the export of cropping and finishing machinery, and John Jones of Leeds was denied a license.[61]

A Coherent Licensing Policy

Thus, there is some consistency in the apparently conflicting permissions and denials of the first ten years of the system. Missionary societies operating in remote regions hardly posed a threat to Britain's manufactures. Likewise a solitary emigrant worker taking a few tools with him no longer represented a critical leakage of technology. Thus, A. McKnight of Tarbolton was allowed to take two stocking looms to Matilda township, Upper Canada, in 1829, and Benjamin Levi took engraving tools with him to the United States in 1835. In each case it was emphasized that the tools were necessary for their owner to carry on his trade. On the other hand, in the opinion of Nottingham manufacturers the single lace machinery maker represented a major leakage of technology, and the Board did all in its power to keep his tools and machinery from leaving the country.

The practice of making gifts of machinery to heads of state was followed only in accordance with the interests of British manufacturers. Egypt, like India, lacked the complex of factors conducive to economic growth found chiefly at this time in Europe and the United States and represented no threat to British manufactures. Nevertheless there were limits. The Turkish ambassador in 1836 was confined to choosing one model of each kind of absolutely prohibited machine up to a total of ten models. But if machinery from a major branch of the textile trade was intended for a potential manufacturing rival, the chances were that no amount of diplomatic intervention could pry a license from the Board. Both the Russian ambassador in 1827

59 B.T. 5/35, p. 155. Italics added.
60 B.T. 5/36, pp. 78–80.
61 B.T. 5/41, pp. 86, 109.

and the Swedish chargé d'affaires in 1830 failed to gain export licenses for cotton machinery.[62]

Not only the future destination and ownership served to indicate whether export would injure British manufacturing interests, but also the type of textile machinery concerned. Between 1825 and 1841 three classes of textile equipment seem to have had no difficulty in gaining export licenses. The first was obsolete equipment. Pre-industrial implements like spinning wheels or wool combs and their broaches were regularly exported, as were textile printing blocks, and also some old if not obsolete industrial machinery. In 1833 "some old defective machinery which had originally been applied to the spinning of worsted by hand" was passed for export to Rotterdam.[63]

Most equipment for preparing fibers for manufacture, such as cotton picking, baling, willowing, batting, scutching, blowing, and spreading machinery was allowed to leave the country, but not carding, drawing, roving, or spinning equipment. In the woolen industry, balers and pickers or devils and also rag or shoddy pickers (developed in the 1830s) and carding engines that did not divide up the woolen lap (i.e., that lacked Goulding's condenser or a piecing attachment) went out, but not billies, jennies, or mules or combing machinery. In the silk industry, preparatory equipment meant machines for reeling or winding the raw material and preparing waste silk, but not for throwing or twisting it. In the flax and hemp industries, preparatory machinery was understood to include breaking and heckling devices. But in the case of tow or coarse bast-fibre processing, for a number of years spinning machinery was allowed abroad, presumably on the grounds that Britain's fine flax spinners would not be threatened. Weaving and knitting machinery was rarely licensed, and lace-making equipment, never. Accessory textile equipment such as card grinders or card–clothing machines were not licensed for export, though card clothing frequently was. Weaving accessories were also sometimes allowed.[64]

The pre-1825 protectionist principle of allowing the export of machinery for initial processing in order to expedite the major manufacturing stages carried on in British mills was at the base of these policies. But two problems developed with this process-based criterion. New preparation machines were developed that revolutionized some branches of textile manufacturing in the 1830s. The rag picker, (on which the shoddy trade was based), waste silk machinery, and

[62] B.T. 5/38, p. 179; 39, pp. 319–320; 42, pp. 189, 357; 43, p. 440.
[63] B.T. 5/41, p. 489.
[64] B.T. 6/151, *passim*.

cotton spreaders, for example, tended to reclaim previously lost raw materials and, by increasing productivity and raising yarn quality – notably in the cotton industry – threatened British manufacturers in overseas markets. Tow preparing and spinning machinery, moreover, was so similar to that used in fine flax spinning that it could easily be adapted for this purpose; after 1835, therefore, its export was consistently prohibited.[65]

In the third group of textile machine types generally licensed for export were various finishing machines. Engraved copper rollers, calico printing machines, and calenders for cotton and linen cloth; cloth washing, drying and pressing (but not fulling, raising, or shearing) equipment for woolen cloth working; and fabric printing pieces in the silk industry were most frequently released. One curious discrimination was made in the case of calico printing machines. Their export to the United States was prohibited for a while, probably because New England cotton manufacturers were turning to cotton printing in the second half of the 1820s, but by 1830 the Board recognized that printing machines were licensed for Continental ports and then re-exported to America.[66]

Over two-thirds of the 2098 applications for licenses to export machinery between 1825 and 1843 involved textile machinery, a proportion that rose from 50 per cent in the 1820s to over 70 per cent after 1838. In the vast majority of cases the recorded destinations were European ports. About 10 per cent of the applications were denied licenses, and over 90 per cent of these refusals covered textile machinery.*

THE OUTFLOW OF MACHINE TOOLS

Since the Board of Trade developed its licensing policy primarily as a reaction to pressure from British manufacturers, the absence of such pressure from machine tool makers must have been responsible for inattention to the outflow of such items. Export of rollers, slitters, presses having screws over 1½ inches in diameter, cutting-out presses, and lathes for "plain, round and engine turning" was prohibited under the 1825 and 1833 acts. Yet between 1830 and 1840 over 1300 metal working rollers, for iron, copper, tinplate, and zinc; 41 lathes, including planing, boring, and screw-cutting lathes; and 20 cutting engines including gear cutters and punching presses, were licensed

65 B.T. 5/43, pp. 103–104, 151.
66 B.T. 5/39, p. 272.
* The author has detailed data on destinations, fibers, and processing stages for the machinery involved in these applications, which he will be glad to supply on request.

for export. Inasmuch as these tools would have incorporated screws larger than 1½ inches in diameter, the total number of machine tools exported must have been much larger. The swiftness with which Americans became self-sufficient in machine tools is reflected in destination data, which reveal that most of this British technology was leaking to continental European rivals.[67]

It was not merely the failure of manufacturers to understand the central role of machine tools in the new technology, however, that accounts for the approval of increasing numbers of applications to export metalworking equipment and machine tools. There is evidence that the Board, a haven for free trade intellectuals in the 1830s, biased machinery export licensing decisions towards free trade in these years. Petitions from the Birmingham Chamber of Commerce in 1827 and the Manchester Chamber of Commerce in 1828 and 1834 to strengthen the prohibitory laws met with the Board's refusal or silence. And in 1827 the Board recorded that it did "not think it would be advisable to introduce a system of more general prohibition."[68] The Board's free trade influence was also revealed in its relations with the Treasury. In the case of tow–spinning machinery Treasury overruled the Board's recommendation of free export. However, when Goole Customs officers confiscated bobbins of yarns exported by Manchester merchants and Treasury sided with Customs, the Board brusquely informed the Treasury that "there is no ground whatever for classing those little wooden rollers with machinery in the cotton manufacture."[69]

Ineffectiveness of Government

The attempt to safeguard Britain's textile technology from foreign rivals was defied in many ways. The licensing system itself, for example, might be used to cover illegal exports. Sometimes would-be exporters applied for "precautionary" licenses, in case their machinery should after all be covered by the law. At first, makers had attempted to export illegally by means of misrepresentation in applications, but after 1835 the Board considered applications defective if their specifications inadequately described the machinery they purported to cover.[70]

A piece of machinery might be broken up and exported in parts in the hope that Customs would be unable to identify it. Of the 51

67 B.T. 6/151, *passim.*

68 B.T. 5/36, pp. 133–135; 37, p. 262; 41, p. 554. These petitions from Birmingham and Manchester were evidently inspired by Manchester Chamber of Commerce. See Musson, "'Manchester School'," 36–37.

69 B.T. 5/42, p. 306 to B.T. 5/43, p. 65 *passim.*

70 B.T. 5/41, pp. 320, 493, 498, 518; 42, p. 383.

seizures made by Customs at Liverpool 1830–1839, indeed, some nine were of unidentified machinery parts, but at London this proportion was halved to nine out of the 121 seizures.[71] Customs evidently followed the policy of detaining unidentifiable parts and placing upon their owners the burden of establishing their true nature. Related to this problem of machinery identification was the ambiguity implied in the use of discretionary powers to waive or enforce the statutory prohibition, a weakness to which Deacon Hume drew attention in 1841.

Smuggling, or totally by-passing the licensing system, was facilitated by the coming of steamboat service across the English Channel. Operating by published timetables, the cross-Channel steamers began loading at a fixed time and cleared two hours later. The smuggler's ruse was to bring cases of machinery down to the dock at the last moment in hopes of avoiding Customs inspection. Or small items of machinery might be taken aboard by passengers in their baggage, which Customs men never examined. Smugglers could send machinery to the Continent by means of the coastal trade, depending upon the inadequacy of the fleet of Revenue cruisers, of which there were in 1839 only 49, plus 21 tenders, manned by 6183 sailors, to monitor the 7000-mile coastline of England, Scotland, and Ireland.[72]

Whatever the success of efforts to keep machinery at home, drawings of them continued to offer a major loophole in the licensing system. When the Nottingham bobbin net manufacturers petitioned for tougher enforcement of the law, the Board observed that "the transmission of drawings cannot be prevented." Eight years later, in 1841, even verbal descriptions were reckoned sufficient for a machine maker to copy a design. The artisan, meanwhile, continued to threaten the retention of technology in two ways. Where technology was centered in the artisan, as in the Nottingham lace trade, manufacturers still pressed the Board to restrain workers from smuggling parts and plans abroad. The artisan or mechanic might also, as agent for a foreign firm or government, engage directly in industrial espionage: in 1826 the Polish government had an agent in Britain pursuing the clandestine export of plans and machinery, while foreign apprentices were often deliberately sent to serve part of their time in British machine shops.[73]

[71] Customs House Archives, Customs 37/56, 'Seizures, 1825–1856," ff. 131–152. I wish to thank the staff of H.M. Customs Library and Archives for facilitating my work with their records, which were inspected by kind permission of H.M. Commissioners of Customs & Excise.

[72] See A.J.D., "Historical Notes on the Coastguard Service" (London, H.M.S.O., 1907), 23–24.

[73] B.T. 5/35, p. 282; 41, pp. 82, 417; 42, p. 273.

II

What might be done to seal these leakages? Before 1841 all that the Board did was to tighten its application procedures, to urge Customs to greater vigilance, and to contemplate more intensive Customs inspection of machinery presented for export clearance. The Manchester Chamber of Commerce assisted Customs, apparently at Liverpool, London, and Hull, in the late 1820s in identifying and arresting illegal exports of machinery.[74] That Customs kept very busy trying to prevent illegal exports is clear from seizure and detentions data, compiled by Customs. The Second Report of the Select Committee on the Export of Machinery in 1841 contains Customs data on detentions and seizures for the period January 1, 1824 through May 15, 1840. A Customs volume of "Returns Relating to Seizures" covers only seizures for the period 1830–1839 inclusive, but, unlike the Committee Report, it identifies types of machinery.[75]

By its very nature, however, the total amount of machinery smuggled abroad is unknown, so we have no means of precisely estimating the efficiency of British Customs houses. From the 1830–1839 return, the London officers were most busy, making 121 seizures in the period while Hull and Liverpool made less than half this number, 59 and 51 respectively. Then came Dover with 23 seizures, and Goole with 12. No other Customs house managed more than three seizures over the decade. The published return of 1841 likewise showed London most active with 84 detentions and 93 seizures. Hull followed with 13 detentions and 76 seizures; then Liverpool with 9 detentions and 70 seizures; Dover with 35 seizures; and Goole with 7 detentions and 5 seizures. The distinction between detentions and seizures was not clarified in the two returns, but presumably the 1830–1839 report included detentions because, on this basis, the figures in the two returns would seem to agree. The seizures in the 1830–1839 return, moreover, included numerous cases of mill work, machine tools, and large rollers being returned to owners.

These returns also suggest that the risk of losing machinery through confiscation was no real deterrent. Both the evidence presented to the Select Committee in 1841 and the 1830–1839 return of seizures give the strong impression that the illegal export of machinery was boldly conducted at major ports, whose well-built wharves could accommodate heavy machinery, and not on remote moonlit beaches, the romanticized setting for smuggling. Unscrupulous exporters appear to have had a low regard for the efficiency of the Customs,

74 Musson, "'Manchester School'," 36.
75 Second Report, 1841, p. 107 (see note 53); for Customs seizures see note 71.

which was shared by the companies that offered them insurance against detection!

IV. The Ending of the Prohibitory Laws

In the early 1840s pressures for repeal mounted. When the severity of the 1825–1826 depression was repeated in the textile industry in 1841–1842, the cotton manufacturers no longer presented a united front.[76] Manufacturers, who supported the Anti-Corn Law League in the interest of lower food prices, moved logically towards a concomitant policy of free machinery exports; and free traders rapidly advanced their cause in the late 1830s in Lancashire, taking over the Directorate of the Manchester Chamber of Commerce early in 1839.[77] By the spring of 1841, the manufacturers were at the end of their tether. In the opinion of a protectionist Manchester cotton spinner, Holland Hoole, "the general expression of feeling is, that the legislature may do whatever they please, they cannot make things worse."[78] The engineering industry in Manchester, moreover, had expanded so far that machine makers rivalled manufacturers as a major industrial pressure group. Furthermore, there is evidence that provincial men were becoming increasingly involved in the textile machinery export trade in the 1830s and 1840s. A quadrennial sampling of applications for textile machinery export licenses, 1826–1842, shows that the percentage of London applicants fell from 78 per cent in 1826 to 33 per cent in 1842.[79]

Mark Philips, one of the two Manchester M.P.s, successfully moved, on February 16, 1841, for a Commons Select Committee to investigate the export of machinery.[80] In a masterly synthesis of the evidence, the Committee's Report summarized the objections to the mixed system of law and license, and surveyed the major changes that had occurred since 1825. Among the former were the difficulty in applying the 1838 principle, the insufficient expertise of Customs officers in both law and mechanical engineering, confusion and uncertainty arising from the conflict between the statutory prohibitions and the Board of Trade's licensing principles, and the sheer impossibility of preventing smuggling.

76 R. C. O. Matthews, *A Study in Trade Cycle History: Economic Fluctuations in Great Britain, 1833–1842* (Cambridge, 1954), 127–151.

77 Norman McCord, *The Anti-Corn Law League, 1838–1846* (2nd ed. London, 1968), 39. See also Musson, "'Manchester School'," for diminution of the activities of Manchester Chamber of Commerce in the 1830s.

78 Second Report, 1841, p. 52 (see note 53).

79 Musson, "'Manchester School'," p. 41 and Table 1. Author's data based on London directories.

80 *Hansard* 3rd ser. 56, cols. 670–692 (February 16, 1841).

II

A series of developments had produced a new situation since 1825: the free emigration of artisans; the almost free export of coal; the revolution in machine tools, also freely exportable; important improvements in flax, waste silk, and cotton preparatory machinery; the rise of machine-making industries on the Continent and the United States; the easy transmission of new inventions through models, drawings, letters, and oral descriptions; active efforts to acquire British technology on the part of foreign governments; and the discriminatory effects of the domestic patent system against British inventors. To these considerations, which substantiated the futility and injustice of protection, were added familiar free trade arguments.

The protectionists could only urge the repeal of the Corn Laws and reductions in foreign duties before, rather than after, machinery export was freed. To maintain efficient protection for machinery exports, they suggested the registration of machine makers, stamping makers' names on machinery, and the funnelling of machinery exports through five or six ports for inspection by competent Customs officers. Meanwhile, manufacturers and machine makers alike faced up to the irritating revelation by the Select Committee's First Report, that more types of machinery than were generally assumed lay outside the export prohibition. Evidently export agents and smugglers had exploited the ignorance generated by the licensing system to rack up large profits, mostly at the expense of foreign purchasers.[81]

In July 1841 the Board allowed free export of spinning machinery to the colonies, if bonded against re-export, and in August 1842 it made a more sweeping relaxation. Most types of cotton and woolen machinery were released to foreign markets through the three ports of London, Liverpool, and Hull where specially competent officers examined the machinery, as Sharp, Roberts & Co., leading Manchester machine makers, had urged in 1827.[82] Though the Select Committee voted eight to one in favor of repealing the prohibitory laws, Parliament tarried for another two years. Gladstone, as President of the Board of Trade, at last took up the question of repeal in 1843. The delay, he claimed, stemmed from the new Tory government's initial preoccupation with reform of the tariff schedule chiefly in 1841–1842.[83] In the summer of 1843 he acted swiftly. His Machinery Exportation bill was reported on July 10 and by August 22 was on

81 See note 53.
82 B.T. 6/151 and 152 *passim;* B.T. 5/36, p. 246.
83 *Hansard* 3rd ser. 70, col. 830 (July 10, 1843).

the statute book. After nearly one hundred and fifty years, all types of British machinery were at last freely exportable.[84]

V. Conclusion

The most important fact about laws to prevent export of technology is that, despite spasmodic yet determined efforts, they were administratively impracticable. In addition to communication difficulties between the half dozen government departments involved, intractable problems at the ports made it difficult to filter emigrants and exports. Discriminating between skilled and unskilled artisans raised practical questions about definition, detection, and deterrence, while hitting upon positive measures to reverse emigration proved impossible. In wartime, the government tried simply to discourage all emigration, but even wartime pressures could not induce it to override the public's faith in the rights of the freeborn Englishman and its fear of the Census apparatus that a passport system would have required.[85] The tide of emigration continued to rise, and it could only be a matter of time before law followed practice. Free emigration was enacted soon after the post-war repression was replaced by a more liberal political and economic climate.

Illegal machinery exports were also very difficult to contain. Not only were machines readily transmogrified into parts, models, plans, specifications, patents, and mental images, but also some machines were legally released because their export was thought beneficial to British manufactures, as with primary processing machinery, or because they apparently held little threat for British manufacturers, as with steam engines, mint machinery, and obsolete equipment. Such relaxations and those dictated by diplomacy helped to open the door to free export. That this movement was aborted for a decade and a half in 1824–1825, when free emigration began, resulted from the manufacturers' failure to perceive the shift in the locus of technology from men to machines.

Administratively the licensing system of 1825–1843 was likewise difficult to implement. The success of machinery smugglers can be surmised from the ease with which foreign rivals acquired British models, the availability of insurance cover for illegal exports, and the infrequency of Customs machinery seizures, averaging less than thirty a year in the 1830s. The principles guiding the award of export

[84] *Hansard* 3rd ser. 71, cols., 493–517 (August 10, 1843), cols. 545–547 (August 11, 1843); 6 & 7 Victoria, c. 84.

[85] See David Glass, *Numbering the People: the Eighteenth–Century Population Controversy and the Development of Census and Vital Statistics in Britain* (Farnborough, Hants, 1973), 19–20. I am obliged to Dr. Charlotte Erickson for this reference.

licenses were never closely and satisfactorily defined, while the avowed principle – the interests of the manufacturers – was too vague to be consistently applied to complex and changing industrial technologies, especially those of machine building and textile manufacturing.

While quantification is impossible, it is likely that laws prohibiting the diffusion of technology did have some effect. They surely discouraged some potential emigrants from setting out on the international tramp; they certainly kept many emigrants from returning home. But despite the laws, thousands of artisans did go abroad during the French Wars, taking their machine building, operating, and managing skills with them. The licensing system likewise reduced the volume of machinery going abroad, but by 1841 no knowledgeable contemporaries believed that new technology would long remain safe from the hands of enterprising foreigners, especially as drawings and machine tools made copying extremely easy. British manufacturers unwittingly made their rivals' depredations easier still. Not comprehending the shift in the locus of technology, they persuaded the Board of Trade to shape the whole licensing system into an engine directed against textile manufacturing equipment: processing machines were halted, but not the machines that made processing machines.

The dominance of textile machinery in applications for export licenses confirms the familiar impression of Britain's early lead in textile manufacturing. The official demand pattern, however, is not readily quantifiable, because machinery values are irregularly, and technical data never, recorded in the Board of Trade papers. The only firm conclusion to be drawn is that northern Europe showed the greatest interest in Britain's textile machinery, especially in 1834 and 1838; expectedly, the Board's licensing policies noticeably weighted official demand toward preparatory machinery.

Britain prohibitory laws thus failed signally to stem the flood of technological information spreading abroad, either via men or machines, in this early industrial period. Administering and policing the sort of protection envisaged by the laws required Draconian measures that public opinion would not tolerate and internal economic and social conditions could not support. After 1825 British manufacturers made an inadequate system of regulated protection more ineffectual still by biasing it away from machine tools and towards manufacturing machinery. In these circumstances the flood could hardly be dammed.

III

British and American Yarn Count Systems: An Historical Analysis

Mr. Jeremy traces since the Middle Ages the interaction of technology and business organization in shaping the measuring systems for yarn in the British and later the American textile industries. Despite the attendant confusion and difficulties, a remarkable motley of such systems arose and has persisted.

A yarn count or numbering system is an elementary formula expressing the fineness of yarn. Most frequently it represents the number of length units contained in one weight unit; alternatively, it indicates the number of weight units per length unit. The former is known as the inverse method of numbering because the count increases as the thickness of the yarn decreases. The latter is described as the direct method because the opposite occurs. More conveniently, they are identified as the fixed weight and fixed length systems respectively. By the early nineteenth century a wide variety of length units and a less extensive collection of weight units had appeared in Europe. Some of the multitude of resultant numbering systems (which differed with country, region, and fiber), especially the British ones, passed to America with the emigration of artisans.[1]

Knowledge of these systems has been, and still is, essential to spinners, weavers, mill managers, and merchants both before and after industrialization. The more uniform the yarn diameter, the more even the consequent fabric. Irregular yarn thickness (size) could be concealed in a heavily fulled and napped cloth but not at all in linen or in most cottons and worsteds in which warp (chain) and weft (filling) were exposed after finishing. Furthermore, a weaver neglecting to order a fairly accurate length and size of yarn for chain and another for filling would have met costing difficulties on one hand and the inconveniences of needless shortages or ex-

Business History Review, Vol. XLV, No. 3 (Autumn, 1971). Copyright © The President and Fellows of Harvard College.

[1] For the opportunity to prepare the groundwork for this article, I am very grateful to the Smithsonian Institution which awarded me a Visiting Research Associateship at its National Museum of History and Technology (1969-1970) where the Division of Manufacturing, headed by Dr. Philip W. Bishop, generously sponsored my work.

cesses of yarn on the other, plus an irregular cloth. Such impediments may have been tolerable in household clothmaking but not in production for a market.[2] Likewise, when a spinner began selling her yarn, it was necessary to ascertain the fineness of her skeins, partly to meet the weavers' demands and partly to ensure that she was not underpaid. With the production of finer cloth, demands for more consistent spinning increased: having established the yarn fineness requisite for his chain, the weaver had to calculate precisely the total amount of yarn needed to dress his loom with a warp of given length using a reed (which regulated the spacing of warp yarns across the loom) of given sett (or fineness) over a given breadth of the loom, and harnesses matching the reed. The amount of filling would vary with the weave chosen.

Spinning of completely consistent yarns came with mechanization during the Industrial Revolution. As yarn size was thereafter determined by carding machines and, with cotton, by roving frames as well as spinning frames, operators at these earlier stages of processing necessarily gauged the size of their card ends and rovings by the spinner's yarn numbering system, or some variation of it.

All this is of great interest to the technological historian concerned with measuring mechanical performance or understanding early manufacturing techniques. But it is also very important for the business historian. He can safely cite ranges of counts produced by one particular firm for long-term internal assessments of economic activity. When several firms or areas are involved however, he could, if unaware of local or fiber differences in yarn counts, develop gross errors in making production comparisons either regionally or internationally. The transmission of count systems also raises provoking historical problems in the transfer of knowledge, nomenclature, etymology, and mensuration, as well as textile technology.

For the non-specialist business historian, the history of yarn numbering systems has a surprising application: it supplies some confirmation for a theory of the history of organizations most recently propounded by sociologist Arthur L. Stinchcombe.[3] He and others

[2] David Ramsey, *The Weaver and Housewife's Pocket-Book; Containing Rules for the Right Making of Linen Cloth* (Edinburgh, 1750), 8: "all along thro' this Book, it is supposed the Yarn is just reeled, and right telled; for in an exact Calculation of this kind, there is no allowance for Deficiency, either of bad Tell or short Reels."

[3] Arthur L. Stinchcombe. "Social Structure and Organizations," in James G. March (ed.), *Handbook of Organizations* (Chicago, 1965), 142–193. I wish to thank the editors of the *Business History Review* for drawing my attention to this article and to the wider implications of the history of numbering systems; Glenn Porter has been especially helpful in persuading me to emphasize organizational as much as technological aspects and in providing constructive criticism of the shape of this essay.

have observed that in the past organizations have been founded in spurts, that those formed at one time have a typically different social structure from those formed at another time, that new bursts of formation indicate organizational restructurings, and that certain structural characteristics of an organizational type are remarkably stable over time. A yarn numbering system, while not an organization, is certainly a technological and economic instrument indispensable to one organization, the textile industry. Further, these systems constitute a set of commercial and technological practices which may be thought of as an economic institution. Such systems appeared before the end of the Middle Ages with the rise of capitalism and were a means of achieving control, control over spinners as well as quantity and quality control over the yarn. What is perhaps most interesting of all is the "power of persistence" of yarn count systems. Of fifteen or so still in use, nine were established during or before the late seventeenth century. Others died out and yet more arose in the nineteenth century, indicative of the organizational restructuring of the textile industry then in progress. Though Stinchcombe has presented arguments explaining, in abstract terms, the above phenomena in the history of organizations, I shall try only to isolate specific economic and technological reasons for the rise and the persistence or disappearance of yarn numbering systems.

Technological and Commercial Origins

A glance at the appended tables (of forty British and American count methods) will show that the most puzzling aspect of yarn numbering systems, and the root cause of their mystifying variety, is the diversity of their linear units. Theoretically, this diversity could be almost infinite; in practice it has been limited by technological and, in England, legal factors.

Because linear yarn measurement was performed on the reel, reel size became the fundamental determinant of these numbering length units. The earliest reel, used domestically from Classical times through the twentieth century, was the conveniently mobile cross or hand reel, often called the niddy-noddy (see illustration 1).[4]

[4] For further illustrations and discussions of the niddy-noddy see A. Schwarz, "The Reel," *Ciba Review*, LIX (August, 1947), 2130–33; Charles H. Livingston, *Skein Winding Reels: Studies in Word History and Etymology* (Ann Arbor, 1957), 1–14; Laurence A. Johnson, "The Niddy-Noddy," *The Chronicle of the Early American Industries Association, Inc.*, XI (March, 1958), 4–7; Marion L. Channing, "From Spindle to Loom," *Spinning Wheel*, XXIII (December, 1967), 24–25, 50.

As an anonymous writer observed in 1733, the use of hand reels is "a Method of making up Yarn so uncertain and precarious, that no Persons who use them, can possibly be

Yarn slippage made it much less accurate than the faster rotary reel (see illustration 2) which appeared in Europe in the twelfth or thirteenth centuries.[5]

Reel size was not so arbitrarily devised as one might at first suppose. First, the range of sizes was probably limited by the dimensions of the human anatomy. In handicraft production, the skein of yarn formed on the reel was removed for further processing (washing, bleaching, dyeing, etc.) and then unwound and fed onto bobbins or spools for warping the chain or onto quills for the shuttle. Because, in a labor-intensive situation (like the early medieval subsistence economy), skein unwinding was most easily accomplished by two persons, the widest diameter of the elongated skein could not be greater than the distance between the palms of the outstretched hands of the person holding the skein. This seems to be confirmed by the fact that no extant records or artifacts I have seen include a reel with a circumference of more than three yards.

The diversity of reel sizes (of any circumference up to about three yards) is attributable to regional custom: the size that "hath been there usual of ancient time," as a seventeenth century English law declared.[6] But in medieval England these sizes did have a rational basis, namely the legal unit of length, the 36 inch yard. Official since the twelfth century at least, it was used in the statutory specifications of cloth sizes in the late Middle Ages. Probably following trade practice, the law began to cite cloth dimensions in quarter yards or quarters of 9 inches in 1350.[7] Consequently amounts of chain and filling would be calculated in quarters. Thus skeins of yarn came to be formed on reels whose circumferences were multiples of the quarter yard: 36 inches, 45 inches, and so on up to 108 inches.

Yarn strength could be another consideration in the choice of reel size. This may explain the variety of reels permitted in the pre-industrial worsted trades. Low tensile strength certainly compelled

---

exact in their Numbers of Threads." Quoted in John Duncan, *Practical and Descriptive Essays on the Art of Weaving* (Glasgow, 1807), 243.

[5] Schwarz, "The Reel," 2132–2133, 2138–2139; Robert J. Forbes, *Studies in Ancient Technology* (11 vols., Leiden, 1955–), IV, 165–167. However, evidence for a cranked, rotary reel earlier than 1405 cannot be found according to Lynn White, Jr., *Medieval Technology and Social Change* (Oxford, 1962), 111. See also Livingston, *Skein Winding Reels*, 1–14. Because they were more accurate than hand-reels, ten thousand check reels were distributed by the Board of Trustees in Ireland during the early eighteenth century. See Duncan, *Essays*, 245.

[6] 7 Jac. 1, c. 7 (1609). The text of British statutes has been taken from Danby Pickering's edition while dates are those given in Great Britain, *Chronological Table of the Statutes Covering the Period from 1235 to the End of 1969* (London, 1970).

[7] 25 Edw. 3, stat. 4, c. 1 (1350).

the use of a smaller reel and cut (skein) in the nineteenth century Irish fine linen industry.[8]

Granted these limitations on reel size, what factors controlled the length of the skein built on the reel? One may have been the amount of yarn that would fill the hand spindle or the spindle or bobbin on the two types of spinning wheel. Such spindles could hold eighty to 200 yards of woolen yarn, depending on coarseness, and bobbins somewhat more.[9]

One strong influence on skein length was yarn diameter. Beyond the point at which the reel cross arms were completely covered with yarn, distortion of the skein length would begin because the outer circuits of yarn would grow longer and longer.[10] To reduce this distortion and simultaneously to increase skein length, a larger reel circumference or longer cross arms could be used. Larger reels were employed with the thicker yarns: woolen, linen and jute. Most cross arms average about 6 inches so that, with spinning wheel and reel in stationary positions, there is no need to guide the yarn as it moves from end to end of the spindle or bobbin in unwinding.[11] These cross arms could hold up to 200–300 strands of yarn, depending on their thickness. Much longer cross arms would give the reel, at first a domestic implement, unacceptably cumbersome dimensions.[12]

Limiting the number of reel rotations composing a skein to within a range of between forty and 200 had a very practical advantage: it minimized the time lost when a slip demanded a recount. This concern for exactness, due to obvious technological and commercial reasons, is reflected by the introduction (during or shortly before the seventeenth century) of the audible or visual or audible-visual check reel (see illustration 3).

So the choice of a linear unit lay between the skein length and the multiple of the skein length (usually called a hank). What is not clear is why one or the other was selected. Advancing technology evidently favored hanks. Mechanized spinning brought consis-

[8] Leslie C. Marshall, *The Practical Flax Spinner* (London, 1885), 221–22.

[9] These conclusions are drawn from an examination of the sixty spinning wheels in the collections of the Merrimack Valley Textile Museum, North Andover, Mass.

[10] A point made in William S. Murphy (ed.), *The Textile Industries: A Practical Guide to Fibres, Yarns and Fabrics in Every Branch of Textile Manufacture* (8 vols., London, 1910–1911), III, 169.

[11] This is another inference from the wheels in the Merrimack Valley Textile Museum's collections.

[12] For the location of reels in colonial homes (in garret, kitchen, bedroom, or side-room) see Abbott L. Cummings (ed.), *Rural Household Inventories Establishing the Names, Uses and Furnishings of Rooms in the Colonial New England Home, 1675–1775* (Boston, 1964), *passim*. Presumably English practice was similar.

tently even yarns and increasingly large-scale operations. With every reason to gauge, process, and package yarn in larger quantities, long industrial reels, capable of reeling several dozen yarns at once, were installed in factories.[13] In the Aberdeen region, their operation was facilitated by applying the 14,400 yard linen spyndle to a fixed length woolen yarn numbering system.

A general consideration affecting the choice of linear unit was "a convenient relation between the length and the weight." [14] Thus the Wiltshire, Gloucestershire and the Stirling, Alloa, and Kilmarnock systems could be reduced to shorter length units weighing an ounce. Such reductions not only greatly helped the weaver in his warp and filling calculations but also the mill carding, roving and spinning masters who, several times daily, needed to test, on small sample reels, the fineness of the card end, roving or yarn coming off their machines. For this very reason of convenience, the length of the Yorkshire woolen skein was extended during the nineteenth century from 1,520 to 1,536 yards.[15]

So much for the linear units of yarn measuring systems. Weight units were those in common commercial use. In England avoirdupois weights, employed since time immemorial, were the sole legal standards between 1587 and 1824.[16] Despite this, as a Parliamentary commission discovered in 1820, a confusing selection of weights existed locally in agriculture, industry, and commerce in different parts of the kingdom. The English pound could vary between twelve ounces (Westmorland) and twenty-four ounces (North Riding of Yorkshire). In Scotland, where troy weights prevailed, the pound ranged between sixteen ounces avoirdupois (Campbeltown) and twenty-seven ounces avoirdupois (Kirriemuir). The stone of wool likewise fluctuated from county to county between seventeen (Dunbartonshire) and twenty-eight (Galloway) troy pounds.[17] Such perplexing local definitions further explain the

[13] Abraham Rees (ed.), *The Cyclopaedia; or Universal Dictionary of Arts, Sciences and Literature* (45 vols., London, 1802–1819), s.v. "Worsted."

[14] Thomas Woodhouse, *Yarn Counts and Calculations* (London, 1921), 5; also emphasized in William Partridge, *A Practical Treatise on Dying* (New York, 1823), 66.

[15] At some point it was realized that the number of yards in the Yorkshire skein (1,520) was almost equal to the number of drams in the 6 lb. watern (1,536), the weight unit of the Yorkshire woolen count system. As a short cut, counts were calculated as the number of yards weighing one dram. The difference reached a skein's length at no. 95 which was really a no. 96 on the 1,520 yd. scale. Convenience proved stronger than custom and by the 1890's the old 1,520 yd. skein had been all but abandoned in preference for the 1,536 yd. one. See Charles Vickerman, *Woollen Spinning* (London, 1894), 342–344. See text note 46 and tables note 35 *infra* for sample reel references and Duncan, *Essays*, 318 for an example of the convenient relation between warp length and linen reel size.

[16] Algernon E. Berriman, *Historical Metrology* (London, 1953), 146–151.

[17] Great Britain, *Parliamentary Papers (House of Commons)* 1820 (314), VII, "Second Report of the Commissioners . . . [on] Weights and Measures," Appendix A.

peculiar differences between the resultant yarn numbering systems developed in pre-industrial, self contained economic regions.

In taking the short step from recording linear yarn measurements to relating these to weight as an expression of fineness, it was generally more practicable to have the weight unit fixed and the length unit as the variable in the equation. The direct or fixed length method required either that each linear unit be weighed in turn or that samples be weighed. The former, laborious even if a very long linear unit or a minimal weight unit (or both) was chosen, would be very costly and therefore suited only to expensive fine yarns and fabrics like silk. In the pre-industrial period, with most yarns destined for clothing or decorative fabrics (in which a smooth textile was a prime requisite), a very long linear unit was undesirable because the quality of spinning was often very inconsistent. On the other hand, a small weight unit required a sophisticated set of weights or precisely calibrated scales,[18] able to measure drams, deniers, and grains, which could be tedious in operation, inconvenient in calculation and therefore generally uneconomical. Sample weighings would be acceptable in two cases only: when yarn had an even consistency; or when yarn thickness did not have to be precisely maintained. Before industrialization, the most even yarns were those of silk, which had been spun by the silk worm and needed only to be twisted, cleaned and plyed in manufacturing.[19] Exactly consistent thickness was least important in the coarse linen, jute, and hemp industries which manufactured such items as bagging, twine, and rope. For these reasons the fixed length yarn numbering system occurs most frequently, before industrialization, in the silk, tow linen, jute, and hemp branches of the textile industry. During the nineteenth century consistently even spinning by machine together with an increase in scale made direct systems much more practicable.

Prior to the Industrial Revolution the wider use of the inverse yarn numbering system derived from the convenience of assessing fineness by throwing a number of skeins of known length and apparently similar fineness into a scale against a fixed weight. Naturally it demanded some experience to judge which skeins had approximately the same count, for the method only gave the average

[18] For an early design for precisely-calibrated yarn scales see W. Ludlam, "An Account of a Balance of a New Construction Supposed to Be of Use in the Woollen Manufacture," *Philosophical Transactions*, LV (1765), 205-217. Dr. Philip W. Bishop kindly supplied this reference.

[19] For comparative fiber structures see Werner von Bergen and Walter Krauss (eds.), *Textile Fiber Atlas* (2nd ed., New York, 1945) and Herbert R. Mauersberger (ed.), *Matthews' Textile Fibers* (New York, 1947).

count of the particular group of skeins or hanks being weighed. In most sections of the textile industry this compromise between mensural precision and cost reduction, between technological and economic considerations, was sufficient.

Exactly when yarn length and weight were first related in a direct or inverse numbering system is impossible to tell. Weaving technology suggests that the production of evenly textured or unnapped fabrics was inevitably accompanied by greater precision in ascertaining yarn fineness as well as length. This is evidently implied in fifteenth century complaints about Norfolk worsteds.[20] The putting-out system must also have encouraged the relation of length to weight because the clothier checked the work of his spinners by weighing the raw wool before distribution and then the woolen yarn after collection. A loss of one quarter pound in every twelve pounds avoirdupois of raw wool delivered to the spinner was allowed by a 1514 statute[21] but the practice of checking must have arisen earlier with the development of marketing. Before the twentieth century the only reliable and commercial method of assessing yarn quantity and quality was to reel and weigh skeins. As an anonymous Scottish author wrote in 1733, "The Use of Weights and Scales is the only sure Way to prove Yarn, both as to its Quantity and Fineness; every good Weaver uses this Method . . . [so should] the Dealers in Yarn."[22]

Emergence of Regional Standardization in Britain

In a subsistence economy, that of England in the early Middle Ages, the unit of production, the family, would at first have established its own reel size. In time, usage of certain reel sizes would have passed into the community with intermarriage, spinning and reeling being female occupations. The settling of national linear units coupled with the regulation of industrial and commercial practices as a market economy emerged in the twelfth century and guilds multiplied in the thirteenth, would have imposed reel sizes that were multiples of 9 inches, as noted above, perhaps by the fourteenth century. During the same centuries, the introduction of

[20] 23 Hen. 6, c. 4 (1444) and 7 Edw. 4, c. 1 (1467) both complain of "the slaies and thread" of worsted goods "falsly made and wrought" or "not well made and wrought."

[21] 6 Hen. 8, c. 9 (1514).

[22] Quoted in Duncan, *Essays*, 244. In seventeenth century Dutch spinning schools, the infant spinners were seated in tiers, according to their ability to spin fine yarn, and their managers "sort and size all the Threads so, that they can apply them to make equal Cloaths." Andrew Yarranton, *England's Improvement by Sea and Land* (2 vols., London, 1677), I, 46.

the rotary reel would have made yarn measuring more exact in those areas where it was adopted.

What is quite evident by the seventeenth century at least, is that changes in the organization of the industry combined with the requirements of weaving technology to necessitate regional reel standards and thus regional yarn numbering systems. Dishonest practices in the spinning and reeling processes became possible as a result of the organizational changes in the English woolen industry which occurred between the fifteenth and seventeenth centuries. The thirteenth century had seen the rise of the independent clothier as the woolen industry spread out into the countryside with the construction of fulling mills on fast-flowing streams in hilly terrain, especially in the West of England. Away from the regulatory controls of incorporated towns dominated by merchant guilds and operating in the much more liberal economic environment of a decaying manorial system, the country clothiers, in conditions of comparative free enterprise, utilized the labor of widely scattered spinners. By the sixteenth century, larger clothiers, more concerned with the problems of management than the details of manufacture, increased in numbers, particularly in East Anglia and the West of England.[23] In the former, master combers came to control the spinners; in the latter, market spinners and clothiers supplied spinners with wool and paid them for spinning the yarn.[24] Large capitalist employers also dominated the Yorkshire worsted industry.[25] If it is recalled that spinning was almost entirely carried out by women, a prolonged but increasingly obvious change in the social-commercial position of spinsters may be registered. Before the sixteenth century and where the industry was not in the hands of large capitalist clothiers, the wife and daughters supplied yarn directly to the weaver (the husband or father). A close family relationship guaranteed honest spinning that needed no regulation. As the putting-out system came to be concentrated in fewer hands, spinsters directly or indirectly returned their yarn to a distant clothier whom they rarely regarded with anything more than the cool deference imposed by economic necessity. The clothier or his agent then passed the yarn, intermingled with other spinsters' skeins,

[23] Eleanora Carus-Wilson, "The Woollen Industry," *The Cambridge Economic History of Europe*, vol. II, M. M. Postan and E. E. Rich (eds.), *Trade and Industry in the Middle Ages* (Cambridge, 1952), 408–428.

[24] Ephraim Lipson, *The Economic History of England*, vol. II *The Age of Mercantilism* (London, 1961), 13 n; George D. Ramsay, *The Wiltshire Woollen Industry in the Sixteenth and Seventeenth Centuries* (London, 1965), 31–49, 90–99.

[25] Herbert Heaton, *The Yorkshire Woollen and Worsted Industries from the Earliest Times up to the Industrial Revolution* (Oxford, 1965), 296–97.

1. AMERICAN HAND REELS OR NIDDY-NODDIES

Examples from the Merrimack Valley Textile Museum collections (acc. no. 59.1.60 on left and 59.1.65 on right); each has a circumference of approximately 76 inches.

2. AMERICAN CHECK REEL

Of audible or snap type (M.V.T.M. acc. no. 59.1.178), this six-arm reel (of 75 inches circumference) must be rotated forty times before its axle worm completely turns the forty-tooth gear to tension and releases, audibly, the metallic spring, thus signalling a skein's completion.

3. AMERICAN AUDIBLE-VISUAL CHECK REEL

Audible-visual or snap-and-clock reel (M.V.T.M. acc. no. 59.1.72) which is extremely rare because it bears both the date of construction and the maker's name: "1749 Daniel Abbot" ("7" reversed) above the clock face. Acquired from the vicinity of Andover, Mass., it is very likely that the reel was made and used there because the Abbots were an early local family. The circumference of the reel's armature is 73 inches.

to weavers who frequently had no family ties with the spinsters – as many as ten spinsters could have produced the yarn for one chain.[26]

This broad picture of the domestic system (much less clear-cut than a summary suggests) implies that the dominance of capitalist clothiers introduced two new elements to the early spinning situation of relevance to the regional standardization of reels: a social and geographic distance between the spinster and her employer that gave her the motivation and opportunity to embezzle raw materials or yarn; and the wider marketing of the spinster's work which obliged conformity to regional practices in measuring yarn lengths and finenesses.

Not surprisingly, before the advent of machine spinning and reeling and its concentration in factories, there are hints of frequent conflict between capitalist clothiers and their worker spinners.[27] This struggle produced statutory reeling units in the English worsted and the Scottish and Irish linen industries but did not do so in the English woolen, linen, or cotton trades.

The absence of this legislation in the English woolen industry may be broadly attributed to two possible factors. First, as a valuable recent overview of the seventeenth century history of state regulation of the cloth industry has shown,[28] the clothiers were subject to a complex set of economic and political cross-currents and changes. Thus legislative controls might have been regarded as, or found to be, undesirable or difficult to obtain, for reeling at least. Second, such legislation may by this time have been unnecessary. Regional reel sizes were certainly long-established by the seventeenth century in the woolen districts. The clothiers simply had to exert greater control over their workers in order to minimize fraudulent spinning and reeling and to rationalize the flow of yarns to the

[26] Yarranton, *England's Improvement*, I, 46; Great Britain, *Parliamentary Papers* (*House of Commons*) 1802–1803 (95), VII, "Minutes of Evidence Taken before the Committee, to Whom the Bill, Respecting the Laws Relating to the Woollen Trade, Is Committed," 301.

[27] The acts of 4 Edw. 4, c. 1 (1464) and 3 Hen. 8, c. 6 (1511) ordained that textile workers should perform their several duties while 6 Hen. 8, c. 9 (1514), noticed above, legislatively recognized for the first time that dishonest carding and spinning were becoming a problem. A study of medieval guild and corporation records is really needed to confirm that yarn embezzlement was a minor problem until the putting-out system came into the hands of large country clothiers. I have relied on the standard studies by Lipson and Carus-Wilson already cited.

After the defeat of a 1593 draft bill against "divers faultes" of spinners and weavers (including, among the former, "deceitfull spinning their yarne"), the tensions between workers and clothiers mounted in the seventeenth and eighteenth centuries. See R. H. Tawney and Eileen Power, *Tudor Economic Documents* (3 vols., London, 1924), I, 372; Ramsay, *Wiltshire Woollen Industry*, 14, 94; Lipson, *Economic History of England*, II, 48–49; Alfred P. Wadsworth and Julia de L. Mann, *The Cotton Trade and Industrial Lancashire, 1600–1780* (Manchester, 1931), 395–400.

[28] J. P. Cooper, "Economic Regulation and the Cloth Industry in Seventeenth-Century England," *Transactions of the Royal Historical Society*, 5th ser., XX (1970), 73–99.

weavers. Since the problem did not at any one point in time present an overwhelming threat to the clothiers as a group (as did the smuggled export of wool), it was never treated with the concerted urgent action that would have gained a legislative response. Moreover, in the West Riding woolen industry the widespread continuance of the clothier's family as the unit of production tended to preclude fraudulent reeling[29] while avoiding the difficulties of regularizing yarn supply from many spinners.

In two English counties where the linen manufacture survived into the nineteenth century, two linen yarn numbering systems lingered until after 1900: in Dorset, where sailcloth was made at Bridport and Beaminster; and in Yorkshire, where the manufacture became factory-based in the nineteenth century. By 1840 both systems were less popular than the statutory Irish-Scottish linen one. Because it was geographically so scattered, because the Irish and Scottish linen industries were more developed (not least to maintain the supremacy of the English woolen industry) and because mule-spun cotton was strong enough for warps (and thus reduced the demand for linen chains), the English linen industry legislatively gained no more than general protection before mercantilism petered out after the 1820's.[30]

Embezzling and false reeling in the cotton manufacture (initially located in Lancashire) was covered by the general law of 1703 and successive acts. And anyway, the cotton interest, during its infancy when reel, skein, and hank sizes were established, found it legislatively difficult enough to withstand the hostility of the ascendant woolen interest.[31]

Once yarn measuring was known (by the sixteenth century at the latest) to be a vexing problem in the domestic system of manufacture, the rise of new textile industries provided the occasion for regulations which in the older trades had been settled by custom and imposed in the clothiers' struggle for control over their workers.

29 Heaton, *Yorkshire Woollen and Worsted Industries*, 91–101.

30 William G. Rimmer, *Marshall's of Leeds, Flax Spinners, 1788–1886* (Cambridge, 1960), 1–9; Robert and William Scott, *Scott's Practical Cotton Spinner and Manufacturer* (Preston, Lancs., 1840), 337–38; Peter Sharp, *Flax, Tow, and Jute Spinning* (Dundee, 1882), 173; H. R. Carter, *Flax and Its Products* (London, 1920), 85.

31 Paul Mantoux, *The Industrial Revolution in the Eighteenth Century* (London, 1961), 199–203, though he has probably overstated the effects of the 1721 act which did not prohibit cotton goods for export, as Wadsworth and Mann noted in *Cotton Trade*, 118.

The cotton yarn numbering system apparently did not derive from Indian practice. At Dacca in Bengal, the center of fine hand spinning, reels were not standardized and yarn was measured by the *hāth* (cubit) of 19¾ inches, fineness being expressed as the number of *hāths* weighing one *ruttee* (2 grains troy). Anon, *A Descriptive and Historical Account of the Cotton Manufacture of Dacca, in Bengal by a Former Resident of Dacca* (London, 1851), 16–24.

Thus the first law, of 1609,[32] specifiying reel size seems to have been instigated by clothiers and/or weavers involved in the New Draperies (cloths of worsted chain and sometimes filling, introduced into Essex by immigrant clothiers and artisans from Flanders who had fled before Alba's troops after 1567), judging from the cloths named and the specification of a two-yard reel for Essex. Furthermore, if the experience of the cotton industry was not far different from that of other textile trades, yarn embezzlement most widely occurred either when the demand for yarn exceeded supply or during times of severe depression.[33] The former condition arose in the worsted industry when the appearance of the New Draperies brought a long period of prosperity stretching through most of the seventeenth century. So the reeling law of 1609 was followed by those of 1650 and 1662 which laid down Norfolk worsted reel, lea (skein), and hank sizes.[34] Depression followed in the eighteenth century and a series of regulations, "the Worsted Acts," which included county by county reel, lea, and hank size definitions, were imposed to check embezzling and generally to revive the industry in the areas of decline.[35] One other, more precise economic factor, had a bearing on the regulation of the worsted trades: due to the abuses of aulnagers, the makers of the New Draperies preferred self-regulation through corporations of master-weavers (men as powerful as the West Country clothiers) and they secured, in two of the above mentioned acts, reeling specifications which were designed to check embezzling.[36]

The setting up and state encouragement of the Irish and Scottish linen industries at the end of the seventeenth century also provided the occasion for defining reel sizes, not only to check yarn frauds but also to improve the quality of workmanship of goods destined for export. The statutory Scottish linen reel of 90 inches together with the 300 yard cut (skein) and 3,600 yard hesp (hank), legally established in 1693, were preserved after the Act of Union by a Westminster statute of 1726.[37] Identical linen reel, cut and hank

[32] 7 Jac. 1, c. 7 (1609).

[33] Wadsworth and Mann. *Cotton Trade*, 89, 276.

[34] Peter J. Bowden, *The Wool Trade in Tudor and Stuart England* (London, 1962), xvii–xviii. Henry Scobell (ed.), *The Collection of Acts and Ordinances of General Use Made in the Parliament . . . 1640 . . . to 1656* (London, 1658), 146–48 for c. 36 (November, 1650). 14 Car, 2, c. 5 (1662).

[35] Heaton, *Yorkshire Woollen and Worsted Industries*, 418–437; Wadsworth and Mann, *Cotton Trade*, 395–99.

[36] Cooper, "Economic Regulation and the Cloth Industry," 83.

[37] For the general economic development of the Scottish linen industry see Roy H. Campbell, *Scotland since 1707: the Rise of an Industrial Society* (Oxford, 1965), 58–59.

According to T. Woodhouse and P. Kilgour, *The Jute Industry* (London, 1921), 70, the Convention of Estates had advocated a 90 inch reel in 1665. The 1693 act is found in

sizes for Ireland were prescribed by an Irish Parliament statute of 1705 for the Irish linen industry.[38] Furthermore, as neither the New Draperies of worsted, generally, nor any linens were napped, yarn quality largely determined fabric quality and so it was especially important to regulate the reeling of worsted and linen yarns.

Impact of Industrialization

Why have many of these pre-industrial count systems survived industrialization, even to the present? Why have several systems persisted in one industry (the woolen industry) and why have efforts to impose a single international system for all fibers not yet succeeded? A number of points suggest some answers to these questions.

Spinning and weaving textbooks written during the first half of the nineteenth century sometimes gave tables in which different numbering systems for one fiber were laid out in parallel columns to facilitate commercial and technical computations.[39] Presumably merchants and factors handling yarns numbered by different methods quickly compiled personal conversion tables; some of these were printed and distributed in the latter part of the century.[40] Such accommodations made by textbook authors and middlemen plainly encouraged manufacturers to preserve traditional count practices.

The amount of potential mercantile confusion due to the export of British woolen yarn would have been related to the number of districts with different numbering systems exporting yarn to any one foreign market and the number of markets so affected. Since we do not know how much woolen (as opposed to worsted) yarn was exported from Britain throughout the nineteenth century, it is impossible to be sure about actual mercantile confusion in foreign woolen yarn markets.[41] The export of worsted, linen and cotton yarn

Acts of the Parliaments of Scotland (12 vols., Edinburgh, 1814–1824), IX, 311–313. The English statute confirming its reel size was 13 Geo. 1, c. 26 (1726).

[38] See Conrad Gill, *The Rise of the Irish Linen Industry* (Oxford, 1925), 61–82 for the background to 4 Anne, c. 4 (1705) which is printed in *Statutes at Large Passed in the Parliament Held in Ireland* (11 vols., Dublin, 1760–1769), IV, 73–75. Miss Edith Henderson of the Harvard Law Library kindly supplied me with copies of the Scottish and Irish acts cited in this article.

[39] Scott, *Practical Cotton Spinner* is one example.

[40] M'Lennan, Blair, & Co., Yarn Merchants, Glasgow (computed and arranged), *Comparative Yarn Tables* (Glasgow, 1883), kindly transmitted to me by Mr. James Hunter of the Paisley Museum, is one I have seen.

[41] The best available figures are in Brian R. Mitchell and Phyllis Deane (eds.), *Abstract of British Historical Statistics* (Cambridge, 1962), 195–97, 293–295, 302–306. Contemporary estimates suggest that Britain exported no woolen yarn whatsoever at mid-century. In

was increasingly significant as the century wore on but by 1800 these three branches of the textile industry had standardized count systems (the Scottish and Irish linen ones differing most materially in nomenclature). We need to know how foreign manufacturers handled counts of imported British yarns: did they accept British numbering systems or did they make any attempt to accommodate them (by conversion tables or even re-reeling or measuring during winding) to indigenous practices?[42]

In the British woolen districts, where count disparity was greatest, most yarn was absorbed locally, thus allowing variations in regional systems. This was due to several factors. More expensive to grow as a fiber and then necessarily spun on the less economical intermittent principle (on jennies, jacks, or mules), in the nineteenth century woolen yarn was more costly than cotton or linen yarn. Elaborate finishing techniques put the cost of woolen cloth further ahead that of other fabrics. Thus the demand for pure woolen cloth was quantitatively and relatively (to the population) not high during Victoria's reign – mass demand for heavy cloths being increasingly served by reprocessed wool or shoddy after the 1830's.[43]

One other pressure, difficult to detect and impossible to measure, was at work conserving regional yarn numbering systems: tradition. This was especially strong in the textile industry where a premium was placed on experience rather than scientific knowledge, until the early twentieth century at any rate.[44]

Despite these powerful conservative influences, industrialization brought two counteracting tendencies which assailed the old count systems: the evolution of new ones and endeavors to reduce all such systems to a single universal one. New numbering systems appeared for several reasons. Mechanization enabled a high degree of uniform linear-density in spinning. As a result, direct systems with long linear units now became practicable (the Aberdeen one, for example).[45] Consistently-even yarns came not only from increased

---

1864 Jacob Behrens computed that England exported 31,824,296 lbs. of worsted yarn. See Kenneth G. Ponting (ed.), *Baines's Account of the Woollen Manufacture of England* (New York, 1970), 141. This is equivalent to the total amount of woolen and worsted yarn exported that year by Britain, according to Mitchell and Deane (p. 196).

42 English mercantile agents in Germany, arranging imports of British worsted yarn between the 1840's and 1870's, certainly quoted English worsted numbers to Yorkshire manufacturers. See, for example, Eric M. Sigsworth, *Black Dyke Mills. A History* (Liverpool, 1958), 284–317.

43 Ponting (ed.), *Baines's Account*, 100–105, 110–14.

44 Not until the 1880's was technical education in the textile industry successfully organized and structured. See W. English, "Beginnings and Early Growth of Textile Technical Education in Britain," *The Textile Manufacturer*, September, 1969, 345–49.

45 Though a hand spinner could have spun up to two spyndles (each of 14,400 yards) of yarn in six days, and perhaps more if others carded for her, uneven hand spinning made

control over drafting and twisting in spinning but also from the application of similar, calculated mechanical regulation in the pre-spinning processes of carding, drawing, and roving (to name the cotton ones). At these stages the much coarser ends necessitated (due to bulk and fragility) shorter linear units in testing size from samples during production.[46] Industrialization also brought new techniques in reclamation: old woolen clothing was reprocessed to yield shoddy; waste silk from reeling was retrieved as spun silk; and waste cotton as bump cotton.[47] These yarns needed new count systems either because they were more fragile or more coarse than virgin yarns.

In Scotland, at the beginning of the Industrial Revolution, the statutory reeling standards of the politically revitalized eighteenth century linen industry were taken over by at least three of the regional woolen industries: at Galashiels and Hawick in the Border Country, where a capitalist organization emerged after 1750, the 300 yard cut became the linear unit in fixed weight numbering systems; and in the Aberdeen region where the 14,400 yard linen spindle was applied in a fixed length one.[48]

By 1870, when most western European powers had commenced industrialization, the costly inconveniences of a multitude of yarn numbering systems, evidently increasing with industrialization, had become apparent. In 1873 the first of several international efforts (a congress held at Vienna) was organized to rationalize the fifty or more systems then in use.[49] Although the history of these efforts is beyond the scope of this essay, their probable motivations are of relevance to the theory of organization history previously mentioned. The strongest of these were at first technological and eco-

such a length unit impracticable before mechanization. I therefore assume that the Aberdeen system was a product of industrialization. No evidence of origin has apparently survived locally according to Mr. A. R. Alexander, Designer for J. & J. Crombie Ltd. of Aberdeen. Letter to Aberdeen Public Library, October 14, 1970, in response to mine of October 1, 1970 to Aberdeen Public Library.

[46] James Montgomery, *The Cotton Spinner's Manual; or a Compendium of the Principles of Cotton Spinning* (Glasgow, 1835), 62–64 and W. Leigh, *The Practical Cotton Spinner and Manager's Assistant* (Skipton, 1892), 28–34, 121–27 are two evidences of this practice. See J. Herzfeld, *The Technical Testing of Yarns and Textile Fabrics with Reference to Original Specifications* (Eng. trans., London, 1898), 77–79 for a late nineteenth century discussion of these reels.

[47] For shoddy see Samuel Jubb, *History of the Shoddy Trade* (London, 1860) and Ponting (ed.), *Baines's Account of the Woollen Manufacture*, 100–105, 109, 111, 163; for spun silk, Frank Warner, *The Silk Industry of the United Kingdom. Its Origin and Development* (London, 1921), 390–439; for waste cotton, Joseph Nasmith, *The Students' Cotton Spinning* (Manchester, 1893), 377–387.

[48] Sir John Sinclair (ed.), *The Statistical Account of Scotland* (21 vols., Edinburgh, 1791–1799), II, 308–313, VIII, 529, for Galashiels and Hawick respectively; Mr. William P. H. Sinclair of Galashiels kindly directed my attention to this source.

[49] Herzfeld, *Technical Testing of Yarns*, 71. At least two other congresses, at Paris in 1878 and 1900, were held.

nomic in character. The mixture of two different fiber yarns in one fabric, by the 1830's, and, much later, the plying of different fiber yarns (which created the fancy twisting industry) were two constraints to adopt one uniform system particularly where yarn and cloth manufacturing were not integrated.[50] International trade in yarns within continental Europe and the twentieth century rise of the scientific analysis of fiber, yarn, and fabric (adding the demands of cost accountants, mercantile houses, and scientists to those of fabric designers and mill managers) produced a number of universal count systems. Of these the Metric was the first (1873) and the Tex (first stage adopted 1960) the latest to be formulated.[51]

The Transmission of British Count Systems to America

Largely because the problem of yarn numbering systems has been neglected, a clear picture of the impact of spreading British systems to Amercia and there commingling them, before and after industrialization, is hard to form. However, it is possible to establish the existence of a problem and to submit a preliminary hypothesis.

While many systems, from France, Holland, and Germany as well as Britain, must have been taken to America with the first waves of settlers in the seventeenth century, the earliest hints of confusion date from the early national period, the time of initial industrialization. Visiting the Springfield, New Jersey, carding mill of a Lancashire immigrant in the summer of 1794, the Wiltshire clothier, Henry Wansey, noticed "a large parcel of linen yarn, of very good quality . . . drawn about nineteen skeins to the pound."[52] Since linen yarn was never measured in lengths denominated skeins in Britain, did Wansey refer to an American woolen skein of 560 yards, an unknown skein peculiar to northern New Jersey, a Lancashire cotton skein of 120 yards, a Wiltshire woolen skein of 320 yards or to some other? The potential for confusion is obvious from the ambiguity of his remark. Another West Countryman found it

[50] According to John James, *History of the Worsted Manufacture in England* (London, 1857), 470–76, the use of cotton warps with worsted filling commenced in England in 1826; the combination gained widespread favor after the 1830's. Summaries of late nineteenth century fancy yarn types are in *The Encyclopaedia Britannica* (29 vols., Cambridge, 1910–1911), s.v. "Yarn," and Murphy, *Textile Industries*, III, 162–66.

[51] British Standards Institution, *Specification for a Universal System for Designating Linear Density of Textiles (Tex System). B.S. 947: 1970* (London, 1970); George E. Linton, *The Modern Textile Dictionary* (New York, 1963), 929–938. The numerous compelling reasons for adopting a single universal yarn numbering system have been lucidly set out in Werner von Bergen (ed.), *Wool Handbook* (2 vols., New York, 1963–1970), II, Part I, 230–33.

[52] David J. Jeremy (ed.), *Henry Wansey and His American Journal, 1794* (Philadelphia, 1970), 133.

realized nearly thirty years later. William Partridge, a Gloucestershire dyer who emigrated to the United States in 1808, discovered that biers (reed units normally relating to sleying) formed the basis of one New Jersey factory's yarn numbering system — "so preposterous a mode of doing it," he thought. This kind of mix-up in technique Partridge attributed partly to the managers' and workers' ignorance and partly to the admixture of workers from the three main British woolen regions, the West of England, Yorkshire, and Ireland.[53]

But technical nonconformity was not confined to British immigrants. In 1816 Raphael Duplanty, manager and part-owner of the Duplanty, McCall & Co. cotton mill on the Brandywine, was informed by a contact then in Havana,[54]

> We have had several Parcels from the U. States which were rather too much twisted. This increases the weight, but does not yield so much spun yarn as the English from the same weight. The spinning Machines should be let [set?] loose. The numbers of the American Twist are as follows: 5, 6, 6½, 7, 7½, 8, 8½, 9, 9½, 10, 11, 12, 13, 14, 15. These Numbers are quite different from the English numbers which the Purchasers in Russia are accustomed to, and which are well known & adopted in their Manufactures. The Numbers 14 & 15 of American are equal to No. 20 & 22 of the English . . . . If the American Manufacturers would consent to spin the Cotton Twist after the English way & to which we are accustomed in Russia, a quantity of 20,000 Poods equal to 720/m English lbs could be readily disposed of.

The equivalencies cited would make the American (Brandywine) cotton hank 1,200 yards long. With his cotton thread, Duplanty used a 42 inch reel and a 140 yard hank.[55]

Arguing *a posteriori* from the late nineteenth century American situation, a fairly simple hypothesis may be constructed. By 1900 the eastern mills were numbering cotton by the 840 yard hank, worsted by the 560 yard hank, line linen by the 300 yard cut and wool either by the 300 yard cut (around Philadelphia) or by the 1,600 yard run (in New England). Except for the woolen ones, all these inverse systems were the prevailing ones for their fibers in

[53] Partridge, *Treatise on Dying*, 31, 65–66.

[54] Memo from Bolleu (?), dated May 18, 1816, transmitted in a letter from Anthony Girard to Raphael Duplanty, June 16, 1816, Accession 500, Duplanty, McCall & Co. Papers, Eleutherian Mills Historical Library, Greenville, Delaware.

[55] Raphael Duplanty's Notebook, 11 (my pagination), Longwood MSS, Group 6, Box 2, Eleutherian Mills Historical Library. On p. 14 he wrote "at McKell & R reel for Thread (sewing & floss) 42 Inches round (reel for) Dyers Bleachers & y[arn?] 54 Inches round." Patrick McKell of the Philadelphia firm of McKell and Rickman first appears in the city directories as a thread manufacturer in 1813; in 1818 he was described as a cotton manufacturer. See *Paxton's Philadelphia Directory* for 1813, 1814, 1818, 1819; *Robinson's Philadelphia Directory* for 1816, 1817.

Britain at the end of the eighteenth century. Their adoption by the American textile mills was therefore quite natural: England's mercantilist policies delayed the capitalist organization of American textile manufacturing until after the Revolution when, fortuitously, Britain's pioneering industrialization provided a revolutionary machine technology; this coincidence largely spared America the confusions of a maze of localized yarn numbering systems such as persisted in Europe. A muddle of techniques would occur in the American factory system in two general cases: when there was still a vigorous conflict of yarn numbering systems in the manufacture of a given fiber in Britain; or when British and continental European systems collided. The Brandywine example, above, comes from a factory managed by a Frenchman. Others could no doubt be found. Such ethnic techniques presumably survived longest where they were geographically and/or economically most isolated. In the east, market pressures and the influx of technicians from the British textile industries combined to impose conformity to the prevailing British yarn numbering systems by the 1820's or 1830's.

The former case, of a clash in techniques in America due to a prior conflict in Britain, occurred in the woolen industry. One early allusion suggests that the New England count, based on the 1,600 yard run, was firmly established during the seventeenth century. In 1734 the General Court of Connecticut offered a premium of 2s. a yard for fine linen cloth, well-spun, woven and whitened, a yard wide and made of yarn that was eight runs to the pound.[56] Two inferences from this measure are particularly striking. First, it is evident that the run was then widely used in Connecticut as a linear measure of yarn. Theoretically England's suppressive mercantilist legislation limited textile manufacturing to the colonial household. This in turn should have prevented the appearance of a regional yarn numbering system, each seventeenth century New England settlement being composed of families from many if not all of the English textile counties, each of which had its own customary standards and techniques. The Connecticut settlers' success may have been repeated in other states east of the Hudson and sprang from several broad factors. Since the balance of trade was continually against them, the colonies north of Maryland had a greater incentive to defy British statutes and to encourage manufacturing than those to the south. Seventeenth century male immigrants to New

[56] Rolla M. Tryon, *Household Manufactures in the United States, 1640–1860* (Chicago, 1917), 36 n.

England who had textile skills (sorters, combers, weavers, fullers, clothworkers, dyers, and clothiers) came mainly from the West of England or East Anglia; [57] since there was more clothing wool than combing wool available in America, the West of England textile experience, including the 1,600 yard run, was most likely to predominate – the existence of a seven knot, 560 yard skein in New England in the eighteenth century may be attributed to the influence of worsted workers from East Anglia. Finally, it is very probable that the itinerant weavers, about whom little is known, played an important part in establishing the run as the regional linear yarn unit.[58]

Secondly, the application of the woolen run to linen yarn in this 1734 measure implies that the use of the run was then deeply ingrained as a customary technique in Connecticut and hence that the run system came over with some of the earliest waves of immigrants. Common in the eighteenth century home, the measurement of yarn in runs was easily transferred in the early nineteenth century to the New England factory where its use was reinforced by the arrival of workers from the declining West of England woolen industry.[59]

The reverse of the 1734 Connecticut measure appears to have happened in the Philadelphia area where the statutory Scottish and Irish linen cut of 300 yards was employed in the woolen industry. Though this awaits positive confirmation, it is most likely that the Scots-Irish of Pennsylvania were the agents of transmission and usage.[60]

[57] A helpful summary and bibliography of the English origins of early New England settlers is in Carl Bridenbaugh, *Vexed and Troubled Englishmen, 1590–1642* (New York, 1968), 463–66.

[58] Tryon, *Household Manufactures*, 144, 190–91. The account book of one late seventeenth century weaver which has been carefully studied is not especially revealing about yarn numbering: John Gould of Topsfield, Mass., quantified his yarn in pounds weight and only once referred to a linear amount – a "bottom," a synonym for a skein. Unless Gould, deliberately or carelessly, omitted to list the counts he bought (which is most likely, I think) either he produced very coarse cloth or else he re-reeled the yarn he purchased. Benno M. Forman, "The Account Book of John Gould, Weaver, of Topsfield, Massachusetts, 1697–1724," *Essex Institute Historical Collections*, CV (1969), 36–49; James A. H. Murray and others (eds.), *A New English Dictionary on Historical Principles* (13 vols., Oxford, 1888–1933), s.v. "Bottom" (15).

[59] So the Woodstock factory of the Hamilton Woolen Company, of Southbridge, Mass., numbered its yarn by the run system in the 1820's and 1830's. See Spinning and Weaving Accounts, 1823–1838 (Woodstock Factory), Hamilton Woolen Company Records, LA-1, Baker Library, Harvard University.

[60] Because Pennsylvania was the only sparsely-populated state that guaranteed religious liberty, the Scotch-Irish from northern Ireland flooded into it during the eighteenth century; by 1790, it has been estimated, they composed 80 per cent of the state's population.

In the early nineteenth century one community in western Pennsylvania was applying the Philadelphia cut system to cotton as well as wool. See Wayland F. Dunaway, *The Scotch-Irish of Colonial Pennsylvania* (Chapel Hill, N.C., 1944); Frederick Rapp to Daniel Fetter, August 5, 1815, Letterbook CO 2, 54, Harmony Society Letterbooks, Pennsylvania Archives (Harrisburg), kindly transmitted by Daniel Reibel, Curator, Old Economy.

III

How long did the chaos in numbering techniques last in the early American woolen factories? A general answer cannot presently be given but one surprising instance of longevity may be cited. At the North Andover, Massachusetts, mill of Nathaniel Stevens (1786–1865), yarn production was recorded in spinning- and stock-books until 1854 in quantities of pounds weight and beers. Not until the beginning of 1854 were qualities noted in counts (of New England runs).[61] The use of the beer (bier or bare – a group of dents in the loom reed, usually twenty), which Partridge had been astonished to find applied to yarn grist (fineness) back in 1823, strongly suggests that the confusions appearing in the period 1783–1820 lingered for nearly two generations in one firm which, in the long term, proved to be a successful one.[62]

Conclusion

"Next to the inconvenience of speaking different languages, is that of using different and arbitrary weights and measures," James Madison wrote to James Monroe in 1785.[63] As has been shown above, the variations in British (and hence American) yarn numbering systems were not, in fact, wholly arbitrary. They derived primarily from differences in linear units which in turn were based on reel sizes. These were ultimately limited by ergonomic constraints to within a circumference of about three yards. Under this length, reel size in the early Medieval subsistence economy could have varied, theoretically, from extended family to extended family.

The growth of a market economy with its statutory prescription of cloth dimensions and the multiplication of weavers' guilds further reduced the range of possible reel sizes and therefore yarn count systems. Between the thirteenth and seventeenth centuries the regional standardization of reels and numbering systems developed as a result of a fundamental change in industrial organization: the rise

[61] "Spinning Book, 1838, 1839, 1840" and "Inventory, January 1, 1850–January 1, 1861," Nathaniel Stevens (1786–1865) group, Stevens Papers, MS 60, Merrimack Valley Textile Museum.

[62] Any further clarification of the history of American yarn count systems is most likely to come from written records because the artifact evidence is generally inconclusive. A survey (which I carried out under the auspices of the Division of Manufacturing in the National Museum of History and Technology) of reels in some two hundred British and American museums disclosed only one reel bearing both the maker's name and date of construction (see illustration 3). Not until 1750 and then only in Scotland were makers of reels (and also heckles, spinning wheels, looms and reeds) legally obliged, by 24 Geo. 2, c. 31 (1750), to brand their products with their name and place of residence. Otherwise such inscriptions were motivated by affection for the reel's recipient or by pride in craftsmanship.

[63] Julian P. Boyd and others (eds.), *The Papers of Thomas Jefferson* (Princeton, 1950–), XVI, 603 (Madison to Monroe, April 28, 1785).

and dominance of independent country clothiers (and, to a lesser extent, yarn dealers) employing scattered spinners within as much as a thirty or forty mile radius of their base towns. By firmly imposing a commonly accepted reel size and numbering system throughout a textile district, these capitalists could gauge the productivity and check the honesty of their spinners, extend quality control to yarn production, and meet their weavers' technical requirements for standard grades and lengths of uniform yarns. With these objects in view, count systems were legally settled in later branches of the textile industry (worsted and, in Scotland and Ireland, linen) that had a sufficiently powerful political interest at Westminster. In others (English linen and cotton) lacking this political leverage during their organizationally formative periods, the establishment of count systems rested with the capitalists involved. Mechanization and labor concentration in factories during the Industrial Revolution eliminated the organizational problem of yarn embezzlement, but the demands of yarn marketing and weaving technology remained to preserve many of the old numbering methods. Though about half (eight at least) of those used during the eighteenth century failed to survive the Industrial Revolution, the ensuing mechanization and reorganization produced six or seven more. On the other hand, industrial reorganization, particularly the expansion of regional into national and international markets, the mixture of different fibers in one type of cloth or yarn, and the utilization of science in textile manufacturing, made the rationalization of a multitude of systems ever more logical. Despite several attempts to induce the acceptance of one universal count system appropriate to all fibers, the present persistence of plurality is a final aspect of this economic institution which tends to confirm Stinchcombe's theory of the persistence of organizational and institutional forms.

In the United States the problem was mitigated by several factors. English mercantilist policies combined with the pioneer agricultural economy to prevent the rise of a sophisticated, handicraft textile industry during colonial times. Only three British count systems, reflecting immigrant origins, came to dominate in household manufacturing, then mostly confined to wool and linen. Fortuitously, when the opportunity to industrialize came after the Revolution, a mechanized system of textile manufacturing could be acquired from Britain. Equally fortuitously, by this time in Britain virtually one numbering system prevailed for each fiber except wool, and in the

American woolen industry, where count chaos was potentially greatest, the arrival of many West of England textile workers during the early nineteenth century served to reinforce technical traditions established in colonial household manufacturing by previous generations of immigrant West Countrymen.

This study illustrates the historical interplay of technology and business organization in shaping one economic institution, yarn numbering systems, in the British and American textile industries. Although the requirements of technology put some limits around the range of Medieval systems and the rise of independent clothiers brought standard regional systems by the seventeenth century, a bewildering set of differing systems existed prior to the Industrial Revolution. Neither technological changes nor alterations in business organizations have sufficed to achieve universal acceptance of a single standardized method of yarn fineness measurement. Despite progress towards this goal made over the past two decades by the exertions of governmental agencies responsible for scientific standards in many western nations, the institution of yarn count systems in Britain and the United States remains almost as diverse as it was in the mid-seventeenth century.

APPENDIX

Inverse or Fixed Weight Systems [1]

Fiber	Location of Use	Reel Size	Table	Length Unit	Weight Unit	Yards Per Pound in No. 10	Known to Be in Use
wool	Gloucestershire, West of England	72″	reel=1 thread=2 yds. 40 threads=1 snap=80 yds. 5 snaps=¼ run=400 yds. 20 snaps=1 run=1,600 yds.	run of 1,600 yds.	1 pound avoirdupois	16,000	17th c. — 1820's [2]
wool	Wiltshire and Somerset, West of England	72″	reel=1 thread=2 yds. 40 threads=1 snap=80 yds. 4 snaps=1 skein=320 yds. 5 skeins=1 run=1,600 yds.	320 yds.	1 pound avoirdupois	3,200	17th c. — present [3]
wool	Leeds and Huddersfield, Yorkshire			skein of 1,520 yds.	6 pounds avoirdupois (watern, wartern or quartern)	2,533	18th c. — 1890's [4]
wool	Leeds and Huddersfield, Yorkshire			skein of 1,536 yds. or 256 yds. or 1 yd.	6 pounds avoirdupois (watern) 1 pound 1 dram	2,560	1890's — present [5]

shoddy (reprocessed wool)	Dewsbury, Yorkshire	36″		1 yard	1 ounce	160	19th c. — present [6]
wool	Galashiels, Scotland	90″	reel=1 thread=2½ yds. 120 threads=1 cut=300 yds. 12 cuts=1 slip=3,600 yds. 4 slips=1 spindle or millful=14,400 yds.	cut of 300 yds.	24 ounces avoirdupois	2,000	1790's — present [7]
wool	Hawick, Scotland	90″	reel=1 thread=2½ yds. 120 threads=1 cut=300 yds. 12 cuts=1 slip=3,600 yds. 4 slips=1 spindle or millful=14,400 yds.	cut of 300 yds	26 ounces	1,846	19th c. — present [8]
wool	Stirling, Alloa, and Kilmarnock, Scotland	72″ (?)	reel=1 thread=2 yds.(?) 120 threads (?)=1 cut=240 yds. 48 cuts=1 spyndle=11,520 yds.	spyndle of 11,520 yds.	24 pounds avoirdupois	4,800	19th c. — present [9]
wool	Ireland			1 yard	0.25 ounces	640	19th c. — present [10]
wool	Philadelphia, United States	90″	reel=1 thread=2½ yds. 120 threads=1 cut=300 yds.	cut of 300 yds.	1 pound avoirdupois	3,000	18th c. — present [11]

Inverse or Fixed Weight Systems

Fiber	Location of Use	Reel Size	Table	Length Unit	Weight Unit	Yards Per Pound in No. 10	Known to Be in Use
wool	New England, United States	72″	reel=1 thread=2 yds. 40 threads=1 knot =80 yds. 20 knots=1 run=1,600 yds.	run of 1,600 yds.	1 pound avoirdupois	16,000	17th c. — present [12]
wool	New England (domestic scale), United States	72″	reel=1 thread=2 yds. 40 threads=1 knot =80 yds. 7 knots=1 skein =560 yds.	skein of 560 yds.	1 pound avoirdupois	5,600	18th c. — present [13]
worsted	Norfolk, Yorkshire, Lancashire & Cheshire; Suffolk; Bedfordshire and Huntingdonshire; Northamptonshire, Leicestershire and Rutland; Lincolnshire; Isle of Ely; Great Britain; United States	36″	reel=1 thread=1 yd. 80 threads=1 (wrap), rap or lea=80 yds. 7 leas=1 hank or skein=560 yds.	hank or skein of 560 yds.	1 pound avoirdupois	5,600	17th & 18th c.; 19th c. — present: Great Britain & United States [14]
worsted	Bradford	36″ (?)	Same as for regular worsted areas probably (560 yard skein/hank)	1 yard	12½ grains (troy)	5,600	20th c.[15]
worsted	Essex	45″					early 19th c.[16]

worsted	Non-specific but Essex at least	54″	reel = 1 thread = 1½ yds. 80 threads = 1 skein = 120 yds. 7 skeins = 1 hank = 840 yds.	840 yds.	1 pound avoirdupois	8,400	early 19th c. — early 20th c.[17]
worsted	Norfolk; Suffolk; Bedfordshire and Huntingdonshire; Isle of Ely	63″	reel = 1 thread = 1¾ yds. 80 threads = 1 rap/lea = 140 yds. 6 leas = 1 hank/skein = 840 yds.	hank or skein of 840 yds.	1 pound avoirdupois	8,400	18th c.[18]
worsted	Suffolk; Bedfordshire and Huntingdonshire; Isle of Ely	72″	reel = 1 thread = 2 yds. 80 threads = 1 lea = 160 yds. 6 leas = 1 hank/skein = 960 yds.	960 yds.	1 pound avoirdupois	9,600	18th c.[19]
worsted	Norfolk; Yorkshire, Lancashire and Cheshire; Lincolnshire	72″	reel = 1 thread = 2 yds. 80 threads = 1 lea = 160 yds. 7 leas = 1 hank/skein = 1,120 yds.	1,120 yds.	1 pound avoirdupois	11,200	18th c.[20]
worsted	Essex	72″	reel = 1 thread = 2 yds. 40 threads = 1 skein = 80 yds. 7 skeins = 1 hank = 560 yds.	hank of 560 yds.	1 pound avoirdupois	5,600	17th c.[21]

Inverse or Fixed Weight Systems

Fiber	Location of Use	Reel Size	Table	Length Unit	Weight Unit	Yards Per Pound in No. 10	Known to Be in Use
cotton	Great Britain and United States	54″	reel = 1 thread = 1½ yds. 80 threads = 1 skein = 120 yds. 7 skeins = 1 hank = 840 yds. 18 hanks = 1 spyndle = 15,120 yds.	hank of 840 yds.	1 pound avoirdupois	8,400	18th c. — present [22]
cotton thread	England	54″	reel = 1 thread = 1½ yds. 280 threads = 1 hank = 420 yds.	hank of 420 yds.	1 pound avoirdupois	4,200	early 19th c.[23]
ounce thread	Great Britain	36″	reel = 1 thread = 1 yd. 30 threads = 1 hank = 30 yds.	hank of 30 yds.	1 ounce	4,800	18th c.[24]
cotton thread	Brandywine-Philadelphia region, United States	42″	reel = 1 thread = 1 1/16 yds. 20 threads = 1 knot = 23⅓ yds. 6 knots = 1 hank = 140 yds. 16 hanks = 1 bunch = 2,240 yds.	hank of 140 yds.	1 pound avoirdupois	1,400	early 19th c.[25]

bump cotton	Great Britain and United States			1 yard	1 ounce	160	20th c.[26]
linen	Scotland	90″	reel=1 thread=2½ yds. 120 threads=1 cut =300 yds. 2 cuts=1 heer=600 yds. 6 heers=1 hank =3,600 yds. 4 hanks=1 spyndle =14,400 yds.	cut of 300 yds.	1 pound avoirdupois	3,000	17th c. — present[27]
linen	Ireland	90″	reel=1 thread=2½ yds. 120 threads=1 lea =300 yds. 12 leas=1 hank =3,600 yds.	lea of 300 yds.	1 pound avoirdupois	3,000	18th c. — present[28]
linen	Yorkshire and England	108″	reel=1 thread=3 yds. 100 threads=1 lea =300 yds. 10 leas=1 hank =3,000 yds. 20 hanks=1 bundle =60,000 yds.	lea of 300 yds.	1 pound avoirdupois	3,000	1840's — 20th c.[29]

Inverse or Fixed Weight Systems

Fiber	Location of Use	Reel Size	Table	Length Unit	Weight Unit	Yards Per Pound in No. 10	Known to Be in Use
linen	Ireland and rest of Britain	54″	reel = 1 thread = 1½ yds. 100 threads = 1 lea = 150 yds. England: 10 leas = 1 hank = 1,500 yds.; 40 hanks = 1 bundle = 60,000 yds. Ireland: 12 leas = 1 hank = 1,800 yds.; 33⅓ hanks = 1 bundle = 60,000 yds.	lea of 150 yds.	1 pound avoirdupois	1,500	19th c. — present [30]
silk (spun)	Great Britain and United States	54″	reel = 1 thread = 1½ yds. 80 threads = 1 skein = 120 yds. 7 skeins = 1 hank = 840 yds.	hank of 840 yds.	1 pound avoirdupois	8,400	20th c.[31]

Direct or Fixed Length Systems

Fiber	Location of Use	Reel Size	Table	Length Unit	Weight Unit	Yards Per Pound in No. 10	Known to Be in Use
wool	Aberdeen, Scotland	90″	reel = 1 thread = 2½ yds. 120 threads = 1 cut = 300 yds. 12 cuts = 1 slip = 3,600 yds. 4 slips = 1 spyndle = 14,400 yds.	14,400 yds.	pounds avoirdupois	1,440	19th c. — 20th c.[32]
worsted (and wool?)	Sowerby Bridge or Halifax Rural District, Yorkshire	36″		wrap or hank of 80 yds.	drams	2,048	19th c. — 20th c.[33]
wool	Cumberland			bunch of 3,360 yds.	ounces	5,376	19th c. — 20th c.[34]
wool	Grain Scale, United States			20 yds.	grains troy	14,000	19th c. — 20th c.[35]
jute, dry spun flax, hemp	Scotland (Dundee at least) and Ireland	90″	reel = 1 thread = 2½ yds. 120 threads = 1 cut = 300 yds. 2 cuts = 1 heer = 600 yds. 6 heers = 1 hank = 3,600 yds. 4 hanks/hesps = 1 spindle/spangle = 14,400 yds.	spindle of 14,400 yds.	pounds avoirdupois	1,440	19th c. — present[36]

Direct or Fixed Length Systems

Fiber	Location of Use	Reel Size	Table	Length Unit	Weight Unit	Yards Per Pound in No. 10	Known to Be in Use
linen	Dorset and Somerset, West of England	81″	reel = 1 thread = 2¼ yds. 40 threads = 1 knot = 90 yds. 20 knots = 1 ran = 1,800 yds. 12 rans = 1 dozen = 21,600 yds.	dozen of 21,600 yds.	pounds avoirdupois	2,160	19th–20th c.[37] Possibly used as an inverse system with fine linen yarn
silk (thrown, artificial, and raw)	Britain: denier scale (old titrage of Lyons); and United States (raw)	47″	reel = 1 thread = 47″ 400 threads = 1 skein = 522 yds.	hank or skein of 476 meters (520.56 yds.)	denier of .0531 grams	445,073.89	18th c. — 20th c.[38]
silk (thrown)	Britain: dram scale (Manchester; thrown silk count); and United States (thrown silk)			hank of 1,000 yds.	drams	25,600	19th c. — 20th c.[39]
silk (raw and thrown)	Europe (legal scale approved by Paris Conference of 1900)			skein of 450 meters	0.05 grams	446,849.5	20th c.[40]
silk (thrown and artificial)	Macclesfield, Cheshire			1,000 yds.	denier (17 denier = 1 dram avoirdupois)	435,200	late 19th c.[41]

Notes to Yarn Tables

[1] These tables do not include any systems peculiar to continental Europe, or any post-1870 reformed ones (Grex, Integral, Metric, Typp, and Tex), or any for compound threads (plyed, doubled or folded yarns). For the first, see Horace Doursther, *Dictionnaire Universel des Poids et Measures* (1st ed., 1840; repr. Amsterdam, Holland, 1965), 135–36; Walter S. Bright McLaren, *Spinning Woollen and Worsted* (London, 1884), 180; Herzfeld, *Technical Testing of Yarns*, 57–79. For reformed systems, see Linton, *Modern Textile Dictionary*, 1068–1072. For compound thread numbering, see *The Textile Recorder*, XLVI (no. 546, September 15, 1928), 49–50.

[2] Partridge, *Treatise on Dying*, 13, 62–65.

[3] Report of Thomas Joyce on the manufacture of wool from sheep of "Spanish mixture," *Letters and Papers of the Bath and West of England Society*, IX (1799), 344–46.

[4] William B. Crump (ed.), *The Leeds Woollen Industry, 1780–1820* (Leeds, 1931), 93, 292, 294.

[5] Vickerman, *Woollen Spinning*, 342–44.

[6] H. R. Carter, *Practical Mill and Factory Arithmetic* (London, 1911), 21, 24; also text note 47.

[7] Sinclair (ed.), *Statistical Account*, II, 308; Thomas Oliver, *Weaving and Designing* (Galashiels, 1911), 11, a reference kindly supplied by Mr. William P. H. Sinclair; Carter, *Factory Arithmetic*, 21, 24. For Berwickshire wool pound of 24 ozs. see Great Britain, *Parliamentary Papers* (*Commons*) 1820 (314), VII, Appendix A, 29.

[8] Carter, *Factory Arithmetic*, 21, 24; Oliver, *Weaving and Designing*, 11.

[9] Carter, *Factory Arithmetic*, 21, 24.

[10] British Standards Institution, *Specification*, 9. This system has been in use for over a hundred years in the Irish tweed industry. Letter from Mr. A. H. Porter, Director, Standards Division, Institute for Industrial Research and Standards, Dublin, to the author, January 28, 1971.

[11] Emmanuel A. Posselt, *Textile Calculations* (Philadelphia, 1896), 9; also text note 60.

[12] William C. Bramwell, *The Wool Carder's Vade Mecum* (Boston, 1881), 364; also text note 56.

[13] Mary M. Atwater, *The Shuttle-Craft Book of American Hand-Weaving* (New York, 1928), 67.

[14] Norfolk: Scobell (ed.), *Acts of Parliament, 1640–1656*, 146–48 for c. 36 (November, 1650); 14 Car. 2, c. 5 (1662); 31 Geo. 3, c. 56 (1791). Yorkshire, Lancashire and Cheshire: 17 Geo. 3, c. 11 (1777). Suffolk: 24 Geo. 3, Sess. 2, c. 3 (1784). Other English counties: 25 Geo. 3, c. 40 (1785). Great Britain generally: Carter, *Factory Arithmetric*, 22, 24. United States: Posselt, *Textile Calculations*, 11.

[15] Carter, *Factory Arithmetic*, 22.

[16] Great Britain, *Parliamentary Papers* (*Commons*) 1820 (314), VII, Appendix A, 30.

[17] Rees (ed.), *Cyclopaedia*, s.v. "Worsted."

[18] Norfolk: 31 Geo. 3, c. 56 (1791). Suffolk: 24 Geo. 3, Sess. 2, c. 3 (1784). The remainder: 25 Geo. 3, c. 40 (1785).

[19] 24 Geo. 3, Sess. 2, c. 3 (1784); 25 Geo. 3, c. 40 (1785).

[20] 31 Geo. 3, c. 56 (1791); 17 Geo. 3, c. 11 (1777); 25 Geo. 3, c. 40 (1785).

[21] 7 Jac. 1, c. 7 (1609).

[22] Great Britain: Robert Owen, *The Life of Robert Owen. Written by Himself* (2 vols., London, 1857), I, 32; Duncan, *Essays*, 319; John Murphy, *A Treatise on the Art of Weaving* (Glasgow, 1824), 407; Carter, *Factory Arithmetic*, 17, 24. United States: J. and R. Bronson, *The Domestic Manufacturer's Assistant* (Utica, N.Y., 1817), 13; Daniel W. Snell, *The Manager's Assistant. Being a Condensed Treatise on the Cotton Manufacture* (Hartford, Conn., 1850), 119; Posselt, *Textile Calculations*, 5. These are typical of the many sources that could be cited.

[23] Rees (ed.), *Cyclopaedia*, s.v. "Manufacture of Cotton."

[24] 28 Geo. 3, c. 17 (1788). Before this act prescribed these standard reel and hank sizes (to prevent "frauds ruinous to the free trader"), ounce thread had been made up in 40 yard hanks. *Ibid.*

[25] Source in text note 55.

[26] British Standards Institution, *Specification*, 8; "Yarn Numbering," *The Textile Recorder*, XLVI (no. 544, July 14, 1928), 37; Linton, *Modern Textile Dictionary*, 118.

[27] Sources in text note 37. To the great loss of spinners, eleven and twelve quarter reels (99" and 108") continued in use in some parts of Scotland after 1726. See Duncan, *Essays*, 244; Murphy, *Art of Weaving*, 402.

[28] Source in text note 38; Marshall, *Practical Flax Spinner*, 209.

[29] Scott, *Practical Cotton Spinner*, 337–38; Sharp, *Flax Spinning*, 173; Fred Bradbury, *Calculations in Yarns and Fabrics* (Belfast, n.d. but *ca.* 1920), 21.

[30] Marshall, *Practical Flax Spinner*, 221–22; Sharp, *Flax Spinning*, 173; Woodhouse, *Yarn Counts*, 12; Bradbury, *Calculations in Yarns*, 21.

[31] *Silk Journal* (comp.), *The Artificial Silk Handbook* (London, 1928), 115; Posselt, *Textile Calculations*, 13.

[32] Duncan, *Essays*, 302, 319 mentions the application of the inverse linen system to wool but (p. 321) also notes a 72" machine reel. Though there must be earlier references to this system, the oldest I have seen is M'Lennan, Blair, & Co. (comp.), *Comparative Yarn Tables*, 1.

[33] Carter, *Factory Arithmetic*, 21, 24; Bradbury, *Calculations in Yarns*, 17, 27; Woodhouse, *Yarn Counts*, 17.

[34] Woodhouse, *Yarn Counts*, 17.

[35] *Ibid.*; "Yarn Numbering," *Textile Recorder*, XLVI (1928), 37–38. According to Bradbury, *Calculations in Yarns*, 18, the length unit was 50 yards. It would appear to have derived from the practice of measuring roving size by weighing samples with grain roving scales. Bramwell, *Wool Carders' Vade Mecum*, 362–363; Werner von Bergen and Herbert R. Mauersberger (eds.), *American Wool Handbook* (2nd ed., New York, 1948), 596.

[36] Oliver, *Weaving and Designing*, 11; Bradbury, *Calculations in Yarns*, 22–23, 27.

[37] Scott, *Practical Cotton Spinner*, 337; H. R. Carter, *Flax and Its Products*, (London, 1920), 85.

[38] Andrew Ure, *The Philosophy of Manufactures* (London, 1835), 103; Carter, *Factory Arithmetic*, 23; Woodhouse, *Yarn Counts*, 17; International Correspondence Schools, *Reference Library*, LXXXVI, Section 11, 14.

As late as the 1930's there were several denier units (the Old Milan denier of 0.0511 gram; the Old Turin denier of 0.05336 gram; and the Old Lyons denier of 0.05311 gram, for example) which caused some confusion and differences in equivalencies. See Herbert R. Mauersberger and E. W. K. Schwarz, *Rayon and Synthetic Yarn Handbook* (New York, 1936), 336–339. I have based my calculations on the French meter length of 39.37 inches and the Old Lyons denier of 0.05311 gram.

[39] Carter, *Factory Arithmetic*, 23; Bradbury, *Calculations in Yarns*, 25, 27; I.C.S., *Reference Library*, LXXXVI, Section 11, 15.

[40] Bradbury, *Calculations in Yarns*, 25, 27.

[41] *Silk Journal* (comp.), *Artificial Silk Handbook*, 112.

IV

British and American entrepreneurial values in the early nineteenth century: a parting of the ways?

INTRODUCTION

In embittered old age in 1828 William Radcliffe, one of the Stockport captains of industry who had almost done for weaving what Arkwright had done for spinning, assessed why his achievements had fallen short of his aspirations. Those aspirations may be taken as a starting point for a comparison of British and American business values in the early industrial period, focused on the cotton industry in Lancashire and New England.[1]

Radcliffe had set up in business in 1789, producing warps for sale to Samuel Oldknow the warehouseman-turned-manufacturer of Stockport, and making muslins on a small scale. After an encounter with foreign merchants in 1794 Radcliffe decided never to sell cotton yarns or warps for export: 'I saw with inward grief this most impolitic and undermining traffic of selling our raw material to continental customers.'[2] It was this market threat, the export of cotton yarn, which impelled the Manchester and Stockport manufacturers (organisers of weaving), bleachers, dyers, printers and merchants, meeting separately, to adopt resolutions calling for export restrictions. Only the factory spinners saw advantage in the situation, arguing that if they could produce more yarn than could be absorbed by weavers at home then they had a right to export it. At the meeting of Stockport manufacturers at the Castle Inn in April 1800 Radcliffe suggested that Britain's weaving capacity could be increased by 'a division of labour and the invention of some plan or other in a straight forward process, something like the carding &c introduced by Sir Richard Arkwright in spinning'.[3] He pledged to devote himself to these improvements and if he succeeded the Stockport manufacturers pledged they would combine with their Manchester counterparts to work for the prohibition of cotton twist exports.

The previous year Radcliffe had taken the son of a rich Scottish merchant as business partner so was able to devote resources to the pursuit of technical innovation. Since these resources amounted to £11,000 the two partners decided to concentrate production and bought the Stockport premises of Oldknow and Richard Arkwright the Younger. One key to increasing handloom production was to mechanise the warping process. This Radcliffe succeeded in doing in 1803-4 when, with the help of a talented weaver-mechanic, Thomas Johnson, he built a practicable warping and dressing machine. To speed up weaving Radcliffe and his workmen designed and built the dandyloom, a handloom with the cloth take-up roll activated by the lay or lathe which the weaver pulled after each knock of the shuttle. These devices he patented, selling the rights to build to other British, but not foreign, manufacturers. By 1806 his profits from handloom weaving, most now concentrated in workshops, amounted to £90-£100 a week, and his patent premiums to about £50 a week.[4] All of this increase in scale and investment in innovation had driven Radcliffe into vast expense. By now he had separated from his partner and was raising capital by mortgaging his Mellor estate, his Stockport premises and his patents. He raised a loan of £5,000 from four 'gentlemen'. Still his credit contracted and in March 1807 he was declared bankrupt. Not until 1817 was he able to pay off all his debts and gain a discharge from bankruptcy. In the meantime his patented ideas had been freely borrowed by rivals. His fellow manufacturers never kept their pledge to fight for restrictions on yarn exports. For his troubles Radcliffe blamed the 'Foreign Anglo Junto' centred on Manchester. Nearly two-thirds of his 216-page autobiography is a rambling and tedious tirade against what he saw as the unpatriotic conspiracy between Manchester and foreign merchants to deprive British manufacturers of trade and workers of employment.

Radcliffe could, however, look back on some civic achievements. At Mellor, where his family had farmed since the seventeenth century and where he once had an estate with a rent roll of £350 per annum, he had a reputation as an improver of roads. After he moved his family to Stockport in 1801 he became a civic leader, holding office as property tax commissioner, captain of volunteers, road commissioner and mayor. When St Mary's Parish Church, Stockport, was rebuilt he served as a trustee. All this is well known.[5]

Especially relevant, however, are Radcliffe's aspirations and the questions they raise about business values. He was concerned about social capital, about employment, and about social stability. He pursued technical innovation in order to create employment, preserve his market share and thus maintain his profits. For the same reasons he liberally shared his inventions with fellow manufacturers and lobbied the government to prohibit the export of machine-made yarns. Altruism and patriotism were the values by which he justified himself. It is hard to say how far Radcliffe's self-serving account can be trusted,[6] or, more importantly, how far he was representative of his class in his thinking. We do not have, but badly need, an analysis of the ideology of British manufacturers in the early industrial period. His aspirations and disappointments do, however, suggest a counterfactual question. Taking his claims at their face value, would he as an entrepreneur have fared any better in the United States? The rest of this chapter uses this question as one of several ways to compare entrepreneurial values in the two societies. Before confronting it, two much broader questions are posed in order to explore the directions of entrepreneurial values. How did society in the two countries view entrepreneurs? How did entrepreneurs in the two nations see themselves?

HOW DID SOCIETY IN BRITAIN AND THE USA VIEW ENTREPRENEURS?

This question might be more easily answered if rephrased 'Which social institutions and mechanisms endorsed the activities of entrepreneurs?' (Entrepreneurs are defined as the owner-managers of firms who finance capital, organise production, decide on the rate of output as they perceive market demand, and who bear the risk entailed in committing resources before outcomes are known.)[7] Here three social mechanisms will be briefly considered: church teachings; legal biases; and perceptions of entrepreneurial success.

Church teachings

There is some debate about what the churches in England taught about business. Tawney would have us believe that from the late seventeenth century until his own day there was a great gulf fixed between religion and business: 'trade is one thing and religion is another' as an eighteenth-century dictum had it.[8] But some of the most influential church leaders of the late eighteenth century had much to say about capital accumulation, not all of it favourable to business. Charles Elliott has shown that John Wesley's 'ideal economic man was a small shopkeeper or master craftsman'.[9] Wesley urged the early Methodists to 'Gain all you can, save all you can, give all your can', warning them against covetousness and expecting entrepreneurs among his followers to give away any income above their bare needs. No capital accumulation was anticipated beyond this point. Not until 1840 apparently did Wesleyan ministers relinquish this principle and thereafter endorse investment for unlimited business growth.[10] Ambivalence about capitalism also marked the teachings of Calvinists, who included Anglican Evangelicals, Baptists and Scottish Presbyterians. Quakers, renowned for their business success, were consistently warned 'that none trade beyond their ability nor stretch beyond their compass'.[11] Only the Unitarians, led by Joseph Priestley, clearly taught that increases in national and personal wealth were godly.[12] By the 1830s and 1840s Anglican clergy, in Lancashire at least, could be found on the side of both masters and men, approving capitalism and defending labour, as Tony Howe and Jane Garnett show in their wide-ranging study. In one case the Reverend William Milton denounced his own patron, a cotton master, for turning his employees into serfs by his 'unchecked absolutism'. His boldness, normally protected by parson's freehold, eventually cost him his living. In another the Reverend Hugh Stowell of Salford castigated masters who lowered wages.[13]

In the United States, to which Priestley fled in 1794, church teaching received the structures and values of capitalism much less critically than in Britain. The post-Revolution separation of church and state meant that all sects relied on popular support far more than in Britain. The parson's freehold, and with it the independence from which Wesley-inspired Anglican Evangelicals thundered against covetousness, was not present in the USA. Tocqueville in 1831 heard the clergy 'inveigh against ambition and deceit', against individual vices but not against the communal shortcomings of business activity.[14] As he put it, 'In the United States religion exercises but little influence upon the laws and upon the details of public opinion; but it directs the customs of the community, and, by regulating domestic life, it regulates the states.'[15] The question remains, however, what happened when new industrial structures, the factory or the joint-stock company for example, no longer fitted community customs whether the clergy tried to defend custom and Christian principle or whether they compromised and modified their stand. It deserves a much more thorough answer than the few impressions possible here but it is important in trying to chart divergent attitudes to entrepreneurship in the period.

In New England, where Samuel Slater's arrival in 1790 heralded the onset of industrialisation, Congregationalists, Baptists, Methodists, Episcopalians and Universalists vied with each other for popular support. Among the Quakers, Slater's mentors and backers, Moses Brown found his narrower religious principles being compromised by business entanglements. Since the Quakers had no professional ministry he was thrown back on his conscience and explanations to fellow Quaker leaders.[16] In rural New England Slater and his son George supported a wide range of denominations but apparently made no attempt to use churches as instruments of control as some English mill masters were able to do. Prude in fact found that religious revival, far from inducing docility in mill-workers, strengthened their resolve to leave the mill and seek new opportunities.[17] To this extent, church teaching had the net effect of increasing labour scarcity and reinforcing the directions in which entrepreneurs sought to innovate.

At Lowell, and in northern New England where the Waltham system prevailed (characterised in part by the employment of girls from New England farms lodged near the factories in respectable boarding houses), the clergy seem to have allied with mill managements. The first church at Lowell, St Anne's, was put up by the mill management and an Episcopal form of service adopted. Congregational, Universalist and Baptist churches followed. All appear to have accommodated their teachings to what was then big business. Indeed the Reverend Henry A. Miles, the Unitarian minister, rallied to the defence of the Lowell corporations when in the early 1840s their treatment of the mill girls was criticised by the Boston newspapers.[18] Out of the First Congregational and Second Universalist Churches, the centres of much social and intellectual activity among the mill girls, came the literary magazines for which the Lowell operatives became famous.[19] Theological debate rather than social criticism seems to have preoccupied the Lowell churches in their early days.[20]

In Boston, the metropolitan base of New England capitalism,

Unitarianism, the most rational and flexible of religious alternatives, prevailed among the commercial–industrial elite. It was essentially conservative and reinforced intensified commercial values. Not surprisingly it attracted a significant proportion of Beacon Hill Brahmins.[21] Between 1828 and 1843 half of all rich church members were Unitarians.[22] By 1850 two-thirds of the wealthiest Bostonians were Unitarians. And 31 of the 34 men elected to Harvard Corporation, the university's governing body, between 1805 and 1860, were Unitarians.[23]

If Rockdale in southern Pennsylvania was typical of the industrial communities further south, then the radical social criticism derived from Enlightenment rationalism gave way to evangelical individualism in the 1820s. Enlightenment reached its zenith in support for the Valley Forge commune, an Owenite socialist experiment. Thereafter, the area was characterised by a close identity between evangelical churchmen (Episcopalian, Methodist, Baptist and Quaker) and small-scale manufacturers like the Smiths, Crozers and Riddles.[24]

In the deep South plantation owners used the services of clergymen, Episcopalian or any others, to teach their slaves the duty of obedience.[25] In the North, where the economy had no use for slave labour, religious conviction fed anti-slavery sentiment.

Out on the industrial frontier before the 1850s religion had a mixed impact, but it was rarely critical of the new industrial order. Sometimes it made little impression at all, as at Harper's Ferry.[26] In upstate New York, revivalism produced utopian industrial and social experiments wholly inappropriate to industrial growth.[27] Equally, revivalism was capable of reinforcing new industrial values. Charles G. Finney, the most furious and popular of frontier evangelists along New York's 'psychic highway', recorded in one of his lectures: 'You [a merchant] are bound to be as pious, and aim *as singly* at the glory of God, in selling goods, as he [a minister] is in preaching the Gospel.'[28] In their respective studies of retailers in Rochester, New York, and paper manufacturers in Berkshire County, Massachusetts, Paul Johnson and Judith McGraw have shown that revivalism might bolster the new capitalist values of shopkeepers and manufacturers alike. If Johnson is right, revivalism reinforced individualism with the notion that all persons are free moral agents. By the late 1830s Berkshire County mill-owners were governing churches, exercising their religious positions to uphold

Sabbatarianism and temperance among their workers in the interests of order and discipline, but not among themselves.[29]

From this impressionistic evidence it would seem that church teaching offered fewer checks on entrepreneurs in the antebellum United States than in contemporary Britain. Even from these scattered sources two things are clear. First, the Church of England offered its parsons (though not its curates) an almost impregnable position from which to sound the most uncomfortable of messages to their congregations and the wider world. There is evidence that clergy did not hesitate to rebuke mill-owners and capitalists when these were perceived as exploiting their workers in defiance of Biblical precepts on justice, poverty and weaker neighbours. Nothing quite like this relationship was possible in the newly-settled industrial districts of the United States. In the older seaport cities and the original colonial settlements churches might have accumulated property to perpetuate and protect themselves. Even then, though, their clergy did not have independence from their own congregations. Second, throughout the United States there is evidence that before the Civil War clergymen of all denominations allied themselves to newly-emerging industrial capitalists. In these circumstances American entrepreneurial values were less constrained by religion than in England.

Legal biases

Unlike in England, where the ascendancy of landed aristocratic interests ensured that traditional concepts of common law persisted well into the second half of the nineteenth century, United States common law was soon refashioned to serve the objectives of the mercantile and entrepreneurial classes. Morton Horwitz has provided a classic analysis of these developments.

Common law, was applied in a fixed and functional manner in England but was employed creatively and instrumentally in the United States to facilitate economic development. 'As political and economic power shifted to merchant and entrepreneurial groups in the post-revolutionary period, they began to forge an alliance with the legal profession to advance their own interests through a transformation of the legal system.'[30] Property rights, which might be limited by the liberties of neighbours in England, were transformed in the United States between 1790 and 1840 to support the 'newly-paramount virtues of productive use and development'.[31]

This was clearly evident in the power of eminent domain, a concept by which land was more swiftly redistributed for new economic purposes such as roads, canals and railways. Americans also moved quickly to develop commercial law. The main achievements were numerous, among them were the cases and judgements establishing negotiability of recognised commercial instruments like bills of exchange; an insurance law based on actuarial understanding of risk and chance; and, after 1825, an assault on the usury laws which by the 1850s left the American commercial community in much the same position as the British whose usury laws were repealed in stages in 1833, 1854 and 1872.[32] The greatest obstacle to American entrepreneurs, the diversity of jurisdiction between local laws, was settled in 1842 by the Supreme Court decision in *Swift v Tyson*. This extinguished the possibility of the state's rights supervening over national economic order — as it had, for example, in Kentucky, where insolvent debtors had been protected by local law from out-of-state creditors, causing commercial chaos in the state in the 1820s.[33] The effects of this legal revolution put the United States ahead of Britain as a society in which the law endorsed entrepreneurial values. While the common law buttressed agrarian values in Britain, in the United States it served commercial and developmental values.

Two examples, of relevance to the case of William Radcliffe, will illustrate and sharpen some of these impressions. The first relates to the mobilisation of capital and the protection of risk. In Britain the characteristic business organisation until the mid nineteenth century was the partnership. A few firms with large capitals, for example insurance or railway companies, secured acts of incorporation. They were untypical before the 1850s and the passing of the Companies Acts which facilitated the formation of joint-stock limited-liability companies.[34] Partnerships suffered from two disabilities: their members were limited in number and hence the amount of capital they might pool was limited; and their members' liability for the firm's debt extended to all their private wealth. It was not surprising, therefore, that in growth industries like the North-West's cotton manufacturers where experience was lacking and risks were high, business failure and bankruptcy rates were also high. However, as Julian Hoppitt emphasises, short-term factors played an important part.[35] When manufacturers relied on credit, as they increasingly did, then factors such as war, harvest and the trade cycle could bring down an over-extended business, like that of William Radcliffe.

In the United States the law responded much more quickly to take account of entrepreneurial values. The member states of the new nation readily granted charters, conferring both joint-stock and limited-liability status, to business corporations. Between 1781 and 1800 sixteen states granted 328 charters to business corporations. Two-thirds were for highways, a fifth for banking and insurance firms and a tenth for local public services. Only a tiny number incorporated manufacturing activities. Nearly 60 per cent were awarded by New England states.[36] Between 1800 and 1820 many more factories were incorporated by the various states. A Congressional source listed approximately 489 manufacturing establishments that had been incorporated in this period, their total authorised capital being in the region of $55 million.[37] The great majority were textile establishments. Some 138 of the 489 were chartered by the state of Massachusetts, 190 by New York. Interestingly, the one state which resisted the practice of incorporating factories was Rhode Island, where the structure of manufacturing was closest to that of England. Here insolvency rates greatly exceeded those of Connecticut and Vermont, though were far behind those of New York (in figures unrelated to population).[38]

Early on, however, competition became an important business value endorsed by the American courts, as Horwitz has demonstrated. While the new corporations were hedged by jurists' decisions on the key issues of property relations and the relationship between the state and entrepreneur, they were not shielded from competition. The state resisted unchecked monopolies, in the area of public works at least, as demonstrated by the *Charles River Bridge* case of 1837.[39] This support for competition derived in part from the earlier concept of non-recoverable injury to property. In short, American courts endorsed competition, rather than monopoly, as a norm for the big-business organisations they were in the process of shaping.

The other example illustrating a widening difference between British and American practice was in the area of patent law. In Britain, as the late Harry Dutton showed, the judiciary long remained hostile to inventors, regarding it as their duty to protect the public interest from monopoly power. Not until the 1830s was there much sign of a change.[40] Despite some slight loosening of the patent process then, the British patent system laboured under massive structural inefficiencies. From the point of view of the inventor, patenting was expensive, uncertain and administratively byzantine.[41] Little changed before the 1850s.

Very different was the situation in the United States. Two features of the American patent system, established by two Acts in 1793 and 1800 and in force until 1836, made it more in tune with the entrepreneurial values of innovation and development than the British system. First, the American system was vastly cheaper. Whereas a British patent cost at least £70 in 1815 and by 1829 at least £150, an American patent cost $30 or £6 15s. While the British patent would cost around three times the annual wage of a cotton factory overseer, the American patent cost a fifth of a skilled artisan's wage. Second, an American patent could be obtained much more speedily: between 1 and 15 days, compared to 2 or 3 months in England by 1829.[42] Part of the difference lay in the fact that British patents were examined for novelty while American patents prior to 1836 were simply registered. To that extent, the American system was inefficient as well as democratic in its inducements to innovators. But when the evils of non-examination were removed in 1836 the system remained relatively inexpensive. According to a rough international comparison it was far more effective than Britain's. By 1851, when the United Kingdom's population was 75 per cent that of the USA's population of 24 million, its patent awards (455) were 21 per cent of those (2,200) in the United States.[43]

The law in the United States, as it related to business organisation and technical change, evidently promoted entrepreneurial values more effectively than in Britain. However, the bias of American law should not be misunderstood. It was not inevitably on the side of the corporation rather than the individual entrepreneur — as some of the early patent cases over textile inventions show. The federal courts, adjudicating in inter-state suits, awarded only 1 of 8 actions between the Boston Manufacturing Company and infringers of its patents to the BMC. The evidence, however, is not easy to interpret. When a corporation was the infringer of the small man's invention, as in the case of Aza Arnold of Rhode Island whose differential roving frame was illegally copied by the Proprietors of the Locks & Canals of Lowell (machinery makers for all the Lowell corporations), he received compensation, though nowhere near what he claimed.[44]

Perceptions of entrepreneurial success

Numerous recent studies have attacked the 'rags to riches' myth. Demonstrably in the early nineteenth century few rose from the working classes to become captains of industry either in Britain or the United States.[45] Despite that, a myth of the self-made man blossomed in the eighteenth century, reaching new peaks in the 1830s, as Francois Crouzet has shown: 'Almost as soon as industrialists had emerged, the myth embraced them, possibly because their contemporaries, used to a slower tempo of change, were struck by the speed with which some factory masters made their fortunes.'[46] Myth, in the wider sense of public perception of entrepreneurs, also flourished. This section of the chapter offers sidelights on the strength of that myth in the cotton industry in Britain and the USA before the 1840s.

The entrepreneur myth was surely fed in England by the career of Sir Richard Arkwright, the pioneer of factory cotton spinning. The son of a Preston tailor, Sir Richard left real and personal property estimated at close to £500,000 when he died in 1792. His only son, Richard Arkwright the Younger, who died in 1843, left £3,250,000 and was 'almost certainly the richest commoner of the first half of the nineteenth century'. The fortunes of both men were reported in the national press soon after their deaths, the *Morning Herald* erroneously stating that the younger Richard Arkwright left £7 million in personal property alone.[47] No matter that the Arkwrights' wealth latterly derived from real estate and safe government or East India Company stocks: in the public mind Sir Richard's career, from Bolton barber to Matlock magnate, would be for ever linked with a fortune made in cotton spinning. The question is 'How precisely was Sir Richard Arkwright perceived by his contemporaries and the early publicists of the cotton industry?'

Arkwright's entrepreneurial success aroused envy and then denigration. Public expressions of envy began in 1780, just over a decade after Arkwright secured his waterframe patent; defamation of his character almost inevitably ensued in the patent battles of 1781 and 1785. At the last of these, *Rex v Arkwright*, Mr Bearcroft the leading counsel for the crown dramatically pointed to Arkwright and declared, 'there sits the thief'. On the other hand, Arkwright was honoured for his achievements. In Glasgow and Perth he was given the freedom of the city; in 1786 he was knighted and the following year was appointed High Sheriff of Derbyshire. Yet social recognition was limited. A king's man he might be: the ladies of the court smirked behind his back and called him 'little fatty'.[48] Sir Richard was a parvenu, at most a knight of the county, lacking the birth and manners (though not the wealth) to qualify for an hereditary title. He also remained the only cotton knight until the 1840s.

The early chroniclers of the cotton industry were divided in their estimates of Arkwright. On the one side there were Richard Guest, Edward Baines Jr and one or two other detractors. On the other there was what might be called the Scottish or heroic, view. Richard Guest, of Leigh, in 1823 expressed the grievances that had long been nursed by those Lancashire mechanics, tradesmen and manufacturers on whose shoulders Arkwright had raised himself to fame and fortune. To cut Arkwright down to size Guest revived the claims of Thomas Highs, to whom he assigned the invention of both the spinning jenny and the waterframe, 'By borrowing his ingenious inventions the late Sir Richard Arkwright lived to acquire a princely fortune, and died with the reputation of being one of the most eminent of those individuals, who have enlarged the resources of their native country, and made her manufactures and machinery the wonder of surrounding nations.'[49] Guest was not the only sniper at Arkwright's reputation before the 1830s. John Corry, the Lancashire historian, nominated James Hargreaves of Blackburn (patentee of the jenny in 1770) as the inventor (Arkwright as the improver) of the waterframe and argued that Hargreaves' name was 'entitled to the same superiority as that of Columbus over Americas (sic)'.[50] A decade or so later the topographer, James Wheeler of Prestwich, judged that Guest's narrative had 'a considerable degree of plausibility'.[51]

The severest blows to Arkwright's reputation came from Edward Baines Jr. Baines first advanced the claims of Highs in the chapters he wrote for his father's history of Lancashire, published in 1824.[52] In the much more comprehensive history of the cotton industry that he published under his own name a dozen years later he expanded his case. He credited Arkwright with a good measure of ingenuity, especially in 'constructing the factory system — itself a vast and admirable machine, which has been the source of great wealth, both to individuals and the nation'.[53] However, Arkwright was not the great inventor others vowed him to be: 'I have found myself compelled to form a lower estimate of the inventive talents of

Arkwright than most previous writers.' The waterframe and carding inventions, Baines asserted, 'belong in great part to other and much less fortunate men'. This assessment produced another defect in Arkwright's reputation. Engineering limitations were excusable; not so the moral shortcomings with which Arkwright had claimed inventions not his own. These 'left a stain upon his character, which the acknowledged brilliance of his talents cannot efface'. The ' "*unbounded invention*" which he claimed for himself, and which had been too readily accorded to him, — the *creative faculty*, which devised all that admirable mechanism — so entirely new in its principles, and characteristic of the first order of mechanical genius — which has given a new spring to the industry of the world, and within half a century has reared up the most extensive manufacture ever known — *this* did *not* belong to Arkwright'. He was responsible merely for 'perfecting the details'.[54] Baines, later Sir Edward Baines (1800-1890), was a Leeds journalist, and the son of one;[55] perhaps this gave him a sceptcism denied to the Scots.

One of the first cotton-industry historians with Scottish connections was James Thomson (1779-1850), a calico printer of Clitheroe and contributor of the article on 'Cotton' (published 1808) to Abraham Rees's monumental *Cyclopaedia*. He had been a student in Glasgow.[56] Echoing the self-made man myth, Thomson saw Arkwright as 'this distinguished character, whose perseverance and invention raised him from one of the most humble occupations in society to affluence and honour'. He brushed aside the awkward claims of Highs, seeing in them Arkwright's 'good fortune' in securing 'the co-operation of some persons who saw the merit of the invention, and were willing to assist his endeavours' and as another aspect of Arkwright's 'reiterated contests with the rival manufacturers of Lancashire'. He cleverly rationalised the undeniable defeat of Arkwright in the carding machine patent trial of 1785: the crank and comb had existed 'in a much less efficacious and approved manner' before Arkwright but he had spotted how to improve it for the carding machine, only to have others anticipate him.[57] Thomson's view was echoed by James Butterworth, postmaster of Oldham and historian of the cotton trade of Manchester, who borrowed from Rees's *Cyclopaedia* in his work.[58]

In chronological sequence the Scots who wrote in defence of Arkwright between 1815 and 1840 were Kennedy, Bannatyne, McCulloch, Montgomery and Ure. Glaswegian Dugald Bannatyne, wrote in a supplement to the *Encyclopaedia Britannica.*[59] John Kennedy (1769-1855) and James Montgomery (1794-1880), both cotton spinners, are considered in the next section. Kennedy, however, was allegedly the link between the Arkwright family and the Scottish economist John Ramsay McCulloch (1789-1864), author of a lengthy review article in the *Edinburgh Review* in 1827.[60] There McCulloch dismissed Guest's book as a 'very meagre, prejudiced and superficial book'. Arkwright was unquestionably the heroic inventor who had 'triumphed over every opposition. The same engenuity (sic), skill, and good sense which had originally enabled him to invent his machine and get it introduced, enabled him to overcome the various combinations and difficulties with which he had subsequently to contend'.[61]

An even more ecstatic promotion of Arkwright's reputation was made by Dr Andrew Ure (1778-1857), former chemistry professor at Anderson's College, Glasgow, and in the 1830s a commercial chemist and prolific apologist of the factory system.[62] Ure hailed Arkwright as the supreme figure in the British cotton industry: 'Arkwright alone had the sagacity to discern, and boldness to predict in glowing language, how vastly productive human industry would become, when no longer proportioned in its results to muscular effort.' The factory system, as a machine and as a work discipline was:

> the Herculean enterprise, the noble achievement of Arkwright. . . . It required, in fact, a man of a Napoleon nerve and ambition to subdue the refractory tempers of work-people accustomed to irregular paroxysms of diligence, and to urge on his multifarious and intricate constructions in the face of prejudice, passion, and envy. Such was Arkwright, who, suffering nothing to stay or turn aside his progress, arrived gloriously at the goal, and has for ever affixed his name to a great era in the annals of mankind.[63]

In his two-volume *The Cotton Manufacture of Great Britain* published the following year, Ure re-examined the printed evidence of the patent trials to defend Arkwright's reputation against the slurs cast by Guest and Baines on Arkwright's reputation. In thirty-six pages he argued the claims he previously made for Arkwright.[64] The claims of Highs he dismissed as 'not only futile but ludicrous'.[65] Arkwright indeed was like Columbus, in the preparations he made to develop his system; comparable to Newton, in standing on the labours of others; akin to Luther, Calvin and

Knox, as an instrument of Providence, reproving 'the unprofitable, frivolous and corrupt practices' of the day. Ure even managed to transfigure Arkwright's pot belly, captured in the portrait by Joseph Wright of Derby: under Sir Richard's obesity Ure perceived a frame 'athletic in form and power' but worn down by 'factory exertions' and asthmatic ailments.[66]

Thus by the late 1830s Sir Richard Arkwright had two conflicting public images, as the archetype of the self-made heroic entrepreneur and as the plagiarist of other men's ideas. Even Samuel Smiles, publicist of the creed of self-made men, noted the mixed reputation but subdued his doubts about Arkwright's prowess as an inventor by observing that there had also been doubts about the inventiveness of Stephenson and Watt with whom Smiles ranked Arkwright.[67] On balance Smiles heeded Arkwright's Scottish defenders rather than his English detractors. Perhaps this was not surprising: Smiles too was a Scot.[68]

What then does the making of Sir Richard Arkwright's public image tell us about the entrepreneur myth in Britain? First, the self-made man was a substantial component of the myth. There was no denying that Arkwright moved from humble beginnings to great wealth as a result of entrepreneurial success. All writers recognised the social distance business success carried him. Second, however, none of those studied here suggested that his example was one to emulate. Guest and Baines obviously felt there were aspects of his career best avoided. Thomson, Bannatyne, McCulloch, Montgomery and Ure presumably felt it was sufficient to set forth Arkwright's achievements for all to admire and that would be enough to inspire pursuers of entrepreneurial renown. Most believed that his spinning system and the complementary one developed by Crompton (using the spinning mule) would suffice for the cotton industry. Arkwright's system could be refined as much as his technology but none imagined that the machine-based factory system would be superseded.

Further, Arkwright's reputation was perceived in terms of his claims as an inventor, rather than his skills as a businessman. Given that the world of work was being transformed by an avalanche of mechanical inventions, it is hardly surprising that Arkwright's reputation was tied chiefly to machines: they had a visual fascination which Ure and, earlier, the contributors to Rees's *Cyclopaedia*, busily relayed to the public. The management techniques behind the machines were much more difficult to describe (though Arkwright was hailed in more general terms as the architect of the factory system as well as an inventor of genius). The evidence on which Arkwright's image-makers divided, namely the reports of the patent trials of the 1780s, is still so confused that it remains impossible to say who was correct. Bias rather than certainty about Arkwright's inventive talents must therefore explain why some chose to destroy, some to build, Arkwright's reputation.

This leads to the last point about the entrepreneur myth as illustrated by Arkwright's early reputation: the identity of its creators. The builders of the myth in his case were Scots. Quoting an unknown source, Montgomery threw into his account of the history of the cotton industry a telling remark: 'He [Arkwright] was particularly friendly to Scotchmen, and gave them free access to his establishments.'[69] They remembered Arkwright with enormous gratitude and enthusiasm. He had introduced his system to Scotland very soon after perfecting his carding machine, by which he had the technical means to commence factory production. He licensed his first Scottish mill in 1776-7, concurrent with the building of a second mill at Cromford.[70] While English landlords were clearing the Highlands, Arkwright planted the seeds of a new industry which by the 1820s turned Glasgow into a major cotton district in Britain. In essence, the myth of Arkwright's success was promoted in Britain by outsiders, not in terms of practical experience (Thomson and Montgomery were both mill managers) but in terms of industrial perspective. Baines was an outsider too but chose to diminish rather than magnify Arkwright; possibly this reflected his journalist's outlook. Guest spoke for all those Lancastrians Arkwright had injured.

If Arkwright's career was projected with very mixed views and a lack of enthusiasm in some quarters in Britain, how would it have been projected in the United States? Would he have fared very differently had he been responsible for taking his machines and his system across the Atlantic? A good idea of what might have happened to him is suggested by the career of Samuel Slater who did just that.

In a sense Slater treated Arkwright much as Arkwright allegedly treated Highs: he borrowed his inventions and used them for personal gain. The difference was that Slater put the Atlantic and national boundaries between himself and Arkwright. Samuel Slater is usually credited with being the first to transfer British cotton spinning technology from Britain to America after he emigrated there in 1789. In fact, he was moving on a crowded stage and was neither the first nor the only one to build Arkwright's waterframe in the USA. He was, however, the first successful manager of an Arkwright-type mill in the United States; his relatives and subordinates carried the new technology through New England; and he became one of the longest-surviving entrepreneurs from those early days.[71] After his death in 1835 at the age of sixty-seven he was the subject of a book-length memoir.[72] A comparison between that biography and the manner in which Arkwright was memorialised is revealing.

First, Slater received a book-length treatment within a year of his death. Remarkably, no book-length study was accorded to Arkwright until nearly two hundred years after his death when Fitton's biography appeared.[73] In fact, White's biography of Slater was a pot-pourri of excerpts from other works, with just a fraction of the volume about Slater. No matter: Slater was given hard covers and Arkwright was not.

Second, Slater's biographer stood as far, perhaps even farther, outside the American cotton industry as the Scots apologists were beyond the English cotton industry. Like his subject he was a British immigrant; unlike Slater he was a professional man, a clergyman. The Reverend George Savage White (1784-1852), born in Bath and the son of a clergyman, had trained at Cheshunt College and been ordained in the Presbyterian ministry in his twenties. In 1811 he emigrated to the USA with several friends and members of his family and settled at Tiverton, Rhode Island. About a year later he was ordained into the Episcopal Church by Bishop Alexander V. Griswold (to whom he dedicated the Slater memoir) but by 1814 had given up the active ministry. Much of the remainder of his life was spent at Canterbury, Connecticut, where he took up various controversial causes including the struggle against slavery. Canterbury was close to the Rhode Island line and about 25 miles from Providence and its northern suburb Pawtucket where Samuel Slater had begun his American career.[74]

Third, despite the illegality involved in transferring British technology to the United States, Slater's achievement was unquestionably hallowed by his biographer. As much as Arkwright (assuming he was to some extent a borrower of others' ideas), Slater stole other men's inventions. By 1789 none was covered by patent restrictions so he was not infringing laws on invention. However, in going to the USA with British technology he was, in British eyes, perpetrating an even more serious offence. He violated Britain's prohibitory laws and thereby betrayed his country's technical secrets to erstwhile colonists and rebels. White knew all this but conveniently passed it over.[75] In America Slater was assimilated into the culture of his choice: he married the daughter of a Pawtucket blacksmith, his brother from Derbyshire joined him in Rhode Island and the two became leading manufacturers in the state.

After his death though, there was the chance that Slater's reputation might suffer like that of Arkwright. White was thoroughly familiar with the work of Baines and Ure, published in the year of Slater's death, (and quoted them) and was well aware of their disagreements over Arkwright's reputation. As he remarked in his Preface:

> without correct information the liability is, for the whole account of their [the cotton manufactures] rise and progress at some future period to run into fiction and fable; and the man who was most instrumental in introducing them, instead of being viewed as a plain practical mechanic, using honest means for his own benefit, and at the same time promoting the best interests of this country, to be ranked among fictitious characters, and to have his name and fame some way mysteriously associated with the business which he has permanently established.[76]

Can we detect behind these phrases White's fear that Slater — who had betrayed the country of his birth to benefit the country of his adoption — would suffer the same fate as Arkwright? If so, we can understand the speed with which White set about the task of cementing together a record that would preserve an acceptable public image for Slater. Yet White had mixed qualifications for the task. He called Slater his friend but did not write as if he knew him intimately.[77] His understanding of cotton manufacturing was second-hand and gleaned by relatively little study so that 'what appeared complex and obscure to me at first, now appears pleasing and easy to be understood'.[79] Originally he 'only intended a memoir of my friend' but 'those whose opinions had weight with me, said the public needed an historical essay on the rise and progress of manufactures'.[79] In this context it was significant that White travelled to Philadelphia where Dr James Mease, a well-known physician and propagandist for the extension of science and industry, lent him assistance.[80] Together they assembled a collection of documents that defended industrialism and the factory system and slanted American textile history towards the Providence–Philadelphia axis and away from the Boston-Lowell axis.

Slater was almost lost under this weight of evidence. Long chunks from a mixture of sources about the British and American cotton industries — mainly Hamilton, *Niles's Register*, the *Edinburgh Encyclopaedia*, Montgomery, Baines, Ure, Babbage, and Zechariah Allen — obscured him. At the Rhode Island Historical Society White gained access to a few papers which Slater had bequeathed relating to his emigration to America. They comprise the small core of original materials in the biography that remain of value to the modern historian. In effect, White succeeded in burying his subject. His book contains a pathetically thin study of Slater which, removed from the volume, is probably not much longer than Ure's treatment of Arkwright. In a sense, though, White's presumed objective was attained. Slater has come down to posterity as a pioneering entrepreneur with his reputation intact and unblemished. All that readers would recall about Slater, and all that White wanted them to remember about him, was the sub-title of the biography: 'Samuel Slater, the Father of American Manufactures'. The imperfections of White's account attracted some critical comment from contemporaries but Slater's reputation remained unsullied.[81]

Of course, one or two case studies do not necessarily prove the point. In America, before the Civil War at least, the kind of adulation accorded Slater posthumously was typical of that lavished on the most successful of departed business figures like Stephen Girard and John Jacob Astor, as Sigmund Diamond found.[82] Arkwright and Slater, however, were the giants among the crowded communities of entrepreneurs in the early industrial periods of Britain and the United States. The differing ways in which their reputations were publicly treated illustrate differing perceptions of entrepreneurial values in the two societies. While rather more investigation is needed, in Britain posterity apparently subjected entrepreneurs' reputations to a bloodier mauling than in the United States. Why, will be hazarded in the next section.

HOW DID ENTREPRENEURS IN BRITAIN AND THE USA SEE THEMSELVES?

What follows is brief, impressionistic and tentative, rather than exhaustive and definitive. Here this critical question is approached from four directions. How did entrepreneurs see themselves as employers of labour? How did they perceive their utilisation of capital? How did they see themselves in the future? How did they understand their past?

In Britain the most enlightened entrepreneurs adopted the stances of humane paternalists. The Strutts at Belper, the Gregs at Quarry Bank, Styal, the Ashtons at Hyde, the Ashworths at Egerton near Bolton: these were the models Ure recorded in his defence of the factory system.[83] These employers consciously sought improved physical, educational, moral and religious conditions and facilities for their employees. Despite assaults of left-wing historians, who have interpreted the devices of paternalism as typically instruments of worker control,[84] the fact remains that for their day these model paternalists offered the best working conditions available within the factory form of organisation. Regrettably, these model paternalists were a small minority of all employers. At most they numbered 40 or 50 out of 4,800 cotton and woollen mill owners. The vast mass of these presented 'a picture of neglect, abuse, oppression, cruelty and indifference'.[85]

As paternalists, these new industrialists inherited two codes. From much earlier times came the assumptions that society was authoritarian and stratified, with many spheres of authority each with its own hierarchy. In this world view the wealthy, privileged and powerful faced the principal duties of ruling, guiding and helping their dependent subordinates.[86] The second was the Puritan code, the work ethic underpinned by a sense of transcendant purpose and a belief that temporal behaviour had eternal consequences.[87] Model captains of industry in Britain regarded themselves as responsible not only for recruiting, training, paying and disciplining their employees but also for providing housing, education, health, leisure and religious opportunity for them and their families. Model industrial paternalism in Britain (if Ure's reports are to be trusted) was all-sufficient, all-enveloping and rural. Its object was to protect the employee from himself and from exploitation by the world outside the model factory community, and to fit him or her for the world to come. It did not envisage much independence on the part of the individual or much social mobility, and it rested on the values of rural rather than urban communities.

Rather different was the view of labour which American entrepreneurs entertained. They faced a dilemma not shared with British counterparts: how could industrial paternalism be compatible with democracy? Jefferson, in his *Notes on the State of Virginia* (1788) revealed the fear that manufactures would corrupt American morals and destroy the independence of the citizens of the new republic producing 'The mobs of great cities [who] add just so much to the support of pure government, as sores do to the strength of the human body'.[88] Later, however, more moderately, if not always readily, Americans came to see that manufacturing independence could complement political independence and might be checked when it threatened non-manufacturing community interests.[89] Yet into the 1820s and 1830s an abhorrence of some forms of industrialisation, as exhibited in Europe, persisted. Zachariah Allen, the Rhode Island woollen manufacturer, observed, 'God forbid, however fondly the patriot may cherish the hope of increasing the resources of his country by opening and enlarging the channels of national industry, that there ever may arise a counterpart of Manchester in the New World.'[90]

If Americans resolved the dilemma, they did so in one of three ways. Some manufacturers copied English models of rural factory paternalism. This was the route chosen by Slater and his followers in the so-called Rhode Island system.[91] As in Britain, whole families were hired to work the mills. Francis Cabot Lowell and his fellow Boston merchants found a different way to solve their labour scarcity and labour discipline problems without incurring the destructive forces of industrialism then afflicting British cities. They recruited their unskilled workforce from among the daughters of New England farming families, housing the girls in boarding houses leased to reputable matrons: a solution that was replicated through northern New England and with greatest force at Lowell.[92] If Slater, in Barbara Tucker's words, 'tried to construct a bridge to the past',[93] then Francis Cabot Lowell and the Boston Associates, in relying so heavily on a female workforce, built a bridge to the future. Thirdly, in Philadelphia a large handicraft weaving industry complemented the factories by offering production flexibility and thereby survived well beyond the Civil War.[94]

In contrast to the Philadelphia model the Waltham–Lowell and the Rhode Island systems originally rejected an urban industrialism of European dimensions. Both presupposed a measure of paternalism and a rural environment. Both expected moral and educational 'uplift' (Prude's word)[95] to flow from the factory into the countryside. When the underlying pattern of economic relations was disturbed accusations of tyranny might develop into strike action, as happened in the 1830s. At Lowell, as Dublin found, the reasons why 800 women went on strike in 1834 in response to a 12.5% wage reduction revealed a great deal about how their employers treated them. They turned out because the cut threatened social 'bondage' and undermined their economic 'independence'.[96] Independence, indeed, was popularly regarded as one of the prime characteristics of the Lowell mill girls.[97] Under the Waltham system employers regarded independence, however illusory, as proof that industrialism could be reconciled to democracy.[98] Such an evaluation had moved them some distance from British paternalists.

There was another departure at Waltham. This is apparent from the way in which entrepreneurs viewed themselves as deployers of capital. In Britain, as already noted, there were legal constraints on the pooling of capital. Eventually this inhibited industrial growth and in the 1850s they were swept away. No such inhibitions operated in the United States and the Boston Associates took full advantage of joint-stock, limited-liability at Waltham, Lowell, Lawrence and elsewhere in New England. The concentration and large scale of capital investments was not the only pattern laid down at Waltham.

Technical solutions were sought to meet the pre-existing imbalance between America's productive resources. The price of labour in relation to capital equipment costs appeared high in the United States compared to Britain, and its supply less elastic. This was particularly true of unskilled labour in the early nineteenth century.[99] On the other hand the United States had enormous natural resources. In response to this situation Waltham system engineers developed a technology that was power- and capital-intensive.[100] More to the point of this general discussion, their successes fed on and strengthened the widespread faith in technical innovation, the belief that machine shops could master nature. Unlike Arkwright's admirers, American manufacturers and mechanics did not believe that the factory system was already perfected, technically or organisationally. The immigrant mill manager James Montgomery in 1840 fully expected Americans to improve technology and management to such an extent that they would be able to compete successfully against the British.[101] As Joseph Whitworth put it after visiting America in 1854, 'The [American] workmen hail with satisfaction all mechanical improvements, the importance and value of which, as releasing them from the drudgery of unskilled labour, they are enabled by education to understand and appreciate.'[102] Such a view was as much held by entrepreneurs as by ordinary workmen. It led Americans into the interchangeable system of manufacture and, eventually, mass production.[103] In contrast, as William Radcliffe's concern illustrated, some early nineteenth-century British entrepreneurs preferred innovations that used labour. The attitude derived in part from seventeenth- and early eighteenth-century anxieties about unemployment and opposition to labour-saving innovations.[104]

The ways in which British and American entrepreneurs saw their future, in the sense of encouraging their children to follow them, were markedly diverging. The wealthiest entrepreneurs in both countries sent their sons to the best schools. In England this meant one of the Nonconformist academies (or similar institutions), which offered a liberal and vocational education,[105] or else one of the public schools, which offered a classical one. Richard Arkwright the Younger sent all five of his sons to Eton and most of them to Trinity College, Cambridge.[106] The chances of such sons following their fathers into business seem to have diminished in the first half of the nineteenth century. In the period 1825-50 23 per cent of fathers of boys attending Merchant Taylors' were men in business compared to 6 per cent of their sons who went into business. At Harrow on the other hand the proportions were 0.5 and around 5 per cent respectively.[107] In 1865 only 10 per cent of steel manufacturers are known to have attended public school.[108] The public school pattern is significant. Nineteenth-century public schools were designed to produce Christian gentlemen. The essence of the gentleman was that he never soiled his hands with hard work. The acquisition and maintenance of manners, land, mansion and title were his concern. Consequently, it is held, there was a haemorrhage of entrepreneurial talent: the younger generation were subverted by the gentlemanly ideal and the aspirations of gentrification.[109]

No such haemorrhage occurred in New England, as recent work has shown. Quantifying the education of the New England elite, Peter Dobkin Hall found that by the last four decades of the eighteenth century 50 per cent of the sons of nine Boston merchant families attended college.[110] Ronald Story's fine study of Harvard tells us a great deal about the men who patronised and the young men who attended that institution (though he, like Dalzell, does not sample the business community to see what proportions of sons were not sent to Harvard).[111] Quite clearly Boston's business community fed and fed on a privileged educational system crowned by Harvard College. At this small institution,[112] students whose fathers were in business rose from about 25 per cent before 1825 to around 50 per cent by 1860. Of 22 antebellum Boston millionaires who had sons, all but 5 sent them to Harvard.[113] Unlike their British counterparts, Boston capitalists secured a high degree of control over their educational engine. After 1830 they dominated Harvard Corporation, tightening administrative and financial procedures and ensuring that the college produced men who were both 'gentlemen of liberal accomplishments' and fitted for careers in 'business, teaching, government, law and medicine as well as the ancient standby, the ministry'.[114] When divinity professors became infected by ideas of atheism and jacobinism in the late 1830s a manufacturer-member of the corporation was urged to have the Divinity School sold and turned into a steam-driven stocking mill. While this seemed rather extreme, business benefactors certainly switched their philanthropy from theology to science, from the 1840s.[115]

The effects of capitalist-controlled higher education emerged in the occupational distribution of Harvard graduates. Those entering business rose from 17 per cent in the period 1798-1830 to 24 per cent in 1835-60.[116] While Harvard patrons, professors and students looked to Oxford and Cambridge, as well as continental universities, for models of gentlemanly, elitist education, they did not seek slavish imitation. As Story argues, in Harvard's perceptions of England 'The real factor was comparability of social function: the extent to which Harvard would produce "heroes and statesmen" as Oxford and Cambridge produced the "great men to whom English liberty and English literature owe their support" '.[117] Social eminence in England derived from wealth based on lands and trade, status from an aristocratic honours system. In New England the Boston Associates, heirs of a mercantile elite, preserved their social eminence by squeezing dividends out of their manufacturing companies to fuel private grandeur and the philanthropic enterprises that satisfied Unitarian and Episcopalian consciences and upheld the social balance of power. Urbane manners, a Harvard education, a fine house (on Beacon Hill), membership of an exclusive club (like the Boston Athenaeum), trusteeship of a great charity (like the Massachusetts General Hospital) and support for the politically conservative Whig party: the prizes of hard-won capitalist endeavour adequately marked their status.[118] In such a world gentrification, excluded by a Puritan and Republican inheritance, did not need to be an option for their sons.

If British and American entrepreneurs' perceptions of the future for their sons and grandsons clearly differed, their treatment of their own industrial pasts sharply diverged. In both countries it took between forty and fifty years for survivors of the first generation of industrialists to awake to the importance of making an historical record of their own past. In Britain this meant a flurry of memoir collecting, discussion, publication and controversy in the decades between 1800 and 1840. In the United States a similar consciousness developed between the late 1850s and the 1890s. Similarity stopped there. Whereas a string of cotton entrepreneurs and one cotton mill manager in Britain wrote up their recollections, in the United States only two or three did so. The first industrialist author in the USA to record the history of the cotton manufacture was none other than James Montgomery, the mill manager who had formerly worked and written in Glasgow. He produced his history in 1840 but it derived largely from White's biography of Slater.[119]

The American authors, Nathan Appleton and Samuel Batchelder,[120] both Boston Associates, showed no interest in disputing the claims of Samuel Slater as the pioneer of cotton spinning in the USA. Appleton's study, written at the urging of the treasurers of the Lowell corporations,[121] when he was in advanced age, totally ignored Rhode Island. Instead, Appleton concentrated exclusively on Lowell, the man and the place and the achievements of both. Batchelder carefully assembled records of early attempts at cotton manufacturing in the 1780s and then added, 'This was the situation of cotton manufacturing in Rhode Island when Mr Samuel Slater arrived in this country; — then all this imperfect machinery was thrown aside, and machinery more perfect built under his direction.'[122] The American industrialists in their historical moods are objective, fair, modest, almost flat in their accounts. They are clearly aware of the controversy that raged around Arkwright. Batchelder, in the context of the waterframe, and Arkwright, observed, 'It is not always easy to decide to whom we ought to award the merit of many inventions, which may have been the study of various ingenious mechanics for years without success.' He surmised that Arkwright had become the target of 'all the unsuccessful schemers' who had tried before him to solve the problems he unravelled.[123] We know what Appleton presumably knew, that other hands besides those of Francis Cabot Lowell, Patrick Tracy Jackson, Kirk Boott and Paul Moody were involved at Waltham in the development of the Waltham systems inventions. Yet Appleton made no mention of obscure mechanics nor did those mechanics find a champion to contradict him.[124] In Rhode Island industrialists decided in 1859 to assemble evidence on the early history of the cotton industry in the state. Nothing much happened, however, until 1861 when the Society for the Encouragement of Domestic Industry began publishing recollections in its *Transactions.*[125] These faded into the obituaries of the correspondents themselves and by the early 1870s the topic ceased to excite interest. There were confusions in the Rhode Islanders memories but nothing like the controversy over Arkwright ensued. And Slater's reputation remained intact.

Perhaps American industrialists had in mind the vicious squalls that had carried Arkwright's reputation to and fro among British cotton manufacturers. On one side James Thomson, John Kennedy and James Montgomery, on the other Richard Guest; the great Arkwright debate had raged between 1808 and the early 1830s. At one level it was a debate about the self-made man. Yet it was more than this. At bottom it was a controversy about the uses of history. One use was the defence of the whole industry. John Kennedy, the Scottish-born fine cotton spinner of Manchester, had this in mind. His account was intended 'to show the changes that have taken place from time to time to the present day' in order to 'appreciate the advantages or disadvantages, as well as the causes of these changes'.[126] Kennedy's concern (which later became Ure's) was to defend the manufacturing system as a whole. From it had flowed national wealth, the rise of an industry, but also new exploitations.

It needed an historical perspective to appreciate the sources, pace, direction and merits of these upheavals. History would retrieve the cotton manufacturing system from the calumnies of radicals and idealists.[127] Montgomery also tried to take a broad national standpoint in his treatment of the history of the cotton industry in Britain, though he was much less explicit about his intentions in providing an historical sketch.

It was Richard Guest's attack on Arkwright in 1823 which opened up another use for industrial history. This was the defence of local interests. Guest was born at Leigh[128] whence also came Thomas Highs (or Hayes), the reedmaker, whom Guest came to regard as the true inventor of the waterframe and carding machine. Highs, in Guest's mind, became the anti-self-made man: while 'Highs' proper arena was in his garret, among his wheels and machines', Arkwright was currying favour with 'men of property to engage with him to a large amount' willingly pursuing 'the humiliation of soliciting partnerships or patronage'.[129] Arkwright became the crafty, toadying entrepreneur, Highs the unsullied creative inventor. But there was a deeply prejudiced strain in Guest's perspective. A slighting remark betrayed it when he compared 'the bright, penetrating, shrewd, intelligent Jurors from the south of the county [in which Leigh was located], with the stupidity and utter ignorance of those from its northern parts'.[130] Guest later argued the case for his local hero at great length in a 230-page book ostensibly about the history of the cotton industry.[131] In short, Guest was a Lancashire parochialist *par excellence*.

Kennedy's account of Samuel Crompton, inventor of the spinning mule, was much more dispassionate and impartial than Guest's of Highs and Arkwright. 'It would be vain', he noted, 'to enumerate all the little additions to Crompton's original machine; also, as they arose so much out of one another, it is impossible to give to every claimant, what is exactly his due for improvements.'[132] He did his best, though, listing eight significant modifications and identifying their inventors. Radcliffe ran in the other direction, using his historical account chiefly to advance a chauvinistic, protectionist (mercantilist) point of view.

What is the explanation for the divergence between British and American industrialists in their views of their own industrial pasts? A superficial reading of the situation suggests that it may partly derive from differences in industrial scale and structure. The American cotton industry was clearly divided between the Waltham and Rhode Island systems. Each had its own founders and heroes whose careers had not overlapped. Hence there was little room for historiographical conflict. This does not, however, explain why within each system there was little controversy over the industrial past. Was it because Americans were essentially practical men from start to finish, with little time for history and not needing it to defend, realise or destroy the position of capitalist manufacturers or the ideal of gentrification? Was it because the core of cotton technology was derived from England anyway?

In Lancashire, on the other hand, the enormous controversy generated over Sir Richard Arkwright and the differing uses of history are redolent of the county's localised communities. Fierce local independence was part of the Lancashire tradition. Local figures, perhaps not always as wealthy or loquacious as Richard Guest, could always be found ready to spring forward to rehearse or defend the local achievement.

Thus in cotton textiles, the technically most advanced industry of the early stages of industrialisation, entrepreneurs' views on labour, on capital, on family hopes for the future, and on the interpretation of the recent industrial past, and the application of that interpretation, were markedly diverging between Britain and the USA in the period 1790-1840.

CONCLUSION

We can now return to the original case and the original question. Would a British entrepreneur like William Radcliffe have fared any better had he emigrated to the United States in the 1820s? That is, would he have found business values in America any more conducive than those in Britain? The answer first depends on what part of America he would have chosen to settle. Assuming he settled anywhere in a manufacturing district in the eastern states between Pennsylvania and New Hampshire, he would have found the churches much less critical of entrepreneurship and big business than in England. He would have discovered that the legal system promoted the interests of economic development. He might have avoided bankruptcy by forming a company and obtaining a charter of incorporation in any state outside Rhode Island. Likewise he could have secured patents for his inventions at a much lower price than in England. In the United States he would have encountered much less social hostility towards business success; and gentrification would not seduce the next generation. In England there lurked not only the impediment of Luddism and the irritant of aristocratic snobbery but also the unpredictable views of the opinion makers. The posthumous treatments accorded to Sir Richard Arkwright on one hand and Samuel Slater on the other illustrated how myth might tarnish the entrepreneur's reputation in Britain but add varnish to it in the United States. Surprisingly, the solidarity of class counted for more in the United States than in Britain. Parochialism, exemplified in Richard Guest's mauling of Arkwright's reputation, could cut across the class interests of Britain's new capitalists. In the United States industrialists, in textiles at least, sang their own praises less but when they did they all sang in tune. In one respect Radcliffe's entrepreneurial values would have been deeply uncomfortable in the United States. In his dandyloom he saw the chance to marry an obsolete weaving technology with the new spinning innovations of the factory system. In only one place in the United States would that vision have been welcomed in the late 1820s: among the handloom weavers of Philadelphia.[133] Everywhere else American entrepreneurship meant a commitment to capital-intensive, labour-saving technology, to values that were prodigal with materials and intolerant of craft skills. The United States' factor endowment as well as its capitalist classes by the 1830s had promoted a distinctive if not vastly different set of business values from those prevailing in contemporary Britain.

ACKNOWLEDGEMENTS

I am very grateful to Bob Burchell for setting me a difficult task which, if not completed to satisfaction, has started me on a larger investigation of the comparative development of business values. The essay has also benefited from his editorial skills. Professor Charlotte Erickson of Cambridge, my former supervisor, has generously read and commented in detail on the first draft of this essay. To her I owe numerous improvements, especially in updating my acquaintance with the American literature, and the encouragement to proceed with publication. What remains is of course entirely my responsibility. Finally, I am grateful to the staffs of Manchester Polytechnic Library and the Central Reference Library, Manchester, for their professional services. Dr Geoffrey Tweedale and Bob Burchell have kindly provided indirect access to a number of volumes in the Manchester University Library.

NOTES

1 It has been argued that the middle Atlantic states, not New England, were the base for American industrialisation. Thomas C. Cochran, *Frontiers of Change: Early Industrialism in America*, New York, 1981.
2 William Radcliffe, *Origin of the New System of Manufacture Commonly Called Power-Loom Weaving Fully Explained in a Narrative Containing William Radcliffe's Struggles through Life*, Stockport, 1828, p. 11.
3 Radcliffe, *Origin*, p. 13.
4 Radcliffe, *Origin*, p. 42.
5 Radcliffe, *Origin*; George Unwin, *Samuel Oldknow and the Arkwrights: the Industrial Revolution in Marple and Stockport*, Manchester, 1924; repr. 1968, esp. Appendix H; Duncan Bythell, *The Handloom Weaver: a Study in the English Cotton Industry during the Industrial Revolution*, Cambridge, 1969.
6 Unwin and Bythell, previous note, show that he exaggerated wage rates, for example.
7 A workable definition derived from Graham Bannock et al., *The Penguin Dictionary of Economics*, 3rd edn, Harmondsworth, 1984.
8 R. H. Tawney, *Religion and the Rise of Capitalism*, London, 1926; 1936 edn, p. 192.
9 Charles M. Elliott, 'The Ideology of Economic Growth: a Case Study', in E. L. Jones and G. E. Mingay, eds, *Land, Labour and Population in the Industrial Revolution: Essays Presented to J. D. Chambers*, London, 1967, p. 78.
10 Elliott, 'Ideology', p. 80.
11 Humphrey Lloyd, *The Quaker Lloyds in the Industrial Revolution*, London, 1975, p. 59; Maurice W. Kirby, 'The Failure of a Quaker Business Dynasty: the Peases of Darlington, 1830-1902', in David J. Jeremy, ed., *Business and Religion in Britain*, Aldershot, 1988, p. 158.
12 Elliott, 'Ideology', p. 91.
13 Anthony Howe, *The Cotton Masters, 1830-1860*, Oxford, 1984, pp. 64-6;

Jane Garnett and A. C. Howe, 'Churchmen and Cotton Masters in Victorian England', in Jeremy, ed., Business and Religion, pp. 72-94.

14 Alexis de Tocqueville, *Democracy in America*, 2 vols, New York, 1945, 1, p. 320.

15 Tocqueville, *Democracy*, 1, pp. 314-15.

16 Mack Thompson, *Moses Brown: Reluctant Reformer*, Chapel Hill, NC, 1962, pp. 251-2.

17 Jonathan Prude, *The Coming of Industrial Order: Town and Factory Life in Rural Massachusetts, 1810-1860*, Cambridge, 1983, pp. 125-6, 153-4.

18 Henry A. Miles, *Lowell as It Was and as It Is*, Lowell, 1845.

19 Helena Wright, 'The Uncommon Mill Girls of Lowell', *History Today*, 23 (1973), p. 17.

20 Harriet H. Robinson, *Loom and Spindle, or Life among the Early Mill Girls*, 1898; repr. Kailua, Hawaii, 1976, pp. 13-14.

21 Robert F. Dalzell, Jr, *Enterprising Elite: the Boston Associates and the World They Made*, Cambridge, MA, 1987, pp. 146, 177-8.

22 William H. Pease and Jane H. Pease, *The Web of Progress: Private Values and Public Styles in Boston and Charleston, 1828-1843*, New York, 1985, pp. 130, 135, 301 n22.

23 Ronald Story, *The Forging of an Aristocracy: Harvard and the Boston Upper Class, 1800-1870*, Middletown, CT, 1980, pp. 7, 34-5.

24 Anthony F. C. Wallace, *Rockdale: the Growth of an American Village in the Early Industrial Revolution*, New York, 1978, pp. 243-397.

25 Robert S. Starobin, *Industrial Slavery in the Old South*, Oxford, 1970, pp. 91-3.

26 Merritt Roe Smith, *Harper's Ferry Armory: the Challenge of Change*, Ithaca, N.Y., 1977, pp. 330-31.

27 Whitney R. Cross, *The Burned-Over District: the Social and Intellectual History of Enthusiastic Religion in Western New York, 1800-1850*, Ithaca, N.Y., 1950, pp. 322-52; Alice F. Tyler, *Freedom's Ferment: Phases of American Social History from the Colonial Period to the Outbreak, of the Civil War*, Minneapolis, 1944, pp. 47-226.

28 Charles G. Finney, *Revivals of Religion: Lectures by Charles Grandison Finney*, 1835; repr. London 1913, p. 277.

29 Paul Johnson, *A Shopkeeper's Millennium: Society and Revivals in Rochester, New York, 1815-1837*, New York, 1978; Judith A. McGaw, *Most Wonderful Machine: Mechanization and Social Change in Berkshire Paper Making*, Princeton, 1987, pp. 82-8, 257-63.

30 Morton J. Horwitz, *The Transformation of American Law, 1780-1860*, Cambridge, MA, 1977, p. 253.

31 Horwitz, *Transformation*, p. 31.

32 Sidney G. Checkland, *The Rise of Industrial Society in England, 1815-1885*, London, 1964, pp. 344, 359.

33 Tony A. Freyer, *Forums of Order: the Federal Courts and Business in American History*, Greenwich CT, 1979: a thorough treatment of *Swift v Tyson*.

34 Philip L. Cottrell, *Industrial Finance, 1830-1914: the Finance and Organisation of English Manufacturing Industry*, London, 1980, pp. 39-42; Clive Trebilcock, *Phoenix Assurance and the Development of British Insurance:*

volume I: 1782-1870, Cambridge, 1985, pp. 69-77; T. C. Barker and C. I. Savage, *An Economic History of Transport in Britain*, 3rd end, London, 1974, pp. 72-3.

35 Julian Hoppitt, *Risk and Failure in English Business, 1700-1800*, Cambridge, 1987, pp. 104-39.

36 Joseph S. Davis, *Essays in the Earlier History of American Corporations*, 2 vols, Cambridge, MA, 1917; repr. New York, 1965, 2, pp. 21-8.

37 United States, Congress, Senate, *Report of the Secretary of State of Such Articles Manufactured in the United States as Would Be Liable to Duties if Imported from Foreign Countries . . . with a Schedule of Factories Incorporated by State Laws from 1800 to 1820 Inclusive*. Senate Report no. 45, 1st Session, 18th Congress, Washington, DC, 1824. Not all these authorised capitals were raised, nor were all the companies incorporated successful.

38 Peter J. Coleman, *Debtors and Creditors in America: Insolvency, Imprisonment for Debt, and Bankruptcy, 1607-1900*, Madison, WI, 1974, pp. 71, 81, 96, 125.

39 Horwitz, *Transformation*, pp. 127-39.

40 H. I. Dutton, *The Patent System and Inventive Activity during the Industrial Revolution, 1750-1852*, Manchester, 1984, pp. 69-85.

41 Christine MacLeod, *Inventing the Industrial Revolution: the English Patent System, 1660-1800*, Cambridge, 1988, pp. 40-57. See also David J. Jeremy, *Transatlantic Industrial Revolution: the Diffusion of Textile Technologies between Britain and America, 1790-1830s*, Cambridge, MA, 1981, pp. 45-9.

42 David J. Jeremy, 'Invention in American Textile Technology during the Early Nineteenth Century, 1790-1830', *Working Papers from the Regional Economic History Research Centre, Hagley Museum*, 5, no. 4, 1982. This draws on GB, *Parliamentary Papers (Commons)*, 1829 (332) 3, 'Report from the Select Committee on the Law Relative to Patents for Inventions'.

43 Dutton, *Patent System*, p. 2; Robert C. Post, ' "Liberalizers" versus "Scientific Men" in the Antebellum Patent Office', *Technology and Culture*, 17, (1976), p. 30.

44 Jeremy, *Transatlantic Industrial Revolution*, pp. 185-6, 215-16.

45 For Britain see Katrina Honeyman, *Origins of Enterprise: Business Leadership in the Industrial Revolution*, Manchester, 1982; Howe, *Cotton Masters*, pp. 50-54. For the USA see Edward Pessen, *Jacksonian America: Society, Personality and Politics*, rev. edn, Homewood, 1978, ch. 5; also his 'Wealth in America before 1865' in W. D. Rubinstein, ed., *Wealth and the Wealthy in the Modern World*, London, 1980.

46 Francois Crouzet, *The First Industrialists: the Problem of Origins*, Cambridge, 1985, p. 38.

47 R. S. Fitton, *The Arkwrights: Spinners of Fortune*, Manchester, 1989, pp. 219, 225, 294-6.

48 *Ibid.*, pp. 72, 75, 183, 210-11.

49 Richard Guest, *Compendious History of the Cotton Manufacture: with a Disproval of the Claim of Sir Richard Arkwright to the Invention of Its Ingenious Machinery*, Manchester, 1823, p. 5.

50 John Corry, *The History of Lancashire*, 2 vols, London, 1825, 1, p. 280.

51 James Wheeler, *Manchester: Its Political, Social and Commercial History,*

Ancient and Modern, London, 1836, p. 158.
52 Edward Baines, *History, Directory and Gazetteer of the County Palatine of Lancaster*, 2 vols, Liverpool, 1824-5, 1, p. 115. For authorship see W. H. Chaloner's introduction to the 1966 reprint of Baines's *Cotton Manufacture*.
53 Edward Baines, Jr, *History of the Cotton Manufacture in Great Britain*, London, 1835; repr. 1966, p. 147.
54 Baines, *Cotton Manufacture*, pp. 194-5.
55 *Dictionary of National Biography*.
56 Negley B. Harte, 'On Rees's *Cyclopaedia* as a Source for the History of the Textile Industries in the Early Nineteenth Century', *Textile History*, 5 (1974), p. 121.
57 'Cotton' in Abraham Rees, ed., *The Cyclopaedia; or Universal Dictionary of Arts, Sciences and Literature*, 45 vols, London, 1802-20.
58 James Butterworth, *The Antiquities of the Town and a Complete History of the Trade of Manchester*, Manchester, 1822, pp. 46-7, 81-92. This was reprinted the following year as *A Complete History of the Cotton Trade*. For the author see *DNB* and Giles Shaw, 'James Butterworth of Oldham', *Transactions of the Lancashire and Cheshire Antiquarian Society*, 26 (1908), pp. 124-32.
59 *Supplement to the Encyclopaedia Britannica*, vol. 3, (1824).
60 *The Edinburgh Review*, 46 (June, 1827), pp. 1-39; Richard Guest, *The British Cotton Manufactures, and a Reply to an Article on the Spinning Machinery Contained in a Recent Number of the 'Edinburgh Review'*, Manchester, 1828, p. 98.
61 *The Edinburgh Review*, 46 (June, 1827), pp. 6, 14.
62 *Dictionary of National Biography*.
63 Andrew Ure, *The Philosophy of Manufactures*, London, 1835, pp. 15-16.
64 Andrew Ure, *The Cotton Manufacture of Great Britain*, 2 vols, London, 1836, 1, pp. 220-56.
65 Ure, *Cotton Manufacture*, 1, p. 247.
66 Ure, *Cotton Manufacture*, 1, pp. 250-52.
67 Samuel Smiles, *Self-Help with Illustrations of Conduct and Perseverance*, 4th edn, 1859; repr. 1908, pp. 38-44.
68 Asa Briggs, *Victorian People*, Harmondsworth, 1965, pp. 124-47.
69 James Montgomery, *Carding and Spinning Master's Assistant*, Glasgow, 1832, p. 238.
70 Fitton, *Arkwrights*, p. 50.
71 Jeremy, *Transatlantic Industrial Revolution*, pp. 76-91; Barbara M. Tucker, *Samuel Slater and the Origins of the American Textile Industry, 1790-1860*, Ithaca, NY, 1984; Prude, *Coming of Industrial Order*.
72 George S. White, *Memoir of Samuel Slater, the Father of American Manufactures*, Philadelphia, 1836.
73 This was Fitton, *Arkwrights*.
74 Notices of White's career are in S. Austin Allibone, *A Critical Dictionary of English Literature and British and American Authors, volume 3*, Philadelphia, 1871, and the *National Cyclopaedia of American Biography, volume 4*, New York, 1897, p. 319. Unfortunately the details in these two notices do not always agree. White's death date is that given in the Canterbury tombstone

inscriptions and the Hartford newspapers. I am grateful to Carolyn Picciano of the Connecticut State Library and Pat Markey of the Museum of American Textile History for supplying these sources.

75 White, *Slater*, p. 37; David Jeremy, 'Damming the Flood: British Government Efforts to Check the Outflow of Technicians and Machinery, 1780-1843', *Business History Review*, 51 (1977).

76 White, *Slater*, p. 10.

77 *Ibid.*, p. 11.

78 *Ibid.*, p. 11.

79 *Ibid.*, p. 11.

80 *Ibid.*, advertisement following title page.

81 David J. Jeremy, *Technology and Power in the Early American Cotton Industry: James Montgomery, the Second Edition of His 'Cotton Manufacture' (1840), and the 'Justitia' Controversy about Relative Power Costs*, Philadelphia, 1990, pp. 285-6.

82 Sigmund Diamond, *The Reputation of American Businessmen*, Cambridge, MA, 1955.

83 Ure, *Philosophy*, pp. 343-52.

84 For example, Sidney Pollard, *The Genesis of Modern Management: a Study of the Industrial Revolution in Great Britain*, London, 1965, pp. 160-208; Patrick Joyce, *Work, Society and Politics: the Culture of the Factory in Later Victorian England*, Brighton, 1980, pp. 90-200 *passim.*

85 David Roberts, *Paternalism in Early Victorian England*, Rutgers, 1979, pp. 180-3. See also Pollard, *The Genesis of Modern Management*,pp. 160-208.

86 See Roberts, *Paternalism*, pp. 1-63.

87 S. G. Checkland, 'Cultural Factors and British Business Men, 1815-1914', in K. Nakagawa, ed., *Social Order and Entrepreneurship*, Tokyo, 1977.

88 Quoted in Michael Brewster Folsom and Steven D. Lubar, eds, *The Philosophy of Manufactures: Debates over Industrialisation in the United States*, Cambridge, MA, 1982, p. 17.

89 Prude, *Coming of Industrial Order*, pp. 110-11, 158-80, 244-7.

90 Allen's *Practical Tourist*, 1832, quoted in Folsom and Lubar, *Philosophy*, p. 342.

91 Tucker, *Samuel Slater*, pp. 139-85.

92 Dalzell, *Enterprising Elite*, pp. 26-44.

93 Tucker, *Samuel Slater*, p. 139.

94 Philip Scranton, *Proprietary Capitalism: the Textile Manufacture at Philadelphia, 1800-1885*, Cambridge, 1983.

95 Prude, *Coming of Industrial Order*, p. 115.

96 Thomas Dublin, *Women at Work: the Transformation of Work and Community in Lowell, Massachusetts, 1826-1860*, New York, 1979, pp. 92-3.

97 David Crockett, *An Account of Colonel Crockett's Tour to the North and Down East*, Philadelphia, 1835, p. 92. Charles Dickens, *American Notes for General Circulation*, London, 1842, ch. 4.

98 The element of illusion is noted in Dalzell, *Enterprising Elite*, pp. 67-8.

99 This is the Habakkuk thesis. For that debate see S. B. Saul, ed., *Technological Change: the United States and Britain in the Nineteenth Century*, London, 1970.

100 Jeremy, *Transatlantic Industrial Revolution*, pp. 92-117, 180-203.
101 James Montgomery, *A Practical Detail of the Cotton Manufacture of the United States of America*, Glasgow, 1840, p. 131.
102 Nathan Rosenberg, ed., *The American System of Manufactures*, Edinburgh, 1969, p. 388.
103 David A. Hounshell, *From the American System to Mass Production, 1800-1932: the Development of Manufacturing Technology in the United States*, Baltimore, 1984.
104 MacLeod, *Inventing the Industrial Revolution*, pp. 159-73.
105 'Nottingham's trading families of the first half of the nineteenth century patronised schools which specifically trained their sons for business careers.' Charlotte Erickson, *British Industrialists: Steel and Hosiery, 1850-1950*, Cambridge, 1959, p. 111.
106 Fitton, *Arkwrights*, pp. 254-70.
107 Edward A. Allen, 'Public School Elites in Early Victorian England: the Boys at Harrow and Merchant Taylors' Schools from 1825 to 1850', *Journal of British Studies*, 21 (1982), p. 109.
108 Erickson, *British Industrialists*, p. 33.
109 Peter L. Payne, *British Entrepreneurship in the Nineteenth Century*, 2nd edn, London, 1988, pp. 22-3.
110 Peter Dobkin Hall, *The Organisation of American Culture, 1700-1900: Private Institutions, Elites, and the Origins of American Nationality*, New York, 1982, p. 71. Pease and Pease, *Web of Progress*, pp. 112-14, note the high commitment to education in Boston in the 1830s.
111 Story, *Forging of an Aristocracy*; Dalzell, *Enterprising Elite*.
112 Between 1800 and 1850 the average number graduating annually rose from 47 to 55: Story, *Forging of an Aristocracy*, p. 90.
113 *Ibid.*, p. 92.
114 *Ibid.*, p. 59.
115 *Ibid.*, p. 79.
116 *Ibid.*, p. 95.
117 *Ibid.*, p. 132.
118 Dalzell, *Enterprising Elite* explores these dimensions thoroughly.
119 Montgomery, *Practical Detail*, p. 141. For Montgomery see Jeremy, *Technology and Power*.
120 Nathan Appleton, *The Introduction of the Power Loom and Origin of Lowell*, Lowell, MA, 1858; Samuel Batchelder, *Introduction and Early Progress of the Cotton Manufacture in the United States*, Boston, 1863.
121 Frances Gregory, *Nathan Appleton: Merchant and Entrepreneur, 1779-1861*, Charlotteville, VA, 1975, p. 298.
122 Batchelder, *Introduction*, p. 39.
123 *Ibid.*, pp. 11-12.
124 Jeremy, *Transatlantic Industrial Revolution*, pp. 186-96.
125 *Transactions of the Rhode Island Society for the Encouragement of Domestic Industry*, 1859, p. 37; 1861, pp. 18, 30-33, 76-125.
126 John Kennedy, 'Observations on the Rise and Progress of the Cotton Trade in Great Britain, particularly in Lancashire and the Adjoining Counties', *Memoirs of the Literary and Philosophical Society of Manchester*, 2nd ser., 3

(1819), pp. 115-16.

127 John Kennedy, *Miscellaneous Papers, on Subjects Connected with the Manufactures of Lancashire*, Manchester, 1849, p. 5. Kennedy rather veiled the criticisms of the injustices and miseries of the new industrial system, referring to them as 'frequent complaints. . . against the manufacturing system'.

128 The only note on him I have seen tells little more than his birthplace, that he was in the cotton trade and that he was a quarrelsome individual. Thomas R. Dootson, *Leigh and District Authors: Old and New*, Leigh, 1927, p. 27.

129 Guest, *Compendious History*, pp. 20, 28.

130 *Ibid.*, p. 38.

131 *Ibid.*

132 John Kennedy, 'A Brief Memoir of Samuel Crompton', *Memoirs of the Manchester Literary and Philosophical Society*, 2nd ser., 5 (1831), p. 336.

133 Scranton, *Proprietary Capitalism*, pp. 129-34.

V

British Textile Technology Transmission to the United States: The Philadelphia Region Experience, 1770-1820*

¶ *This study examines the complex process of the transfer of technology in the early stages of industrialization. The manner and timing of the selective acquisition of portions of Britain's new textile technology by entrepreneurs in the Philadelphia region were determined by the subtle interplay of market forces, technological constraints, and active efforts to encourage or discourage the transfer.*

Before the end of the eighteenth century, Britain's astonishing economic progress had demonstrated the central importance of a machine technology in the process of industrialization. Henceforth, the acquisition of such a technology would be indispensable for any society seeking economic parity with Britain. The United States, eager to realize economic as well as political independence, fortuitously possessed a number of strong assets which predisposed it towards the ready acceptance of an industrial machine technology. Among these were the expanding urban markets of a growing population; the legacy of England's handicraft techniques, value system, and language; and, before 1815, spasmodic commercial protection afforded by fluctuations in Anglo-American relations. Furthermore, American society exhibited rising educational levels; the absence of deeprooted, impeditive traditions; and the promise of boundless opportunities for immigrants. But most of all, the relatively low proportion of labor to land, in America's factor endowment, impelled the application of the labor-saving, capital

Business History Review, Vol. XLVII, No. 1 (Spring, 1973).

* In the preparation of this article, the author is indebted to the Eleutherian Mills Historical Library for a research grant used in the summer of 1969; to Dr. Whitfield J. Bell, Jr., Librarian of the American Philosophical Society, for the invitation to read an earlier draft at a session of the History and Philosophy of Science Section of the American Association for the Advancement of Science on December 27, 1971; and to the Merrimack Valley Textile Museum for the support of its research facilities.

In reshaping the paper for publication in the *Business History Review*, I am grateful to its editor and editorial readers (particularly Eric Robinson) for their constructive criticisms. Glenn Porter, especially, in persuading me to reorganize my material, diverted my attention to several of the more provocative questions which are raised here.

intensive techniques of a machine technology in the exploitation of the country's vast natural resources.[1]

In the early Industrial Revolution, however, foreign entrepreneurs trying to obtain Britain's new textile technology met problems arising from its nature, namely its embodiment in men and machines. Until 1812, when comprehensive verbal descriptions and good representative drawings began to be published, the design of the improved textile equipment was known only to those with firsthand experience of building or running it. Likewise, until 1832, when a reliable set of carding and spinning rules was first published, the empirically-developed operating procedures for the new machinery were known only to those initiated in textile mill practice. Thus prior to 1812 at least, the new textile technology had a non-verbal character which compelled foreign rivals to learn about it either through contacting experienced artisans or through possessing the new machines, on which they might (with imaginable hazards) undertake "learning by doing."

Even before the development of an industrial technology placed a growing emphasis on machines, Britain, like other European powers, had protected her staple industries with a mass of mercantilist legislation. Some of this evinced a clear awareness of the loci of technology by aiming to curb the emigration of artisans and the export of tools and implements over a wide range of trades.[2]

Textiles early enjoyed special commercial safeguards against leaks in the national reservoir of manufacturing technology, whether embodied in the "living" or "dead instruments of trade," as Adam Smith called them.[3] Under an act of 1718, an emigrant woolen worker could suffer loss of all his English property as well as his citizenship, while those enticing him abroad were subject to a £100 fine and three months' imprisonment. Even those suspected of preparing to infringe the law were required to give surety to

[1] For a perceptive, stimulating, and lucid analysis of the social and economic mechanisms controlling the American acquisition and modification of European technology see Nathan Rosenberg, *Technology and American Economic Growth* (New York, 1972), 25–86.

The whole question of technology transfer was the subject of an international conference organized by ICOHTEC (International Cooperation in History of Technology Committee) under the title "The Acquisition of Technology by Non-Initiating Countries," held at Pont-à-Mousson, Nancy, France on June 29–July 4, 1970. For this see *Technology and Culture*, XII (1971), 217–227.

[2] For the general mercantilist background to protection see Eli F. Heckscher, *Mercantilism* (2 vols., London, 1955), II, 112–172.

The ten major laws passed between 1695 and 1799 against the emigration of artisans and the export of machinery specifically covered the metal, clock, glass, pottery, harness, mining, and certain machine-making trades as well as textiles. The 1749 act, however, made the enticement to emigrate of "any other Manufacturer, Workman or Artificer of or in any other of the Manufactures of *Great Britain* or *Ireland*, of what Nature or Kind soever," a criminal act.

[3] Adam Smith, *An Inquiry into the Nature and Causes of the Wealth of Nations*, edited by Edwin Cannan (New York, 1937), 624.

the local Justices of the Peace pending determination of their cases.[4] Penalties for seducing away Britain's woolen workmen were increased in 1749 to a £500 fine and a twelve months' prison sentence. The same act extended the earlier law to include all persons engaged in the cotton, silk, and mohair trades,[5] and an act of 1782 added cotton and linen printing workers.[6]

Though the export of machinery received attention as early as 1695, when the export of stocking frames was prohibited, broader restrictions came in the second half of the eighteenth century.[7] The 1749 ban on the export of all woolen and silk manufacturing tools and utensils was widened in 1774 to cover all cotton and linen equipment. At the same time, preparing to export woolen and silk implements became open to investigation by local magistrates. In all cases a £200 fine and forfeiture of the impounded tools constituted the statutory penalty.[8] Reflecting the changing pace of technological innovation, an act of 1781 prohibited the export of any "Machine, Engine, Tool, Press, Paper, Utensil or Implement" and any "Model or Plan . . . Part or Parts thereof" used in the cotton, linen, woolen, or silk manufactures of the kingdom.[9] In 1782 the export of engraved copper plates for calico printing became punishable with a £500 fine as well as forfeiture of equipment.[10]

After Ireland gained some constitutional independence from Britain in 1782 (with the repeal of the Declaratory Act), the core provisions of these prohibitory English laws were reapplied to Ireland by an Irish statute of 1785.[11] Furthermore, nearly all these acts reserved specific penalties for ships' captains or Customs officers implicated in violations. Although these laws failed to halt the efflux either of men or machines and were mostly repealed in 1824–1825,[12] they posed something of a deterrent to the transfer of industrial technology.

[4] 5 Geo. 1, c. 27 (1718). Danby Pickering's edition has been used for the text of British statutes, and dates are taken from Great Britain, *Chronological Table of the Statutes Covering the Period from 1235 to the End of 1969* (London, 1970).

[5] 23 Geo. 2, c. 13 (1749).

[6] 22 Geo. 3, c. 60 (1782).

[7] 7 & 8 Wm. 3, c. 20 (1695). The background to the 1695 act is briefly noticed in William Felkin, *A History of the Machine-Wrought Hosiery and Lace Manufactures* (Cambridge, 1867), 71, and Frederick A. Wells, *The British Hosiery Trade: Its History and Organization* (London, 1935), 35–39.

[8] 23 Geo. 2, c. 13 (1749); 14 Geo. 3, c. 71 (1774).

[9] 21 Geo. 3, c. 37 (1781). To reduce unemployment, the export of wool or stock cards and spinners' cards was specifically excluded from these laws by two statutes, 15 Geo. 3, c. 5 (1775) and 26 Geo. 3, c. 76 (1786).

[10] 22 Geo. 3, c. 60 (1782).

[11] Ireland, 25 Geo. 3, c. 17 (1785), which prescribed a £500 fine and twelve months' imprisonment for seducing workers in linen, wool, cotton, silk, or mohair to emigrate, and a £200 fine with forfeiture of equipment for exporting linen, woolen, or silk manufacturing tools.

[12] Many sections of the prohibitory acts were repealed by 5 Geo. 4, c. 97 (1824) and

Certainly they evidenced the jealous regard with which the mercantilist state eyed its own manufacturing secrets – while countenancing the "borrowing" of those of its rivals. As A. E. Musson and Eric Robinson and others have shown, such secretiveness generated a good deal of industrial espionage in Europe both internationally and internally. The trouble to which leading French textile manufacturers, for example, would go to recruit skilled English workmen and, having secured them, to retain their services, underscores the key role of non-verbal technology during the early Industrial Revolution.[13] And the competition for men and machines occurring at the international level was repeated at lower levels between firms.

Once an economy achieved a pool of domestic experience of the skills in question, dependence on immigrant workers diminished. Additionally, once the empirically-based, new technology had been reduced to a set of physical and/or chemical rules (verbally or mathematically expressed) and then published, its acquisition became a much more objective process. As a result, the transmission and adoption of that technology would be greatly facilitated.

This paper treats, within the context of the Philadelphia region (Philadelphia and Delaware counties in Pennsylvania and New Castle county in Delaware), certain aspects of the international transfer of a single machine technology during its early, non-verbal stage of development. Requiring initial consideration are the American initiatives to obtain that technology.

American Efforts to Acquire Britain's New Textile Technology

For the most part, Philadelphia businessmen were limited to strategic schemes in their efforts to procure British textile tech-

6 Geo. 4, c. 105 (1825), but the export of machinery was still controlled under the Customs Regulations Acts of 1825 (6 Geo. 4, c. 107) and 1833 (3 & 4 Wm. 4, c. 52), which were not repealed until 1843 (6 & 7 Vict., c. 84). For these see John Clapham, *An Economic History of Modern Britain* (3 vols., Cambridge, 1926–1938), I, 485–86.

[13] A. E. Musson and Eric Robinson, *Science and Technology in the Industrial Revolution* (Manchester, 1969), 216–230. Other recent works touching this subject are W. O. Henderson, *Britain and Industrial Europe, 1750–1870: Studies in British Influence on the Industrial Revolution in Western Europe* (Leicester, 1965); J. Clayburn La Force, "Technological Diffusion in the 18th Century: the Spanish Textile Industry," *Technology and Culture,* V (1964), 322–343; and J. J. Hémardinquer, "Une Dynastie de Mécaniciens Anglais en France: James, John et Juliana Collier (1791–1847)," *Revue d'Histoire des Sciences*, XVIII (1964), 193–208.

Witnesses before the Hume Committee on Artisans and Machinery in 1824 told of French manufacturers or their agents crossing to England, visiting workshops or factories, probably at meal or closing times, and enticing away workers with offers of high wages. Once in France, immigrants were bound by contracts, indebtedness, and a passport system operated by employers and police. Yet some workers were sent back to England by their French employers to obtain the latest machines and to recruit more artisans. See Great Britain, *Parliamentary Papers (House of Commons)*, 1824 (51), V, "Six Reports from the Select Committee on Artizans and Machinery," 23, 36, 107–108, 278, 554, 579.

nology. At least four kinds of conscious effort are evident: through formal organizations, pressure group activities, state legislation, and individual initiatives.

The first of the formal organizations designed to secure improvements in manufacturing in the Philadelphia region was the United Company of Philadelphia for Promoting American Manufactures, formed in late February 1775 as a corollary to the non-importation austerities invoked in resisting British rule.[14] Modelled on similar societies created in Boston in 1751 and in New York in 1764, it raised capital by issuing an unknown number of £10 shares and entrusted the management of these resources to a board of twelve managers who then acted as a corporate clothier. The board rented a building for their manufactory, purchased equipment and raw materials, put out the flax, hemp, wool, and cotton to nearly 500 spinners, supervised the collection of yarn and the weaving of cloth, and ran the manufactory store. In September 1775 the company's managers conferred about "the employment of three (it is said) complete spinners on the machine and cotton weavers" at the "factory." The machine in question was Christopher Tully's version of James Hargreaves' spinning jenny (patented in Britain in 1770), one of two jennies built in Philadelphia in the winter of 1774–1775. With twenty-four spindles, each could theoretically perform the work of that number of spinners, though the company expected only a tripling of productivity. Perceiving the labor-saving potential of the jenny (which spun a relatively fine, soft, and even yarn on the intermittent principle),[15] the company petitioned the Pennsylvania Assembly in February 1776 to publicize the machine by setting up models through the province at the colony's expense (£7 per machine) in order to hasten its use and so "contribute greatly to the Establishment of *American* Manufacturers." The scheme, in-

[14] William R. Bagnall, *The Textile Industries of the United States, 1639–1810* (Cambridge, Mass., 1893), 63–72 gives the best account of the United Company.

[15] Intermittent spinning, performed with the spindle, required that twisting and drafting be carried out separately from winding, the third component of the spinning operation. Thus when the fibers in the spinning zone were formed into yarn, the twisting and drafting was stopped for the yarn to be wound onto the spindle. Naturally this was more time-consuming than continuous spinning (for which see note 21 below). However it had two distinct advantages over continuous spinning. Spindle drafting allowed yarn evenness to be improved at the spinning stage. Further, intermittent spinning, by imposing less strain on the yarn during its formation, allowed the spinning of finer yarns, more fragile cardings, and less "spinnable" fibers like wool. For this reason, intermittent spinning was the only mechanical method by which woolen (as opposed to worsted) yarns could be spun at all before the twentieth century. Prior to 1850 the most important intermittent spinning machines were the jenny, jack, and mule. For comment on the principles of spinning see W. E. Morton and G. R. Wray, *An Introduction to the Study of Spinning* (London, 1962), 133–175. I am grateful to Dr. John A. Iredale of the School of Textiles, Bradford University, England, for a discussion on some of these points.

stigated by ruptured relations with Britain on the eve of the Revolution, was apparently swept aside by the tide of war.[16]

When the United Company was revived in August 1787 as the Pennsylvania Society for the Encouragement of Manufactures and the Useful Arts, it was clear from the prophetic address Tench Coxe delivered to its inaugural meeting that the pursuit of the Society's goals encompassed borrowing European inventions, utilizing European immigrant skills, and emulating European methods of encouraging manufactories.[17] Coxe even proposed a precise technique for procuring foreign technology: "It might answer an useful purpose, if a committee of this society should have it in charge to visit every ship arriving with passengers from any foreign country, in order to enquire what persons they may have on board, capable of constructing useful machines, qualified to carry on manufactures, or coming among us with a view to that kind of employment." [18]

Whereas the managers of the Pennsylvania Society exercised general supervisory powers, the manufacturing committee was charged with the specific duty of applying a manufacturing fund to the establishment of "factories." The minutes of the committee disclose its frustrations and successes in searching for the improved British machines and immigrants who could operate them.

In September 1787 the committee learned that a carding engine and three spinning machines were in the city. Very shortly after, however, it was discovered that these had been sent back to Europe the previous May. Several committee members mentioned the rumor "that there were one carding engine & two spinning machines lately sold by one Guthrie living in Chester County & put on board a Boat at Wilmington." An *ad hoc* committee made fruitless enquiries to verify this second story of peripatetic textile equipment, so the manufacturing committee decided, in November of that year, to get a carding engine custom built and, later, to have two jennies and a "slubber" (jenny for preliminary spinning)

[16] *Pennsylvania Archives* 8th series (Harrisburg, Pa.), VIII, 7215–16, 7220, 7325–26, 7366, 7390–91, 7504, 7532, 7588; *The Pennsylvania Magazine: or American Monthly Museum*, I (1775), 157–58.

[17] See Bagnall, *Textile Industries*, 77–79 and J. Leander Bishop, *A History of American Manufactures from 1608 to 1860* (2 vols., Philadelphia, 1864), I, 404–409 for the Pennsylvania Society's history; also *The American Museum, or Universal Magazine*, II (1787), 167–69 for the Society's "Plan."

Except where otherwise specified, information on the activities of the manufacturing committee comes from "Minutes of the Manufacturing Committee of the Pennsylvania Society for the Encouragement of Manufactures and the Useful Arts," vol. I (September 11, 1787–November 3, 1788) and vol. II (January 21, 1789–June 3, 1789), MS A M 302, Historical Society of Pennsylvania, Philadelphia. I am indebted to the Historical Society for permission to use these minute books.

[18] *American Museum*, II (1787), 248–255 for Coxe's address; this quotation is from p. 253.

made. A search for machine builders then followed. By late January 1788 the three jennies were near completion in the Society's manufactory, being built by Enoch Richardson, a Philadelphia carpenter, and others. For the carding engine, "calculated to card about 50 lbs. of Cotton p. day," the committee had to go farther afield. One John Hague,[19] an "Artest" of Alexandria, Virginia, was located by committee member Thomas Wood, who contracted with him to build the card "on their joint Accot. & Resque & when finished to be at the option of this Committee to take the same if they see proper." The machine was finished by March 12, 1788 and the committee accepted it.

The manufacturing committee's report of all this progress, issued early in January 1788, attracted other actual and aspiring machine builders. Four days later John Barber, a Scotsman, announced that he had some cotton manufacturing machinery and conveyed his desire to be employed by the committee. When the latter offered only to buy up to £30 worth of yarn from Barber to allow him to complete his carding engine, the Scot, evidently dissatisfied with the slender inducement, left the city. Another machine maker, Robert Lemmon of Baltimore, advised the committee of his intentions of constructing a carding machine (apparently for the committee) and requested "assistance . . . to encourage some Artists to come here." Both proposals were declined on the ground that the Society already had enough machine builders. Of these, Hague benefited most from the Society's initial activities: he and Wood received £100 for his card, which could be hand-operated, and for a jenny he gratuitously supplied, Hague got £14 (the other three jennies built in Philadelphia cost a total of £65. 7s. 3d.). The carding machine, two of the jennies, and two looms, one equipped with the fly shuttle (patented in Britain in 1733), were proudly displayed on the Society's float in the Grand Federal Procession of July 4, 1788 – fruits of the international appropriation of technology.[20]

Other equipment or skills then non-existent or scarce in America of course greatly excited the Society's interest. In July 1788 a woman possessing a twisting mill was employed "on the best Terms." Ten months later the Society's superintendent of cotton manufacturing, James Walsh (very likely a British immigrant in

[19] Whether John Hague was the same man as Joseph Hague (who built the first known jenny in Philadelphia in 1774–1775 and who failed to assemble a spinning mule imported there in 1783) is not clear. Perhaps erring on the side of caution, I have assumed that they were two different persons.

[20] Whitfield J. Bell, Jr. (ed.), *Francis Hopkinson's Account of the Grand Federal Procession, Philadelphia, 1788* (Boston, 1962), 12.

view of his skills and European connections), informed the manufacturing committee that "having receiv'd Letters from friends in Europe desiring to know what encouragement could be given to them should they come over & in conjunction with him set up the perpetual spinning as also Dying Cutting & finishing," he needed advice in replying to his correspondents. "Perpetual spinning" was a contemporary phrase for continuous spinning,[21] first mechanized in the waterframe, patented in Britain by Richard Arkwright in 1769. This the committee was most anxious to acquire, for the manufactory had run into difficulties which compelled the hiring out of some of its looms in April 1789. Possibly the four jennies (totalling 224 spindles) and the two or three hundred household spinners could not produce enough cotton and linen yarn, respectively, to keep pace with the Society's twenty-six handlooms, some of which had fly shuttles.[22] Certainly the lack of bleach yards and the limited bleaching services provided by one English immigrant [23] were causing a bottleneck in finishing.

The committee therefore responded with alacrity to the chance to obtain the waterframe, a machine offering the highest productivity in spinning hard, medium quality yarns, most suitable for cloth warps, and to the opportunity to procure artisans well-qualified in cotton finishing. Two weeks after Walsh's communication, in late May 1789, the committee drafted a letter for their cotton superintendent to forward to his European friends. It explained the determination of Americans to develop their manufactures, sketched the many advantages those manufacturers would find, and assured immigrant manufacturers of encouragement. As proof of their good intentions, the committee gave Walsh a £50 letter of credit "towards defraying the expences of any Artists that may come out to him & that when such People arrive here that sum to be paid." Shortly after this date the committee's *extant* minutes cease, and the outcome of these ventures is unknown. The enterprise met an almost fatal blow when the manufactory was destroyed by fire in March 1790, believed to be the work of an arsonist. Not until 1803 did the Society make an unremarkable revival.[24]

[21] In continuous spinning, performed with the bobbin and flyer or some variation of it, winding occurred simultaneously with twisting and drafting. This method was faster than intermittent spinning (see note 15 above) and was particularly suitable for coarse and medium yarns. The major continuous spinning machines before 1850 were the waterframe, throstle, cap frame, and ring frame.

[22] Bishop, *History of American Manufactures*, I, 407–408.

[23] This was the calico printer John Hewson (d. 1822), active in Philadelphia between 1773 and 1810. See Florence M. Montgomery, *Printed Textiles: English and American Cottons and Linens, 1700–1850* (New York, 1970), 92–98.

[24] Samuel Rezneck, "The Rise and Early Development of Industrial Consciousness in the United States, 1760–1830," *Journal of Economic and Business History*, IV (1932), 795.

V

As Samuel Rezneck has ably demonstrated,[25] a small pressure group, centered around Philadelphia (scene of the 1787 constitutional convention and the nation's capital, 1790–1800), worked in the 1780's and 1790's through press and political connections to mobilize the national interest in industrialization. They argued for economic independence and pointed to the impressive results achieved by Britain's expanding technology. The most important accomplishments of that movement (in addition to the Pennsylvania Society) lay in the creation and dissemination of ideas. This included the publication of the *American Museum* (12 vols., 1787–1792) a periodical of polemical and informative pieces edited by Irish-born publisher Mathew Carey; Alexander Hamilton's national program for industrialization, stated best in his 1791 *Report on Manufactures*; and a number of cogent articles and pamphlets by Federalist merchant Tench Coxe. In addition, the movement led to the founding of a model manufacturing town, the Society for Establishing Useful Manufactures (S.U.M.) at Paterson, New Jersey. As a means of overcoming the chronic American shortage of skilled labor, these proponents of American industrialization recommended the adoption of Britain's new capital-intensive technology, if only this could be applied. As Tench Coxe remarked, "We are beginning to see the great, and to us *peculiar*, value of labour-saving machines. . . . The first judicious european capitalists, who shall take good situations in the United States, and establish manufactories, by labour-saving machines, must rapidly and certainly make fortunes." [26]

A third conscious effort to acquire British technology was made by the Pennsylvania legislature. Both before and after the Revolution it awarded premiums to persons who introduced new machines or techniques of production into the state. For example, the provincial assembly voted £15 each to the two men (one of whom was an immigrant English weaver) who built jennies in Philadelphia in the winter of 1774–1775, as well as £40 to an immigrant English thread-maker, for an improved twisting machine which he had set up in the Philadelphia Bettering House in the fall of 1775. In 1788 the state assembly, advised by the Pennsylvania Society's manufacturing committee, considered petitions from two calico-printers, but it is not clear whether premiums were awarded. Certainly though, John Hague (as mentioned above) was voted £100 for introducing the carding machine that year.[27] The legis-

[25] *Ibid.*, 784–811.
[26] Tench Coxe, *A View of the United States of America* (Philadelphia, 1794), 165–66.
[27] *Pennsylvania Archives* 8th series, VIII, 7418–19, 7428, 7463, 7532; *Proceedings and*

lature further encouraged textile manufacturing and hence, indirectly, the siphoning of foreign technology, by voting in March 1789 to purchase one hundred £10 shares of the Pennsylvania Society's Manufacturing Fund.[28]

The strongest measure taken by the legislature, however, was a protectionist law passed in March 1788. The Pennsylvania Society, furious at the loss of some valuable carding and spinning machinery, petitioned the state assembly in November 1787 to pass legislation to outlaw the destruction or export of any manufacturing hardware as well as efforts to encourage the departure of skilled artisans from the state. In the debate on the bill the following March, all the recorded speeches were made by Philadelphia representatives who displayed no unanimity. Those in favor pointed to "the absolute necessity which existed of calling forth the aid of our own manufacturers and mechanics to counterbalance the pernicious effects of a balance of trade against us;" to manufacturing as a means of slowing down the westward population drift from the state; and to the international political ascendancy of the nation which was the industrial model – Britain. On the other side, a Jeffersonian legislator expressed his belief in the preeminence of agriculture while a merchant articulated mercantile fears for commercial freedom. The resultant act specified a £350 fine and twelve months' imprisonment for those guilty of exporting or destroying any machinery or implements for manufacturing wool, cotton, linen, silk, iron, or steel, as well as a £100 fine and similar prison term for those convicted of seducing artisans in those manufactures to emigrate. It is impossible to gauge the effects of this law, which was not renewed after its first two years.[29] To be sure, these joint endeavors of the Pennsylvania Society and the state assembly did not go unnoticed among British workers.[30]

Debates of the General Assembly of Pennsylvania (1787–1788), as Taken in Short-Hand by Thomas Lloyd (4 vols., Philadelphia, 1787–1788), III, 130, 142, 143, 168, 171, 193, 194, 204, 213, 217, 226, 231; Pennsylvania, *The Statutes at Large of Pennsylvania from 1682 to 1801* (16 vols., Harrisburg, 1896–1911), XIII, 138–39 (act of October 3, 1788 awarding £100 to John Hague).

The two jenny builders were Joseph Hague and Christopher Tully; the thread-maker was John Marshall; and the calico printers were Robert Taylor and John Hewson.

[28] *Ibid.*, 239–240 (act of March 26, 1789).

[29] *Ibid.*, 58–62 (act of March 29, 1788); *Proceedings and Debates of the General Assembly of Pennsylvania 1787–1788*, III, 200–203. The Jeffersonian legislator was George Logan (1753–1821) and the merchant was Thomas Fitzsimons (1741–1811). Wool and cotton cards were exempt from the 1788 act (which had a statute life of two years) because in Philadelphia they were "made in sufficient quantities to serve all the continent" and so formed a major export item, according to Fitzsimons.

[30] Samuel Slater, for example, after reading a newspaper report of the state premium to Hague and state support for the Pennsylvania Society, determined to leave Derbyshire for America. He disembarked at New York from whence, ironically, he moved north to Rhode Island, not south to Philadelphia. See George S. White, *Memoir of Samuel Slater* (Philadelphia, 1836), 37; Mack Thompson, *Moses Brown, Reluctant Reformer* (Chapel Hill, N.C., 1962), 220–233.

Finally, Americans tried to secure British textile technology by numerous individual initiatives. Several of these related to the Philadelphia region. In 1787–1788 Tench Coxe reportedly contracted with an English mechanic residing in Philadelphia to return to Britain, secure brass models of Arkwright machinery, and forward these to the American Minister in Paris, Thomas Jefferson. English Customs officials, however, detected the models and discovered the subterfuge.[31] Slightly more successful was the Philadelphia merchant William Pollard. In spring 1789 he paid several hundred pounds to someone in England for a model, purportedly of Arkwright's waterframe. By October 1789 he had constructed a full scale machine, only to find that it would not spin and that the vendor ignored his letters. Undaunted, Pollard went on to make improvements for which he was awarded a federal patent.[32]

More effective still were the exertions of the disgraced son of a prominent Maryland family, Thomas Digges, who sublimated his kleptomania by pilfering technology. At Belfast, on one of his many forays into British manufacturing districts, Digges recruited William Pearce, an ingenious, unsteady, and intemperate mechanic. Before his American sponsors, Pearce audaciously laid claim to the inventions of Arkwright and Cartwright. Reaching New York in July 1791 armed with Digges's introductions and some machine models, Pearce sent the latter to Tench Coxe and then followed them to Philadelphia. There he set up spinning and weaving machinery which was admired by European weavers and visited by a Presidential inspection party. Through Hamilton's influence, Pearce concurrently prepared models and machines for the S.U.M. at Paterson, to which he moved in July 1792. Quarrelling with the S.U.M. in late 1793, he was hired by the Delaware merchant Jacob Broom to erect an Arkwright-type spinning mill on the Brandy-

[31] Bagnall, *Textile Industries*, 75. This sounds quite uncharacteristic of Jefferson; unfortunately the only letter from Coxe to Jefferson around this date (received by Jefferson on December 13, 1787) has not been found. See Julian P. Boyd and others (eds.), *The Papers of Thomas Jefferson* (Princeton, 1950–), XII, 93.

[32] Petition of William Pollard to the Patent Board, June 29, 1790, Restored Patents, Specifications, I, 2–3, Record Group 241, National Archives, Washington, D.C.

Though I did not see Pollard's original reconstructed patent drawing in the National Archives, a printed black-and-white copy of it exists in the Patent Office. The design, as Pollard claimed, was based on the Arkwright waterframe of 1769. But whereas Arkwright's machine had four pairs of drafting rollers, Pollard's appears to have had only two pairs. This arrangement would seriously limit its drafting capacity. On the other hand, Pollard's frame apparently incorporated two subsequent English improvements: a bobbin lifter rail (patented by Coniah Wood in 1772) and its controlling mechanism, a heart-shaped cam. Together, these devices greatly accelerated the winding operation. In the original waterframe, the yarn had to be shifted manually from hook to hook on the flyer as each level of the bobbin was filled; in frames with the lifter motion, the yarn was distributed up and down the bobbin mechanically.

wine. This was completed in 1795 but was destroyed by fire two years later.[33]

All these various efforts by Americans in the Philadelphia region to secure British textile technology – the activities of formal organizations, pressure groups, the state legislature, and individual persons – were indicators of a strong demand for the new technology. To understand the process of technical transfer, however, we must also examine the supply side. That is, we need to know the nature and extent of the mechanisms by which both the men and the machines embodying the productive innovations were supplied to America. The emigration of artisans, the export of machinery, and the publication of practical information obviously played a vital role in the process of the transfer of British textile technology.

Methods of Transit

Though it has been argued for power and metallurgical technologies that "specific transfers rested more heavily upon copies of drawings, models, patent specifications, or visits to the initiating country" than upon individual mechanics, the reverse pattern developed in Philadelphia's encounter with Britain's new textile technology.[34] The key role here seems to have been played by the transfer of people familiar with the technology rather than the transfer of information in other forms.

Despite British protective legislation, the pushes and pulls of transatlantic migration swept many artisans westward, some of whom were the victims of handicraft obsolescence, while others helped create such obsolescence via the new technology. Two benchmarks indicate the size of this movement. The British returns of emigrants taken in 1773–1776, a high point in emigration, show some 6,000 persons sailing from English ports for North America and the West Indies. Of the 315 engaged in the textile trades, fifty-eight, or nearly 20 per cent, came to Philadelphia. In the United States during the War of 1812 about 7,500 individuals, including 1,300 workers in textile or textile related industries, registered as

[33] Carroll W. Pursell, Jr., "Thomas Digges and William Pearce: an Example of the Transit of Technology," *William and Mary Quarterly*, 3rd series, XXI (1964), 551-560; Lynn H. Parsons, "The Mysterious Mr. Digges," *ibid.*, XXII (1965), 486-492; Harold C. Syrett and others (eds.), *The Papers of Alexander Hamilton* (New York, 1961–), IX, 85-87; X, 345-47; XI, 241-49, 505-506, 566-67; XII, 78, 135-36, 140-42, 298-99, 516-17.

[34] Brooke Hindle, "The Transfer of Power and Metallurgical Technologies to the United States, 1800-1880: Processes of Transfer, with Special Reference to the Role of Mechanics," prepared for the ICOHTEC Conference on "The Acquisition of Technology by Non-Initiating Countries," 1970, typescript, 2.

enemy aliens. Of these textile workers, 314 or 24 per cent located in the city and three counties under consideration; ninety-four in Philadelphia City; eighty-four in Philadelphia county; twelve in Delaware county; and 124 in New Castle county.[35]

Given this stream of skill, it was hardly surprising that some Philadelphia entrepreneurs would simply wait for the ships to berth, contemplating a recruiting scheme no more aggressive than a visit to the docks. Yet there were some sound reasons for relying on immigrant transmission of the new technology.

Though immigrant mechanics have been viewed as vehicles of non-verbal technology,[36] their function calls for a closer analysis in the present context. By "non-verbal technology" I mean that stage at which the technology has not yet been systematically published in written or two-dimensional forms such as verbal descriptions, working drawings, and physical or chemical rules or formulae. How much non-verbal British textile technology was unique in the late eighteenth century and therefore how much would depend upon individual mechanics for international movement?

It is often overlooked that all the new machines preserved the core elements of many of the old devices, so that the basic principles of fiber processing offered few learning difficulties to American workmen. For example, the teasing and stripping actions of hand carding persisted in the Arkwright machine card, the basic seventeenth century spinning mechanisms (spindle and bobbin-and-flyer) appeared in the jenny, waterframe, throstle, and mule, and the essential movements of the handloom remained unchanged in the powerloom. Of course the design, construction, and operation of these fundamental mechanisms were quite familiar to American craftsmen and household producers.

In trying to appreciate the importance of non-verbal technology in textile manufacturing, an understanding of the constructional inputs of textile manufacturing equipment is essential. Constructional inputs included: (1) obtaining raw materials; (2) manipulating these materials; (3) machine designing; (4) construction of machine components; and (5) putting the components together

[35] Emigration Returns for England and Wales, 1773–1776, Great Britain, Public Record Office, T. 47/9–11; these have been generally analyzed by Mildred Campbell, "English Emigration on the Eve of the American Revolution," *American Historical Review*, LXI (1955–1956), 1–20. United States, Marshals' Returns of Enemy Aliens, War of 1812 Papers, Record Groups 45 and 59, National Archives; this source has been broadly surveyed by Herbert Heaton, "The Industrial Immigrant in the United States, 1783–1812," *Proceedings of the American Philosophical Society*, XCV (1951), 519–527. Textile workers in these returns are presently being analyzed by this author for a larger study of the transmission of textile skill and technology between Britain and the United States, 1770–1840.

[36] Hindle, "Transfer of Power and Metallurgical Technologies," 2.

in correct static and dynamic interrelationships. Of these, (1), (2) and (4) offered few problems to American craftsmen in the late eighteenth century; American carpenters, cabinet makers, wheelwrights, ironmasters, blacksmiths, wiredrawers, and clockmakers (for brass gear cutting) could equal most of their British counterparts.[37] The inputs of machine design and component interrelationships, however, involved the radically new elements of the improved textile technology. These critical construction inputs could only become available to American entrepreneurs by three routes: by written (published or handwritten) descriptions, drawings, rules, formulae, etc.; by Americans visiting British industrial districts; or by the immigration of artisans experienced in building the new technology. Machines and models unaccompanied by a skilled mechanic were of very limited utility at this non-verbal stage.[38]

Effective written dissemination remained generally unavailable until October 1812, when the article on "Manufacture of Cotton" * appeared in that mine of systematic technological information, Abraham Rees' *The Cyclopaedia; or Universal Dictionary of Arts, Sciences, and Literature* (39 vols. text and 6 vols. plates, London, 1802–1820). The other important widely available, written con-

[37] Charles F. Hummel, *With Hammer in Hand; the Dominy Craftsmen of East Hampton, New York* (Charlottesville, Va., 1968) records in detail the equipment and business of a Long Island clock and cabinet making firm which, operating between the 1760's and 1840, occasionally turned to the production of household implements like spinning wheels.

For a valuable critical bibliography of other works covering the pre-industrial and early industrial machine making crafts in America see Brooke Hindle, *Technology in Early America: Needs and Opportunities for Study* (Chapel Hill, N.C., 1966), 67–73.

[38] That an informed mechanic was an essential adjunct to the hardware of the new technology at the introductory stage of its transfer, is a point illustrated by the fate of the spinning mule sent from England to Philadelphia in 1783 (see text below). Once a new technological system had been adopted and adapted, it was of course possible to transmit fresh modifications to the receptor economy through any of the inanimate media.

Evidently the transition into the second stage had begun by the 1820's. Presenting evidence before two Parliamentary committees, Britain's leading engineers and manufacturers disputed over the means by which the new textile technology could be replicated abroad, particularly in France (but surely too in America). The London engineers led by Henry Maudslay thought that the French needed only drawings from encyclopaedias in order to copy the improved machines. The manufacturers of Manchester, Glasgow, and Norfolk, such as Peter Ewart, John Kennedy, Henry Houldsworth, Sr., and John Harvey, strongly dissented. In their judgement, even the possession of models was inadequate to permit perfect copies; only a full size example, a "pattern machine," would suffice. Taking a less extreme position, the Manchester millwright, William Fairbairn, believed that though the French had obtained English machines "they have failed in setting them to work."

The divergence in opinion probably reflected two different machine-building experiences. No London engineer at this time had built much textile machinery to speak of, and handicraft work was more prevalent around Manchester. Maudslay spoke from the perspective of his own pioneering achievements, thus probably overestimating the abilities of foreign rivals – until they too acquired the precision tools for metal cutting which he had developed. See Great Britain, *Parliamentary Papers (Commons)*, 1824 (51), V, 160, 161, 251, 255, 300, 343, 346, 380, 546, 548, 567, 570, 572; Great Britain, *Parliamentary Papers (House of Commons)*, 1825 (504), V, "Report from the Select Committee on the Laws Relating to the Export of Tools and Machinery," 11, 33–34; Musson and Robinson, *Science and Technology*, 427–509; L. T. C. Rolt, *Tools for the Job: a Short History of Machine Tools* (London, 1965), 83–91.

* October *should read* November

tributions on textile technology, the *Cyclopaedia's* articles on "Weaving," "Woollen Manufacture," and "Worsted Manufacture," did not appear until summer 1818, delayed by the accident of alphabetization. An edition of Rees for American readers was speedily published at Philadelphia between 1810 and 1822.[39] But even in this monumental record of technological progress, the articles and plates (like most British patents to this date) only occasionally specified dimensions, speeds, or settings, let alone rules by which adjustments in processing could be made. Rees' static descriptions thus resisted easy practical application until at least 1832. That year James Montgomery, a Scottish spinning master, published his *The Carding and Spinning Master's Assistant* in which, "every machine employed in the process is introduced in its order, their use and operations described, and how to adjust and adapt them to suit the various kinds of cotton, and the different sizes of yarn pointed out. Rules for performing all calculations connected with the business, are distinctly laid down and exemplified . . . [together with] drawings of the different machines." [40]

The new British textile technology remained private until 1812 and unreduced until 1832. Theoretically this detailed technical information might have been elicited by correspondence with a British artisan; such was not only a tedious method, but was also unlikely since the mechanic would have to be both literate and extraordinarily motivated to release his hard-won means of livelihood. For him and the American entrepreneur, emigration was the simpler and safer way to remove the new skills to the United States.

Alternatively the American promoter might visit Britain to try his hand at industrial espionage.[41] But here he would meet wide-

[39] Benjamin D. Jackson, *An Attempt to Ascertain the Actual Dates of Publication of the Various Parts of Rees's Cyclopaedia* (London, 1895), 6–7; Ralph R. Shaw and Richard Shoemaker (eds.), *American Bibliography: A Preliminary Checklist for 1810* (New York, 1961), 65–66; Eugene S. Ferguson, *Bibliography of the History of Technology* (Cambridge, Mass., 1968), 55.

[40] James Montgomery, *The Carding and Spinning Master's Assistant; or the Theory and Practice of Cotton Spinning* (Glasgow, 1832), v.

[41] As did the Brandywine manufacturer, Joshua Gilpin, with far-reaching consequences for the American paper industry but none for its textile technology. See Norman B. Wilkinson, "Brandywine Borrowings from European Technology," *Technology and Culture*, IV (1963), 10–12; George H. Gibson, "Fullers, Carders, and Manufacturers of Woolen Goods in Delaware," *Delaware History*, XII (1966), 35.

Shrewd and ambitious Yankees undertook some of the most effective industrial espionage in Britain's early textile mills. Abel Buell recruited workers and secured machine designs in 1789; Francis Cabot Lowell obtained the power loom design in 1810–1811; and Zachariah Allen procured gig mill and steam dresser plans in 1825. See David J. Jeremy (ed.), *Henry Wansey and His American Journal, 1794* (Philadelphia, 1970), 73; George S. Gibb, *The Saco-Lowell Shops: Textile Machinery Building in New England, 1813–1949* (Cambridge, Mass., 1950), 7–14; Zachariah Allen, *The Practical Tourist, or Sketches of the State of the Useful Arts, and of Society, Scenery, &c. &c. in Great-Britain, France and*

spread secretiveness surmountable only by much ingenuity or frequent disbursements.[42] Even if he could obtain an exact copy of a complex machine such as a waterframe, problems would arise later when he needed to make adjustments in processing machinery as a result of changes in raw material supplies, product market demands, or growing competition.

At this point knowledge of how to arrange the machines' components in correct static and dynamic interrelationships became critical. Early textile machine making was not merely a matter of shaping gears, cams, pulleys, rollers, and so on, but of giving them the dimensions, positions, settings, speeds, and surfaces suited to work the cotton or wool fibers into rovings, yarns, and cloths without incurring excessive breakages. This would be most difficult with machines incorporating a new device, such as the cotton drawing frame, roving frame, waterframe, throstle, and mule, all of which utilized the deceptively simple drafting rollers.[43] The American promoter therefore could either discover the correct adjustments by trial and error (which might be expensive in materials), or he could try to locate a competent immigrant mechanic. Those who successfully pursued the first strategy in Philadelphia (such as William Pollard) appear to have been exceptional. The policy of the Pennsylvania Society and of Jacob Broom in seeking out the immigrant artisans, or their pupils, to build the new machinery was the more typical solution.

The case of the transfer of the spinning mule conclusively, if negatively, establishes the importance of the immigrant artisan as the vehicle of technology transmission to the Philadelphia region.

Holland (2 vols., Providence and Boston, 1832), I, 198–200 and Allen's diary of observations on early European industrial technology, his "Journal of European Trip, 1825," in the Zachariah Allen Papers, Rhode Island Historical Society, Providence, Rhode Island.

[42] Allusions to the secretive practices of early manufacturers are scattered through their surviving papers. Eric Robinson has nicely marshalled the evidence of Boulton and Watt's covert, protective practices. One contemporary engineer, however, thought that Boulton and Watt "have always displayed an uncommon degree of mystery." Whatever the truth of this comment, many millowners closed their doors to strangers. Rare must have been the optimistic naivety of the Norfolk manufacturer who opened his factory to foreigners "with caution," believing that "when machinery is peculiarly complicated, you may show it with good effect, I think, because it makes the difficulty of imitation appear greater." Artisans, too, resisted pressure towards "learning foreigners," knowing that sharing their skills would eventually lower their own wages.

The early attitudes of suspicion and secrecy were frequently built into the architecture of the nineteenth century British mill, which often was reminiscent of a medieval castle, surrounded by high walls and entered through a gatehouse. This sort of defensive layout American traveller Zachariah Allen found in Benjamin Gott's woolen mill at Leeds in 1825 and considered it "like those of most other similar establishments." See Musson and Robinson, *Science and Technology*, 216–230; G. B., *Parliamentary Papers* (*Commons*), 1824 (51), V, 20, 159, 277; Allen, *Practical Tourist*, I, 195. *

[43] The point is well demonstrated in the experiments of Lewis Paul and John Wyatt, for which see Richard L. Hills, *Power in the Industrial Revolution* (Manchester, 1970), 38–40, 44–52.

* V, 20, 159, 277 *should read* V, 20, 159, 278

Along with a carding machine and two jennies, a sixty-four spindle "Hallwood" spinner was crated in casks labelled Queens Ware and dispatched from England by one Benjamin H. Phillips in 1783. Phillips, who sent his son ahead to Philadelphia, sailed with the machines from Liverpool. When the ship put in at Cork, however, he fell ill, returned to Liverpool, and died soon after. Meantime the machines went on to Philadelphia, where young Phillips, unable to use them, sold them to Joseph Hague (builder of the first known jenny in Philadelphia) and his friends. The Hague group, however, managed only to assemble the card and smaller jenny. This is hardly surprising because only a competent mechanic already acquainted with the machine could have made sense of a box of mule parts. In frustration, Hague packed the lot up and put them in storage until he was able to sell them in the spring of 1787. They were then shipped back to England, touching off the Pennsylvania Society's demands for protectionist legislation.[44]

How then did the artisans convey the technology across the Atlantic? In evading British Customs officials, disguise and deception were easy for individual artisans but were less feasible for equipment. A fairly simple machine like a jenny, which incorporated only one new mechanism (the carriage clasp) and relied upon the operative for critical regulatory movements, could be memorized without too much difficulty. Indeed the first man known to have built a jenny in Philadelphia (or America) – Derbyshire weaver-manufacturer, Joseph Hague – spontaneously recalled its design. Hague left Liverpool in May 1774 and, on arrival in the United States, found a bottleneck at a previous processing stage jeopardizing the chances of continuing his trade. He succinctly expressed the manner of the jenny's transit in a petition to the Pennsylvania Assembly, presented March 1775: "finding the high Price paid for Spinning a very great Discouragement to his Undertaking, he gave Instructions for making a Machine similar to those lately invented, and used in *England* for that Purpose, whereby the Expence of Labour is exceedingly reduced; that not being furnished with either a Plan or Model of such Machine, he has been at the Expence of much Time, and of some Money in causing one to be constructed."[45] Once its design was known, the jenny could be

[44] For a good treatment of this episode see Joanne L. Neel, *Phineas Bond: a Study in Anglo-American Relations, 1786–1812* (Philadelphia, 1968), 55–61 and 65, n. 13. Though Dr. Neel misses the identification, the "Hallwood" must have been a spinning mule. Named after the early home of its inventor, Samuel Crompton, the mule was first known as the "Hall-i'-th'-Wood Wheel." For this see George W. Daniels, *The Early English Cotton Industry* (Manchester, 1920), 114.

* [45] G.B., P.R.O., T. 47/9, 145, entries for May 17–24, 1774 for Hague's embarkation from Liverpool aboard the packet *Boston; Pennsylvania Archives* 8th series, VIII, 7215–16 for the text of his petition of March 11, 1775.

* 1774 for Hague's *should read* 1774 for John Hague's

built by carpenters, as the success of the Pennsylvania Society demonstrated.

Complicated machines, like mules or waterframes, would either have to be taken over as entire machines or as very good drawings, specifications and models, accompanied by a competent mechanic of course. Phillips' audacious attempt to transport machinery was not often repeated. William Pearce, who in the 1790's built waterframes and mules as well as his own double-loom, depended upon models, key machine parts, and drawings which were smuggled from Dublin to New York aboard a ship other than the one in which he sailed.[46]

By 1810, after three decades of machine building in England, trained mechanics could be so practiced in their trade as to need to take no more than their tools to the United States. Two such were the Manchester brothers, George and Isaac Hodgson, aged twenty-nine and twenty-two respectively when they emigrated in 1811, sending their tools marked as fruit trees by another vessel.[47]

By the outbreak of the War of 1812, 15 per cent or forty-eight of the Philadelphia region's British textile workers were engaged in machine-making: six millwrights, fourteen machine-makers, eight machinists, two engineers, one steam engine builder, one cotton machine-maker, one reed-, one shuttle-, and one spindle-maker, and thirteen wire-drawers. There were several geographical clusters of these workmen. Twelve of the wire-drawers, for example, worked in the Falls of Schuylkill rolling mill and wire factory established soon after 1809 by Josiah White and Erskine Hazard. The factory's products – wire for card teeth – were basic to manual and mechanical textile processing. Within the city of Philadelphia, Thomas Ore, resident in America since 1794, described himself as a cotton machine maker; his neighbor, Joseph Bamford, had built mules since 1805.[48]

But the largest concentration of British machine makers was on the Brandywine, where fourteen were registered. Of this group, the Hodgson family attained the most lasting success. Altogether five brothers emigrated in 1811–12. Between March 1813 and May 1814 they built for the Duplanty, McCall & Co. fine cotton spinning mill, four breaker and six finisher cards of the flat top type so popular with the medium and fine spinners of Glasgow and Lan-

[46] See note 33 above for sources.

[47] Wilkinson, "Brandywine Borrowings," 3; U.S., Marshals' Returns of Enemy Aliens: Delaware, War of 1812 Papers, Record Group 45, National Archives.

[48] War of 1812 Papers quoted in note 34 above; Edwin Iwanicki, "The Village of Falls of Schuylkill," *Pennsylvania Magazine of History and Biography*, XCI (1967), 339; Samuel H. Needles, "The Governor's Mill, and the Globe Mills, Philadelphia," *ibid.*, VIII (1884), 378.

cashire for forty years after Arkwright. For these cards, with iron frames and mahogany-covered main cylinders three feet in diameter and two feet wide, the Hodgsons charged $200 apiece. The March 1813 contract with Duplanty, McCall & Co. also specified one blowing machine ($150), one drawing frame of twelve heads or units of drafting rollers ($456), two roving frames totalling fourteen heads ($434), and two 120-spindle throstles or modified waterframes ($660 each). Though they were not the only machinists to receive contracts from the firm, the Hodgsons continued to build equipment for Brandywine textile mills well into the 1830's. Other English-born machine makers used by Duplanty, McCall & Co. were William Hartley, a Philadelphia reed-maker and James Wagstaff, a Wilmington spindle-maker who turned more than 1,000 spindles for the firm in 1815.[49]

The transfer of British textile technology thus occurred primarily via the movement of men accompanying machines, parts, or models, all encouraged by American efforts, and not very effectively deterred by British attempts to prevent leakage. The immigration of skilled artisans and the smuggling of machinery, rather than the transmission of knowledge via published materials, clearly constituted the methods by which the Philadelphia area secured an ability to imitate Britain's textile industry. That ability, however, was obtained over a period of many years and in pieces, rather than all at once.

Technology Transmission Rates and their Implications

Under close scrutiny, some interesting differences in the rates of transit of the basic British textile machines are apparent. The rotary card, perfected by Arkwright in 1775, was set up in some form at Philadelphia by Joseph Hague in 1783–1784, and then publicly exhibited in the Grand Federal Procession of 1788: a transit period of between eight and thirteen years.[50] Improved jennies, built at Philadelphia in 1774–1775 by Joseph Hague and

[49] Sources quoted in note 47 above; quotation of *ca.* February–March, 1813 and contract of March 22, 1813 between the Hodgson brothers (George, Thomas, Isaac, Henry, and Richard, in sequence of age) and Duplanty, McCall & Co., Longwood MSS, Group 6, Box 2, Eleutherian Mills Historical Library, Greenville, Delaware; Ledger 195, f. 144, Duplanty, McCall & Co. Papers, Accession 500, Eleutherian Mills Historical Library; Evan Leigh, *The Science of Modern Cotton Spinning* (2 vols., Manchester, 1873), I, 104–106; Roy M. Boatman, "The Brandywine Cotton Industry, 1795–1865" (Hagley Research Report, 1957; copy in Eleutherian Mills Historical Library), 95–96; George H. Gibson, "The Delaware Woolen Industry," *Delaware History*, XII (1966), 103.

[50] See Abraham Rees (ed.), *The Cyclopaedia; or Universal Dictionary of Arts, Sciences,* ** and Literature* (45 vols., London, 1819–1820), s.v. "Manufacture of Cotton," and Hills, *Power in the Industrial Revolution*, 73–81 for the carding engine; for its introduction to Philadelphia see sources in notes 17, 20, and 44 above.

* London 1819–1820 *should read* London 1802–1820

Christopher Tully, crossed in under five years.[51] The slubbing billy, invented in 1786, arrived in 1791, brought by William Pearce.[52] Astonishingly, the spinning mule's abortive transit took only three years, though its successful installation waited until the 1790's, when William Pearce may have set up one in 1791–1792 and James Davenport certainly had a number in the Globe Mill in 1796.[53] In contrast, roller printing, developed by Scotsman Thomas Bell in 1783–1785, was delayed over twenty years before the brothers Issachar and James Thorp brought the first copper cylinder calico-printing machine and accessory bleaching and glazing equipment to Bristol Township, Philadelphia county, in 1809.[54] At the extreme end of the scale, the gig mill, used in Gloucestershire continuously though illegally since the sixteenth century, was not found in America until two centuries after its invention; as late as 1813 Dr. James Mease noted that "for the purpose of raising the nap on cloths, handcards are at present universally used in Pennsylvania," though by 1820 the region had at least four gig mills.[55]

Apart from the fortuitous coincidence that the Revolution swept

[51] For the jenny see Rees (ed.), *Cyclopaedia*, s.v. "Manufacture of Cotton" and "Woollen Manufacture;" also Christopher Aspin and Stanley D. Chapman, *James Hargreaves and the Spinning Jenny* (Helmshore, Lancs., 1964). For its introduction to Philadelphia see sources in note 16 above.

The illustration and specification of Tully's jenny in the *Pennsylvania Magazine*, I (April, 1775), 157–58 show a number of improvements over that patented by James Hargreaves on June 12, 1770 (G.B. Patent 962). These include a horizontally axled driving wheel; power transmission to the spindles via a layshaft (a long tin or wooden roller) vastly increasing the potential number of spindles; wheels and an opening mechanism on the carriage-clasp; activation of the faller wire from the carriage-clasp; friction-reducing glass bases under the spindles; tensioning devices for both the driving wheel and the tin roller; and a movable roving box. Some six feet in depth, Tully's machine gave a draw of up to fifty-four inches approximately and a maximum drafting ratio of about 1:7. Apart from modifications to the carriage-clasp and roving box mechanisms and the addition of a clock to record the number of draws (and hence length and fineness of yarn spun), the 1775 Philadelphia jenny came close in design to the most sophisticated jennies found in nineteenth century England.

[52] I have assumed that a slubber was not the same as a slubbing billy. If it was, the slubbing billy reached Philadelphia in 1787, a year after its appearance in England. See above, p. 30. *

See Rees (ed.), *Cyclopaedia*, s.v. "Woollen Manufacture" and Julia de L. Mann, *The Cloth Industry in the West of England from 1640 to 1880* (Oxford, 1971), 288 for the billy; and Syrett and others (eds.), *Papers of Hamilton*, IX, 86, n. 2 for its transmission via Pearce.

[53] For the mule see Rees (ed.), *Cyclopaedia*, s.v. "Manufacture of Cotton;" Harold Catling, *The Spinning Mule* (Newton Abbot, 1970); and Hills, *Power in the Industrial Revolution*, 116–133. For its introduction to America see source in note 44 above; Syrett and others (eds.), *Papers of Hamilton*, IX, 86, n. 2; and Bagnall, *Textile Industries*, 222–24.

[54] See Geoffrey Turnbull, *A History of the Calico Printing Industry of Great Britain* (Altrincham, 1951), 50–98 for the early history of roller printing. For the Thorps see J. Thomas Scharf and Thompson Westcott, *History of Philadelphia, 1609–1884* (3 vols., Philadelphia, 1884), III, 2316–17; U.S. Marshals' Returns of Enemy Aliens: Pennsylvania, War of 1812 Papers, Record Group 59, National Archives; United States, Records of the 1820 Census of Manufactures, Schedules for Pennsylvania (Eastern District), Record Group 29, National Archives.

Wooden rollers for crude surface printing appear to have been used by the Irish calico printer Archibald Hamilton Rowan on the Brandywine in the 1790's; see Montgomery, *Printed Textiles*, 98–103.

[55] Mann, *Cloth Industry*, 141–42; *Archives of Useful Knowledge*, III (1813), 343–44.

* p. 30 *should read* p. 29–30

away England's mercantilist restrictions on colonial manufacturing just when the pace of invention began to accelerate significantly in Britain, there are other partial explanations for these particular time lags. The carding machine, slubbing billy, jenny, and mule were basal processing machines at the bottom of the processing "tree" and so were essential equipment for any manufacturer of cotton or woolen yarn.[56] On the other hand, the printing of cottons or linens was an optional finishing process practiced only by a relatively few specialists in America before the 1820's.[57] Until then, the domestic demand for decorative fabrics was adequately met by these skilled manual workers and, more importantly, by European imports.

Raising or gigging, however, was a necessary operation in the finishing of all woolen (as opposed to worsted) cloth. Several factors may therefore explain the belated arrival of the gig mill in America: the statutory prohibition on its use, effective for most of Britain until the late eighteenth century; [58] the relatively little raising required in the finishing of coarser cloths on which American manufacturers concentrated, in consequence of the American middle and upper class customers' preference for imported finer cloths (a taste induced by England's mercantilist policies); [59] and even, perhaps, the shortage of teasels (a type of thistle used for raising) before 1810.[60]

This does not explain why the waterframe and the power loom, also basal processing machines, required up to two decades to reach Philadelphia, and indeed America. The waterframe, patented by Arkwright in 1769, declaredly came as a model to William Pollard in 1789, but operating machines were apparently first set up on the Brandywine by William Pearce in 1795 and in Philadelphia county in the Globe Mill by James Davenport in 1796.[61] Powerlooms,

[56] Admittedly, the cotton manufacturer of coarse or medium yarns would choose waterframes or throstles instead of billies and jennies; the woolen manufacturer would prefer mules to jennies if he could afford them and if he could find mule spinners.

[57] Montgomery, *Printed Textiles*, 83–103.

[58] Mann, *Cloth Industry*, 141–42.

[59] In the West of England fine cloth might receive as many as twenty-nine courses or treatments with the teasels. In America raising was a less elaborate operation and the early finishing handbooks gave it a correspondingly cursory treatment. Bemiss, for instance, prescribed only "three good nappings" or courses for a superfine Thomas, see *ibid.*, 300; John Wily, *A Treatise on the Propagation of Sheep, the Manufacture of Wool, and the Cultivation and Manufacture of Flax* (Williamsburg, Va., 1765), 26; Asa Ellis, Jr., *The Country Dyer's Assistant* (Brookfield, Mass., 1798), 107, 110–11; Elijah Bemiss, *The Dyer's Companion* (New York, 1815), 274.

[60] Barnes Riznik, "New England Wool-Carding and Finishing Mills, 1790–1840" (Old Sturbridge Village Research Report, 1964), 42.

[61] See Rees (ed.), *Cyclopaedia*, s.v. "Manufacture of Cotton" for the waterframe which, when modified with the addition of a long, horizontal tin roller to drive the spindles, was known as a throstle. For its transmission to Philadelphia see source in note 32 above; Pursell, "Thomas, Digges and William Pearce," 558; Boatman, "Brandywine Cotton Industry," 11; Bagnall, *Textile Industries*, 222, 224For the Pollard story see Anthony F.C. Wallace and David Jeremy, 'William Pollard and the Arkwright Patents', *William and Mary Quaterly* 3rd ser. 34 (1977).

functioning in Britain by 1796, were still at a prototype stage in Philadelphia in 1814, though the region contained forty by 1820.[62] Both were temperamental machines, at first heavily dependent upon the services of experienced mechanics for correctly setting the ratios of roller speeds and drafts in the waterframe, and for ensuring the proper synchronization of shedding, picking, beating-up, let-off, and take-up movements, to say nothing of primitive stop motions, in the early powerlooms. Such technological drawbacks, rather than capital costs or power needs, initially may well have been the main reason for retarded transit in these cases.

In Philadelphia's experience, the new technology came piecemeal (one machine at a time and in no particular sequence) and not, as at Waltham in 1813, as an essentially complete manufacturing system. Even so, the acquisition of single machines created the demand for others at adjacent processing stages as bottlenecks appeared in a promiscuous handicraft-machine production system. Thus, the Pennsylvania Society, employing one carding machine, jennies, household spinners, and fly shuttle looms in 1788–1789, met imbalances at the spinning and finishing stages which impelled the quest for the waterframe (with a higher productivity than the jenny's) and more finishing workers. Whilst a primary piecemeal-exposure to the new technology provoked a demand for other components of the new system, it also meant that, until mechanization was complete, an individual machine's performance was rated only against its corresponding manual technique. Sharper pressures developed in the primary system-exposure experience. Here, in the early nineteenth century, before textile manufacturing was thor-

[62] For the early history of the power loom see Julia de L. Mann, "The Textile Industry: Machinery for Cotton, Flax, Wool, 1760–1850," in C. Singer and others (eds.), *A History of Technology* (5 vols., 1954–1958), IV, 299–304; Walter English, *The Textile Industry: an Account of the Early Inventions of Spinning, Weaving, and Knitting Machines* (London, 1969), 89–97; Duncan Bythell, *The Handloom Weavers: a Study in the English Cotton Industry during the Industrial Revolution* (Cambridge, 1969), 66–93.

When the Brandywine cotton spinning firm of Duplanty, McCall & Co. came to look for power looms in spring 1814 they learned that three were available in Philadelphia, belonging to Walter Jarvis, Robert Miller, and Thomas Siddall. Siddall's, which he patented that July, had the essential warp protector motion which stopped the beater or batten when the shuttle got jammed in the warp shed. An English engineer then in Philadelphia, Daniel Large, identified it as the motion patented by the Scotsman Robert Miller in 1796. Siddall, if not English himself, was much involved with English cousins and with the Thorps, Wagstaffs, and Hodgsons, local immigrant families from Lancashire. He claimed that his loom wove at the rate of two or three yards an hour. See Archibald McCall to Raphael Duplanty, February 17, 23, April 13, 19, 26, May 7, 20, 1814, Duplanty, McCall & Co. Papers, Accession 500, Eleutherian Mills Historical Library; United States, Patent Office, *A List of Patents Granted by the United States from April 10, 1790, to December 31, 1836* (Washington, D.C., 1872), 140 (July 9, 1814); *The Emporium of Arts and Sciences*, V (1813), 459; G.B. Patent 2122 (June 28, 1796) awarded to Robert Miller; Teckla Anne Siddall, *Siddalls in America (the Phineas Line)* (privately printed, n.p., 1962), copy in the Eleutherian Mills Historical Library; U.S., Census of Manufactures, 1820, Schedules for Pennsylvania and Delaware, Record Group 29, National Archives for power looms in 1820 – three more were narrow or ribbon (Dutch) looms.

oughly understood, imbalances between processing stages acted as a powerful innovational incentive, in stimulating efforts to keep productivity levels at each stage in line with one another. This may partly explain the innovational backwardness of the Philadelphia region before 1820, compared to the Boston Manufacturing Company at Waltham.[63]

Problems in Adopting the New Technology

We have seen how the new textile technology crossed the Atlantic. What was necessary for it to be adopted by the businessmen of the greater Philadelphia region? Increasing urbanization, ramifying transportation networks, a system of mercantile credit, and utilitarian educational opportunities all existed to some extent in late eighteenth century America, and the Philadelphia region in particular. Still required, if the British pattern of industrialization tallied, were sectors of steady expansion and the triggers of rapid population growth and labor-saving technology.[64] Between 1790 and 1820 the lower Delaware valley provided one region of industrial growth. Here Philadelphia's population more than doubled up to 63,000, while concentrated cotton and wool machine spindles in the three counties rose from 240 in 1790 to nearly 28,000 in 1820.[65] In adopting the new technology in America, at least three kinds of problems — economic, technological, and political — presented themselves in the Philadelphia area.

For the individual manufacturer, the central economic problem lay in choosing optimum product markets. On the one hand household production and Asian imports supplied the lower and middle quality fabrics; on the other, European manufactures dominated the market in fine goods, and some middle qualities too. Of these two groups of competitors, the household producer in Asia and America was the more vulnerable. The ascendancy of British imports is illustrated by the difficulties attending woolen broadcloth production in America. Against the American consumers' long-conditioned preference for English broadcloth, even immigrant English clothiers failed in attempting this line of business in the United States.[66]

[63] See David J. Jeremy, "Innovation in American Textile Technology during the Early 19th Century," forthcoming in *Technology and Culture*, XIV (1973), for some elaboration of these points in the New England context.

[64] Michael W. Flinn, *Origins of the Industrial Revolution* (London, 1966), 91–104.

[65] U.S., Census of Manufactures, 1820, Schedules for Pennsylvania and Delaware, Record Group 29, National Archives.

[66] One such was Moses Hey (1792–1866), who emigrated about 1816. In one of his obituaries it was observed that "tempted by a protective tariff, he here [Philadelphia]

Urban population growth, however, made the cheap floor covering, oil-cloth, popular in Philadelphia. Introduced by the British carpet manufacturer, Isaac Macauley, two years after he emigrated in 1806, its value was endorsed by several awards from the Franklin Institute in the 1820's.[67]

Most successful in the Philadelphia area was the machine spinning of medium quality cotton yarns. By 1820 cotton claimed over 22,000 of the region's 28,000 cotton and woolen spindles, concentrated in twenty of the fifty-four cotton and woolen mills. Initially supplementing professional hand-weavers, cotton yarn manufacturers directed the new British textile technology against the weakest producers for the urban markets, namely the household spinners, mostly women and children. As a basal component, cotton yarn had numerous applications in urban consumer goods. By 1820 handweavers needed it for plain goods (shirting and sheeting), medium quality fancy cloths (like checks, plaids and ginghams), and "smallwares" (tape, furniture binding, saddle girths, etc.).[68]

As markets were identified, manufacturers adjusted equipment to suit their specialties. For this reason New England cotton yarn manufacturers utilized throstles rather than mules, the throstle having a higher productivity per spindle for coarse and medium yarns. Philadelphia region manufacturers retained a relatively high proportion of jennies and mules (though figures are uneven) before 1820, reflecting both their attempts at fine spinning during the War of 1812 and their production of cheap mixed cotton and woolen goods, such as satinet.[69]

Operational problems remained for American manufacturers who

---

commenced the broad cloth manufacture, but this, in so new a country, has always been attended with difficulties, which, together with ever changing circumstances, more than neutralized that protection, and, *to the present day*, has only proved a lure to Destruction." Newspaper clippings, Moses Hey Papers, Historical Society of Pennsylvania. See Charles Robson, *The Manufactories and Manufacturers of Pennsylvania in the Nineteenth Century* (Philadelphia, 1875), 447–48 for Hey; and Arthur H. Cole, *The American Wool Manufacture* (2 vols., Cambridge, Mass., 1926), I, 195–207 for this problem.

A good comparative study of early American cloth types, prices, and categories of producers is badly needed to establish the effects of market forces on early manufacturing. In this context the cessation of imports of coarse cottons from India and China by 1816 also requires close investigation. See Caroline F. Ware, *The Early New England Cotton Manufacture* (Boston, 1931), 71–72.

[67] Oil cloth was first made, apparently, by Nathan Taylor of Knightsbridge, London, in 1754. See Edward H. Knight, *Knight's American Mechanical Dictionary* (3 vols., Cambridge, Mass., 1876), I, 888; Scharf and Westcott, *History of Philadelphia*, III, 2231–32; U.S., Marshals' Returns of Enemy Aliens: Pennsylvania, War of 1812 Papers, Record Group 59, National Archives; *Journal of the Franklin Institute*, II (1826), 266; VI (1828), 405; VIII (1831), 325.

[68] Sources in note 65 above.

[69] *Ibid.* Of the region's 28,000 spindles in 1820, only 5,728 can be surely identified from the Census. Of these, 3,580 were on intermittent spinning machines (2,760 jenny and only 820 mule spindles) and 2,148 on continuous spinning machines (480 waterframe and 1,668 throstle spindles). When a machine's spindleage was not given, I credited a jenny with sixty spindles and the other three types with 120 spindles each.

had acquired the technology to the extent of constructing a set of equipment. On the machine making and maintenance side, American employers may or may not have been assisted by one significant change in the immigrant labor supply. This was the increase in the division of labor, attendant upon the accelerating progress of Britain's machine technology. Whereas in the 1780's one man undertook the construction of a variety of textile machines, by the 1820's he made but one kind of machine component, which was installed in several different types of machines. This was especially true of the Manchester district, the fountainhead of British cotton machinery innovation. Here, for example, in 1824 at least six different kinds of skilled workers supplied the parts of a spinning frame.[70] Specialization of labor, of course, could improve the quality of the component in question, and this was a very important consideration for critical machine elements, like spindles and rollers. In these cases, American manufacturers surely benefited, as did Duplanty, McCall & Co., who retained the services of a spindle maker from the vicinity of Manchester.[71] On the other hand, immigrant English builders of parts ordinarily made by more general craftsmen in the United States may have been lost to American industry until the domestic demand for textile machinery could support such specialization.

Besides maintenance technicians, operatives, production managers, and artisans skilled in non-mechanized processes all needed to be recruited. The immigrant artisans offered one obvious but fluctuating source. How long would it take for their skills to pass to the indigenous population?

It may be surmised that the time required for the absorption of the new technology into American industry, up to the level of moving from the immigrant to the native artisan, would depend minimally upon the various skills' learning periods. The shortest of these occurred in the most unskilled of the operative functions: supplying a missing mechanical movement with human strength. This ranged between light tasks such as throstle doffing (unloading full bobbins and replacing them with empty ones) and heavy jobs like lifting loom beams, work which could be understood and learned in a matter of minutes or hours. When the operative acted as a mechanical regulator, as in jenny spinning (essentially maintaining a fine balance between drafting and twisting for a large

[70] These were the spindle maker, roller maker, turner or bobbin maker, founder, tinman, and flyer maker. See G.B., *Parliamentary Papers (Commons)*, 1824 (51), V, 545.

[71] James Wagstaff. See Ledger 195, f. 144, Duplanty, McCall & Co. Papers, Accession 500, Eleutherian Mills Historical Library; U.S., Marshals' Returns of Enemy Aliens: Delaware, War of 1812 Papers, Record Group 45, National Archives.

number of ends) or in putting up a mule (winding on the spun yarn in cop formation), the coordination of hand and eye supplemented by experiential knowledge of fiber behavior required a training period of several years. The same would be true of operations involving mathematical ability, such as a weaver employed in his warp calculations. In both these cases, however, there would have been a major carry-over from the pre-industrial trades.[72]

Perhaps production management and skills defying mechanization would take the greatest time to develop in America. On the former, James Montgomery underlined a well-known fact: [73]

> the manager of a Cotton Spinning Factory can only acquire a proper knowledge of his business, by long experience and application in the practical department of the manufacture, and it will depend upon the situation in which he is placed, and the advantages he enjoys, if he ever obtain that correct knowledge of all its details, which is necessary to render him fully qualified for managing a large establishment with satisfaction or profit to the proprietors.

The point was borne out in the Philadelphia region by John Price Crozer, the manufacturer and philanthropist, who remembered an English immigrant surviving most successfully during the 1818–1819 depression: [74]

> In the year 1812, a good many small manufactories had sprung up, chiefly of woolen, but those in Delaware County and almost everywhere in Pennsylvania had proved unfortunate. The factories all closed. . . . One cotton factory — Wagstaff and Englehorn's . . . — continued to operate with partial success. Wagstaff was a practical cotton spinner from England. And now, in 1821, when attention was newly awakened, this factory seemed working to great profit.

The same was true of the larger and more profitable woolen mills on the Brandywine, where English manufacturers served as production managers: Nathaniel H. Clifford in the du Pont woolen mill at Louviers (with a fixed capital of $50,000 in 1832, but somewhat less during the War of 1812 when Clifford organized it); and Isaac Bannister in William Young's Delaware Woolen Factory at Rockland (of a similar fixed capital).[75]

[72] Until the sciences of ergonomics and industrial psychology have been applied to early machinery and techniques, or until someone marshals all the evidence on operative learning problems, we shall remain ignorant of one of the vital aspects of non-verbal technology transfer. I am grateful to Professor Eugene S. Ferguson of the University of Delaware for bibliographic comments in my first effort to explore this area.

[73] Montgomery, *Carding and Spinning Master's Assistant*, i.

[74] John P. Crozer, *Biographical Sketch of John P. Crozer Written by Himself* (Philadelphia, 1861), 54.

[75] Gibson, "Fullers, Carders, and Manufacturers of Woolen Goods in Delaware," 36–41; U.S., Marshals' Returns of Enemy Aliens: Delaware, War of 1812 Papers, Record Group 45, National Archives.

Of the unmechanized and unquantified skills demanding a long apprenticeship, dyeing was the most important. Many of the European dyeing secrets became widely known when the Swiss immigrant John Rauch published his *Receipts on Dyeing* in 1815 and after the Gloucestershire dyer William Partridge published his *A Practical Treatise on Dying* (New York, 1823, 1834, 1847). The value put upon a good dyer is suggested by Partridge's income from the du Pont, Bauduy & Co. woolen mill at Louviers: nearly $1,700 for a year's services in 1812–1813.[76]

Crossing the threshold operationally from dependence on immigrants could thus not be considered completed until after the War of 1812, and perhaps not until the 1830's. For machinery construction, one indicator of the length of this final stage in technology transmission appears in the case of Alfred Jenks. One of the more enduring machine-makers trained under Samuel Slater, he left Pawtucket to establish a textile machine works at Holmesburg, Philadelphia County, Pennsylvania, in 1810, twenty years after the immigrant master-manufacturer's arrival. Though the length of Jenks's training under Slater is unknown, a leading British engineer, William Fairbairn, in 1824 reckoned that it took five to seven years to train a millwright.[77]

Finally, in adopting the new technology, early manufacturers faced political opposition from merchants, Jeffersonian agrarians, and British Loyalists. The last, though least significant, deserves brief notice here as a curious aberration peculiar to the Philadelphia experience and as especially relevant in the present context. Stemming from a determination to defeat American industrialization by upholding British protective legislation against the emigration of artisans and the export of machinery (in effect, maintaining America's old mercantilist relationship with Britain), Loyalist resistance revolved round the person and activities of Phineas Bond, British consul at Philadelphia, 1786–1810. Learning, soon after his arrival, of the influx of new textile machinery and skilled workers from Britain, Bond did all within his power to reverse the flow.

Not only did he give *ex post facto* encouragement to English cotton merchant Thomas Edemsor, who had returned Phillips' machines to Britain in 1787, but he sent Henry Royle, a calico printer

[76] Vol. 37, Journal 1810–1815, 80, du Pont, Bauduy & Co. Papers, Accession 500, Eleutherian Mills Historical Library. Partridge's income may be compared to that of Duplanty, McCall & Co.'s cardingmaster, in charge of half a spinning mill, who earned nearly $500 per annum in 1815. See Wage Sheet for November 1815, Duplanty, McCall & Co. Papers, *ibid.*

[77] G.B., *Parliamentary Papers (Commons)*, 1824 (51), V, 566. For Jenks, see Joseph W. Roe, *English and American Tool Builders* (New Haven, Conn., 1916), 123, 246–47.

(who sold the Phillips machines to Edemsor) back to England. In January 1788 the manufacturing committee of the Pennsylvania Society recorded its considered opinion that the two major obstacles to its work were "the uncertain value of the Paper Currency" and "the interference of British Agents in removing those Machines which they had in prospect for the Use of the Society." So great was the Society's fury that Bond felt obliged to take temporary refuge in New York over Christmas 1787. Later, sometime after 1792, Bond met a Scottish machinist in Philadelphia who had plans of a gig mill and a slubbing billy in his possession. Before the machinist could establish himself, the consul persuaded him to return to England, paying his expenses and providing him with a recommendation to the Board of Trade.[78] Beyond delaying the second transfer of the jenny for a few months, the mule a couple of years, and the gig mill two decades, it is impossible to ascertain the extent and effectiveness of Bond's gadfly tactics. Jeffersonian and mercantile opposition to tariff protection for manufactures were probably far more consequential but lie beyond the scope of this article.

Conclusion

Philadelphia's experience suggests that Britain's new textile technology crossed the Atlantic at the impulse of the immigrant artisans and in response to the initiatives of the American manufacturing interest. The latter's role lay especially in mobilizing the will to industrialize, in advertising the broad industrial opportunities, and in creating the matrices for a few prototype manufacturing systems.

Furthermore, the artisans appear to have been the major vehicles of transit, rather than unaccompanied drawings, models, or machines. This mechanism was the logical one, for Britain's new system of textile manufacturing involved not merely shaping machine parts but giving to them the static and dynamic interrelationships, in terms of dimensions, positions, settings, speeds, and surfaces, which would (relatively) smoothly and predictably perform a given processing operation over a range of adjustments.

Transmission rates seem to have been shortest for basal processes and mechanically straightforward equipment, while piecemeal ac-

[78] Sources in notes 17 and 44 above; also Great Britain, *Parliamentary Papers* (*House of Commons*), 1806 (268), III, "Minutes of Evidence Taken before the Committee on the Woollen Manufacture of England," 423–431. When the machinist, John Douglas,* returned to England, success continued to elude him. He therefore emigrated to France in 1802 and set himself up as a machine maker near Paris.

* John Douglas *should read* William Douglas

quisition stimulated efforts to secure complementary machinery, either by borrowing or inventing – American ingenuity having developed card clothing and rotary shearing machines before 1820.

Several important obstacles stood in the way of the adoption of the new technology. The most significant hurdle lay in identifying optimum product markets. Next in importance was the difficulty in building up a reservoir of the new operational skills, which drew upon immigrant artisans until the War of 1812 at least. In addition, surmounting political resistance to industrialization was also an obstacle, though surely of lesser consequence.

In many ways the experience was similar to that of New England and the Hudson valley down to 1812. These regions witnessed numerous fugitive attempts, but few successes, in establishing the new technology in its factory organization, though Samuel Slater's endurance record eclipsed those of all Philadelphia region firms. After 1814 the potent injection of Boston capital into vertically integrated manufacturing put the Waltham system mills far ahead of mills in Philadelphia, where the continuing investment attractions of commerce and a growing concentration of handweavers contributed to the rise of smaller and less innovative, horizontally specialized firms.

This article underscores the complexity of the general problem of the transfer of technology from one country to another. The manner and the timing of the selective acquisition of portions of Britain's new textile technology by entrepreneurs in the Philadelphia region were determined by the subtle interplay of market forces, technological constraints, and active efforts to encourage or discourage the transfer. This suggests that while some generalizations can surely be made about the transmission process, particular economic needs and opportunities and particular technological and political constraints were operative for different kinds of technologies and for different regions. And, at least in the case of the textile industry, the role of the physical movement of men transferring machines, machine parts, or mechanical concepts was critical in the early stages of industrialization. This was especially true in the non-verbal period of industrial technology, before its reduction to reliable, published machine topographies and processing rules, and also before the development of intermediary institutions capable of effectively marshalling and diffusing technological knowledge.

VI

Immigrant Textile Machine Makers along the Brandywine, 1810–1820

INTRODUCTION

In the transfer of early industrial textile technology from Britain to the USA, a crucial phase in America's industrial development, the migrant artisan was the pre-eminent technology carrier. Between 1790 and the 1820s, a relatively small number — dozens rather than hundreds of British workers — conveyed information about the new manufacturing equipment and techniques across the Atlantic. If the movement is conceptualized in stages, it is clear that immigrant artisans were more important at the introductory stage (bringing information to America) in cotton spinning than at the introductory stage in cotton powerloom weaving, which Americans shared with British inventors in developing. At the prototype factory stage immigrants predominated where lengthy learning periods were required, as in mule spinning and cotton mill management. And at the stage of internal diffusion, immigrants' kin and trade networks, like those centred on Samuel Slater, played a major role before large capital goods firms were set up in the 1820s.[1] This broad picture does, however, need to be filled in, particularly with respect to the role of machine makers. How many immigrant machine makers were active in the respective mill districts in this early period? Were they introducing new equipment from Britain (acting as agents at the introductory stage of transfer)? Did they set up pilot plants (either pilot plants for the region or for the whole of the USA) or were they assuming more importance by loosening the supply of new but familiar equipment and by imparting their skills to native-born Americans? Only a series of local studies can answer these questions and this article is intended as one such study.

The focus of this paper is the Brandywine valley in Delaware, part of the much larger Philadelphia region which was then undergoing significant economic change. The whole commercial region comprised over 40 counties in Pennsylvania, Delaware and southern New Jersey. In Philadelphia and its hinterland, the conditions for industrial textile manufacturing ripened between 1800 and 1830. The city of Philadelphia, whose population doubled up to 80,000 in the three decades (and throughout the period was the second largest city in the USA), offered manufacturers a growing market, shielded from foreign competition by embargo and war between 1807 and 1815 and by some tariff protection in the 1820s. Though exports beyond the whole commercial region did not grow until after the 1820s, a water-borne freight trade was well organized along the Delaware River, while the western hinterland, stretching over 150 miles west of Philadelphia, was penetrated by turnpikes before 1810 and by canals in the 1820s.[2]

The three counties closest to Philadelphia and on the western bank of the Delaware River (Fig. 1) offered numerous waterpower sites for manufacturing. The most developed stream

VI

Immigrant Textile Machine Makers along the Brandywine

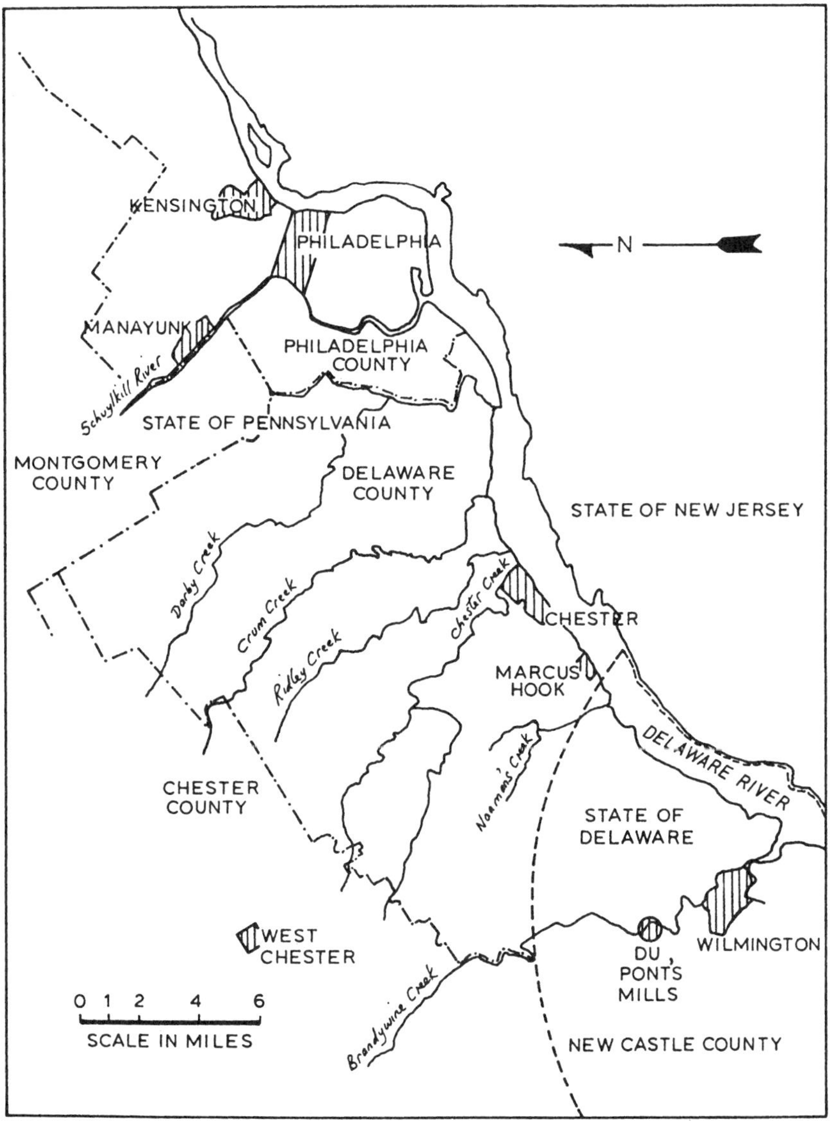

FIG. 1. The Philadelphia–Wilmington area. Adapted from Anthony F. C. Wallace, *Rockdale, The Growth of an American Village in the Early Industrial Revolution* (New York: Alfred A. Knopf, 1978), 2.

TABLE 1. COTTON SPINDLES IN THE VICINITY OF PHILADELPHIA, 1810–1830

	1810[1]	1820[2]	1826[3]	1831[4]
Pennsylvania				
Philadelphia County		14,038		
Philadelphia City		7[a]		
Delaware County		1,680	15,116	
Total		15,725		
Total Pennsylvania	8,849	23,853		120,810
Delaware				
Newcastle County		14,204		
Total Delaware	1,822	14,204		24,806
Pennsylvania and Delaware	10,671	38,057		145,616
Total USA	122,647	342,266 (356,900)[b]		1,246,503

Sources

[1] Tench Coxe, *A Statement of the Arts of the United States of America for the Year 1810* (Philadelphia: A. Cornman, Jr., 1814).
[2] Jeremy, *TIR*, Appendix D.
[3] George C. Leiper and William Martin, *Report* (1826).
[4] Friends of Domestic Industry, *Report*, 16.
[a] Spinning wheels probably.
[b] Highest figure from extrapolation.

was the lower Brandywine, where 14 merchant flour mills, pivoted between the farmlands of Pennsylvania and consumers on both sides of the Atlantic, ground half a million bushels of wheat in 1816.[3] These mills, in which Oliver Evans invented his automatic flour processing system, and other mills and industrial works in Philadelphia, created a reservoir of millwrighting and mechanical experience that would help to set up the new factories. By 1810 Pennsylvania reportedly had 7.2% of the USA's 120,000 cotton spindles and Delaware 1.48% (*see* Table 1). A decade later, Pennsylvania's share of the country's 350,000 spindles had not significantly changed while Delaware's had more than doubled up to about 4.2%. In absolute terms, however, while Pennsylvania's share went up 2.7 times, from 8,849 to 23,853 spindles, Delaware's shot up 7.8 times, from 1,822 to 14,204 spindles.[4] Obviously Delaware's cotton industry in the decade 1810–1820 deserves closer attention. In the 1820s the growth rate of Pennsylvania's cotton industry, measured in spindles, climbed more steeply than Delaware's: the latter's grew by 1.7 times and Pennsylvania's by 5 times. By 1831 Delaware had 24,806 cotton spindles compared to its giant neighbour Pennsylvania with 120,000 spindles.[5] Consideration of the Pennsylvania counties in the 1820s is planned for another article.

MACHINE MAKING AND IMMIGRANT MACHINE MAKERS

Machine making and immigrants

In the establishment of a spinning industry, throughout the USA as much as in the Philadelphia region, immigrant British machine makers played a key role and for several broad reasons. Firstly, of course, Britain was the fountainhead of the new technology associated with factory organization and machine production. At the non-verbal state of that technology, before it was reduced to written or graphic forms and operating rules in the decades 1812–1832, the new knowledge and skill both in building and operating machine spinning systems resided in experienced artisans. While American craftsmen — carpenters, cabinet makers, wheelwrights, ironmasters, blacksmiths, wiredrawers and clockmakers — could obtain and to some extent shape the basic raw materials of wood, iron, steel and brass, only British sources could supply machine designs, constructional information for new components and assembly information for those components. And Britain's prohibitory laws, probably acting more effectively against the export of machinery than the emigration of artisans, increased reliance on skilled workers.[6]

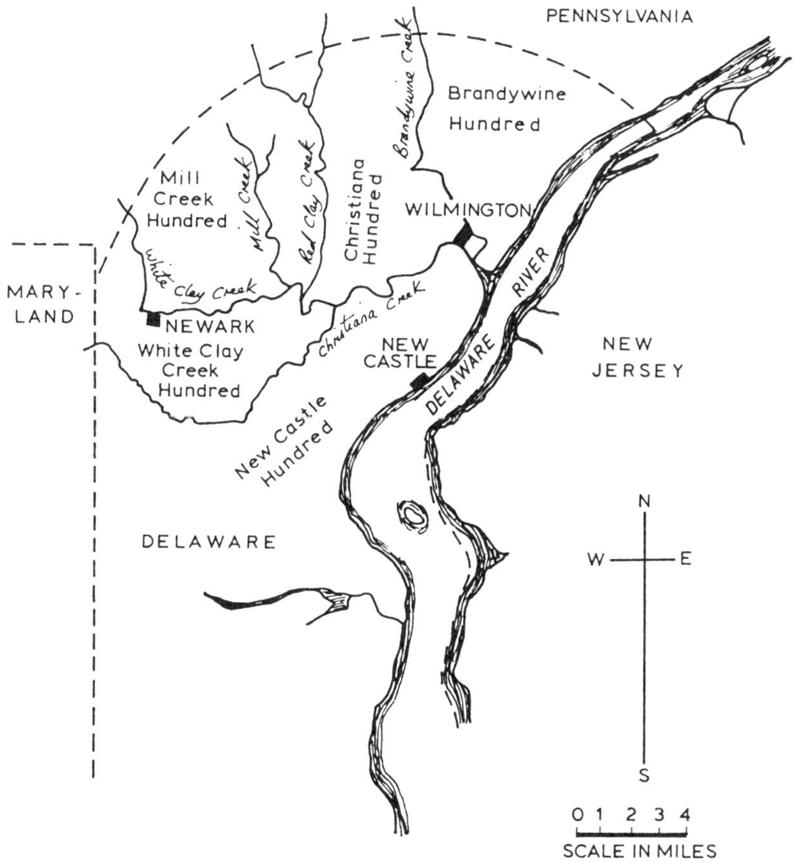

FIG. 2. Northern Hundreds in New Castle County, Delaware.
Source: Daniel G. Beer, *Atlas of The State of Delaware* (Philadelphia, 1868).

By the War of 1812 the nature of textile machine making was changing fast. The first American machines, built in the early 1790s, were of wood, incorporating few metal parts (by quantity); thus two New York cabinet makers could advertise in 1791 that besides household furniture they would make spinning jennies and carding engines.[7] By 1814 iron was replacing wood in the cotton machines of the Philadelphia region mills. And whereas some of these early machines were suitable for home use, the later equipment was intended

TABLE 2. IMMIGRANT MACHINE MAKERS IN THE VICINITY OF PHILADELPHIA DURING THE WAR OF 1812

	Phila. Co.	Phila. City	Del. Co.	Total Pa.	Wilmington, Brandywine Hundred & Christiana H.	Mill Creek and White Clay Creek Hs.	Total De.	Total USA
General								
Millwright	3	1		4	1	1	2	36
Engineer		1		1				2
Machinist					6	2	8	10
Machine maker	3	1		4	10		10	45
Steam engine builder		1		1				
Tinplate workers		1		2				14
Total	6	5		12	17	3	20	125[a]
Textile								
Cotton machine maker		1		1				9
Wire drawer (surrogate for card makers)	11	1		12				15
Spinning wheel maker				1				1
Turner (surrogate for spindle makers)		6		6				20
Spindle maker					1		1	1
Reed maker		1		1				4
Shuttle maker	1			1				2
Total	12	9		22	1		1	60[a]
Grand totals	18	14		34	18	3	21	185

Sources

US National Archives, RG 45, 59, War of 1812 Papers, Marshals' *Returns of Enemy Aliens*.

[a] Including other workers and trades, *see* Jeremy, *TIR*, Appendix C.

for factory production, requiring increased size, greater power, and a longer machine life. In addition, the new machines were part of a total interlocking production system, unlike the isolated attic jenny. A balance between the outputs of the equipment at each stage of processing (carding, drawing, roving and spinning) was therefore essential if mill managers were to avoid bottlenecks. While managers and overseers could regulate production line output to some degree, the parameters were settled by the original design of the machines ordered from the machine maker, who submitted suggestions for the mill manager to consider before the order was placed. A flow production system also intensified the need for a much higher level of mechanical reliability. This requirement was in turn reinforced by inputs of relatively lower operative skill with the Arkwright system of spinning, compared to the mule system. As in England, the changing nature of textile machines altered the character of machine making. Cabinet makers, clockmakers, and blacksmiths were replaced by specialist machine makers, numbers of whom came from Britain where a much larger market and industry supported a much greater division of labour.

This is one feature made clear in Table 2. Together, Pennsylvania and Delaware attracted nearly 30% of all recorded immigrant machine makers in the USA during the War of 1812 — compared to their combined share of 24% of all textile immigrants (operatives, managers, machine makers).[8] Within the two states there was a strong tendency for general and specifically textile machine makers to cluster in Philadelphia City and County and, more strongly, for general machine makers to concentrate on the Brandywine Valley, bounded to the east by Brandywine Hundred and to the west by Christiana Hundred, with the four thousand-strong city of Wilmington at the river's lowest bridging point (Fig. 2).

With these generalizations in mind we can now look at the work of several immigrant machine makers and try to discern the sort of part they played in the mechanization of textile manufacturing in the Philadelphia region during the formative period 1810–1820.

Daniel Large

One of the first new specialist machine makers in the region was Daniel Large who emigrated from England in 1806 at the age of 25.[9] In 1808 he built and installed a 10 horsepower low pressure engine to grind paint at the white-lead works of Wetherill and Brother in Philadelphia. At the instigation of Benjamin Henry Latrobe, engineer of the Philadelphia waterworks and also an Englishman, Large formed a partnership with James Smallman, another immigrant and a former employee of Boulton & Watt. Smallman ran the rolling mill appended to the waterworks where he started building rather poor steam engines in 1804. They came together because Smallman needed Large's improvements, including the Watt parallel motion, and Large needed a loan of $10,000. But the partners fell out and in 1811 Smallman sued to recover his money, forcing Large, not for the last time, into bankruptcy. Large nevertheless continued to build steam engines at his foundry in Otter Street below the German Road and Latrobe regarded him, with Smallman and Oliver Evans, as the leading engine manufacturers of the region.[10] When Large registered as an enemy alien in August 1812 he gave 'Steam engine maker' as his occupation; at this point he lived on North Front Street with a wife and three children.[11] That his early machine building work was not confined to steam engines is clear from a later New England patent case in which one of the witnesses testified that Large had built four-cone double speeders in Philadelphia 14 years earlier, in 1807. Models of this important cotton roving machine, patented in Britain in 1797, must have been among the first of their type built in the USA.[12]

It was therefore to Large that some of the region's cotton manufacturers turned during the Embargo and War of 1812, years when severed commercial and diplomatic relations between the USA and Britain lent indigenous American manufacturers widened prospects in expanding home markets. One of these manufacturers was Duplanty, McCall & Company, a five-man partnership, formed early in 1813, which set up a cotton factory on the Henry Clay mill site on the Brandywine with a modest capital of $30,000. Eventually the firm had a four-storey building measuring 73 × 44 feet containing 1800 spindles. The partners, tempted into the manufacture of relatively fine yarns (Nos 40–50 and later Nos 100–150), were eventually caught out by the inundations of imported English and Scottish yarn, thread and cloth in 1815–1816. Army contracts kept the firm going a few more years but on 12 August 1819 operations ceased with the two McCalls and the two du Ponts estimating their collective losses at around $50,000 (though the company's tangled financial affairs were not wound up until 1839).[13]

While the partners had previously associated in gunpowder and woollen manufacturing, none of them had any practical experience of cotton manufacture. Archibald McCall (1767–1843), the firm's general agent and a leading merchant in Philadelphia, sought out Large because he had 'made the machinery for Craig's Cotton Factory'[15] — located in the Globe Mill just north of the Philadelphia city boundary.[15] 'Large is unquestionably the best mechanician in this country, and his knowledge on all those subjects far before Coxan's', McCall told Raphael Duplanty, manager of the factory to be set up in the Brandywine mill.[16] McCall repeated the opinion of Samuel Wetherill (1736–1816), the Quaker chemical manufacturer who had earlier bought a steam engine from Large.[17] Wetherill rated Large as 'a very modest man', but thought that Coxan, an alternative British steam engine and machine maker in Philadelphia, was a 'visionary and scheming and always promises more than he can do, from not perfectly understanding the principles of his business'. Resting on Wetherill's opinion, for he regarded Wetherill as 'a man of sound judgement and one who would not deceive one', McCall urged Duplanty to employ Large. Duplanty had already made contracts, which will be examined later, with local machine makers for the mill's first set of cotton spinning machinery.[18] Fortunately, as it turned out, Large could only be hired on the fitting up side of the factory.

A week later Large, evidently preparing estimates for the mill's power transmission equipment, was asking for details of the projected spinning machinery:

> Whether the works were to be calculated for mules or trussels [throstles]? If of each kind, what to be the proportion of mules to the trussels? How many spindles you wish to each trussel? If the trussels are to drive in the middle or at the ends? Whether with rims or without? If the mules are to be double or single? He says there is a material difference in the expence and therefore it is necessary for him to know which you wish.[19]

This was no salesman's bluff: much more power was needed to drive throstle spindles (about 100 per horsepower) than to drive mule spindles (700–1000 per horsepower).[20] Consequently the size and cost of gear wheels and shafts varied with the combination of spinning machines chosen — though no one had yet devised a means of accurately measuring an individual machine's power consumption. McCall, in discussions about these matters with Large, was especially heartened to learn that he 'has been a cotton spinner himself in England and appears to know a good deal of the business'.[21]

Duplanty and McCall decided to employ Large to build 'the pier head, water wheel and gearing' if his prices were reasonable. In this way McCall forecast that Large would also have to impart

> all the information he possesses relative to a plan for the Factory. But, having once obtained his ideas, I mistake if you and Mr. Dupont are not much more capable of drafting a proper plan for a Factory than he is. However, of this you will be better able to judge when You converse further with him.[22]

McCall advised Duplanty to contract at fixed prices for items Large himself would make with a commission on the rest at a rate which depended on the total amount of the bills.[23]

The war years brought numerous demands for Large's services, and McCall could not get hold of him for days at a time. Eventually, on 15 April 1813, three weeks after the factory's spinning machinery had been ordered, Large was on his way to Wilmington and the Brandywine. Soon after, he signed a contract to build the Duplanty McCall & Co. factory's water wheel, gearing and shafting, and its heating system.[24]

In the months, and years, that followed, McCall and Duplanty's plans to use and exploit Daniel Large turned out far differently than they expected. Large rarely and reluctantly visited the Brandywine. He himself made only the more profitable lighter castings and did not closely superintend the heavy work which he subcontracted to the Weymouth Ironworks in New Jersey. A fly wheel and a crown wheel were mislaid in transit to the factory site.[25] And not until mid-January 1815, almost two years after the contract was settled, were the water wheel and its power transmission system ready to run. By this time McCall was exhausted and exasperated by what he regarded as Large's incompetence. He had discovered that Large dealt similarly with a Philadelphia merchant, Francis Makoe, when installing a steam engine in a saw mill. McCall concluded

> I believe the fellow is an imposter who has undertaken things he knows nothing about and endeavours to pick up his knowledge as he proceeds and to learn his business at the expence of his employers. Such a man deserves to be punished.[26]

Harsh words! Were they really deserved?

From Large's correspondence it seems that his water wheel design was not unsound. Indeed its iron construction and its diameter are reminiscent of the wheels in the Belper North Mill, Derbyshire, belonging to the Strutts, which were well known and described in Rees' *Cyclopaedia* in November 1812.[27] Whether Large was personally familiar with Belper is unknown but not unlikely. Large specified a breast or undershot (which is unclear in the record) wheel 18 feet in diameter to sit in a head and tail race 10 feet wide and 4 feet deep. The shaft of the water wheel was to turn 11 feet above the bottom of the race or 9 feet over the usual water level. Enclosed within the factory (to reduce the hazard of icing in winter), the water wheel was located at one end of the building so that machinery in an extension could be easily powered.[28]

If his design was sound, Large was surely at fault in his choice of construction materials. Some of his shafts were of cast iron but some, including the main shaft on which the cast-iron water wheel turned, were of wood, unlike the Belper wheels which were cast in sections together with their axles. With a weight of 17,569 pounds of cast iron resting on a 25 foot long shaft, Large's arrangement proved to be a serious mistake. Over the years the shaft wore and warped and produced desultory attempts by local millwrights to tighten and repair it and an unnecessary irritant for the mill managers.[29] The company bitterly quarrelled with Large over his water wheel installation and delayed settlement of his bills. Eventually Large brought an action against Duplanty which came to a head in 1819. In a deposition, James Blackie, Large's pattern maker, testified that his employer 'was very particular in giving directions about the said models [wooden patterns for castings]' when they were being made in 1814. Further, that Duplanty and McCall visited Large's workshop to monitor the progress of these patterns. The defendants countered with testimony from other millwrights, William Murphy and Englishman Thomas Oakes, arguing that Large 'to increase his percentage, [made] all the Casting . . . three or four times as Heavy as it should have been' so that the wooden members of the water wheel were overstrained.[30] Whether Large designedly or unwittingly erred remains a matter of doubt.

Some months before the final fracture and his total condemnation of Large, McCall attributed technical difficulties to Large's ignorance of America's natural resources:

> I do not think Mr Large has been long enough in this country to be well acquainted with the qualities of our Timber. I have not seen him, but without knowing his opinion I can inform you [he told Duplanty], that Spanish Oak is not proper for a Mill shaft.[31]

Ship carpenters would not use this sort of oak if it was to be exposed alternately to wet and dry conditions; they preferred white oak or, if this was unobtainable, white pine. Several years after Duplanty, McCall & Co. stopped manufacturing, the surviving partners scrapped Large's cast-iron water wheel and replaced it in 1823 with a wooden one, signalling a reversion to a traditional but better known technology.

In fire-proof construction Large's proposals were progressive in concept, if not in execution, and were derived from British best-practice. The upper factory floors were to be supported by iron pillars, and piped water raised by a forcing pump was to be used both to flush water closets (his suggestion for the form of 'the necessaries') and to 'overflow the different floors of the factory in case of fire'. Duplanty feared that temperature variations would cause the metal pillars to expand and buckle the floors but Large assured him that they had been used in Manchester for about eight years with no such ill-effects.[32] In fact iron pillars had been employed in factory architecture as early as 1793–1795, at Belper, a feature also reported in the Rees' *Cyclopaedia* article of November 1812.[33] Duplanty evidently modified Large's ideas in some particulars. A staircase to the upper floors was placed outside the factory building instead of inside. And Alexander McCall offered another alteration: 'if the Necessaries could be over the water, I should suppose it would be very desirable & prevent much of the offensive smell arising.'[34] Whatever the extent of the modifications made by the mill managers and their Philadelphia agent, it is fairly clear that Large was not giving wholly bad advice. Sound in design principles, it was defective and damaging because it was poorly executed. And it was poorly executed chiefly because Large was too busy with other orders to go down to the Brandywine or to chase up subcontractors and supervise his share of the work in person and in detail.

Little is mentioned in the correspondence received by Duplanty about the heating system ordered from Large. In fact, Duplanty turned for advice on this matter to a New England manufacturer who recommended using Russian stoves.[35] In spring and summer 1814 the Brandywine manager was considering two rival plans for conveying heat from stoves around the factory — one using horizontal hot air pipes, the other vertical hot air pipes, both embedded in the mill walls — but Large was associated with neither.[36]

Though he made no machinery for the Brandywine company, Large certainly supplied advice about the equipment ordered and installed by Duplanty, McCall & Co. On the difficult but crucial question of balancing machine capacities, he told McCall that

> to correspond with the 6 finishers, 24 In. in the wire & the main Cylinder 36 In. diameter, it will require 2 drawing frames of four heads each & 3 roving frames of 12 Cans each.[37]

Whether the recommendations, expressed with the precision associated with accuracy, were sound we cannot tell for the data concealed design details. But since he had the contract for the factory transmission system and other machine makers the contract for the manufacturing equipment, Large would have had to be extraordinarily devious to give deliberately bad advice at this point. Although he did not get the Duplanty, McCall & Co. machinery contract (because it was settled before he appeared on the scene), Large was evidently a competent machine maker as well as steam engine manufacturer. He built cotton machinery during the War of 1812, if not after 1815 when capital goods' demand suddenly fell. In spring 1818, when mills and machine makers were going out of business, Archibald McCall was searching for bargains in second-hand equipment. After failing to snap up Joseph Bamford's stock — Bamford, an English machine maker operated in Philadelphia from 1805 until he failed (and went to jail) in 1818 — McCall found a line of Large's machines at Jenkin Town, 12 miles north of Philadelphia. It consisted of a picker, 2 cards, a drawing frame (4 heads), a roving frame (6 heads), 2 throstles (108 and 132 spindles), a grinding cylinder, 3 bobbin winders, a warping mill and 3 handlooms. The equipment, costing $3000, had run for a few months before the peace (December 1814). Now, three years later, it was nearly as good as new. The vendor, who seemed fairly impartial, rated the throstles as excellent. They had cast iron frames and brass stands. The cards were good, though their clothing was damaged. The picker was small and no longer the latest design while the roving frame, like most, was 'apt to get out of order.' The lot was available for $1200 in 1818 and McCall, short of cash and credit, was negotiating to pay for them with a small house belonging to his mother.[38] This appraisal of Large's machines in the absence of their builder sufficiently demonstrates that Large was technically competent as a machine maker and that his problems arose less from his technical failings and more from his commercial ineptitude, his failure to organize his business activities efficiently. Essentially he was over-stretched and apparently lacked competent partners who, by their injection of skill and capital, could help his business to expand and meet growing wartime demands. Apparently Large went bankrupt again, sometime between 1815 and 1820, if John Price Crozer's memory was good.[39] What happened to him thereafter is unclear. His name appears in the Philadelphia directories until 1840 but in 1821 he was reported at Eastport, Maine — serving a customer or conveniently on the Canadian border far from his Philadelphia creditors?[40]

The Hodgson Brothers

Another firm of immigrant British machine makers active in the Philadelphia region during the formative period in cotton factory spinning, the Hodgson family, presented various contrasts to Large in their migration and activities on the Brandywine. These five brothers, sons of John Hodgson, a turner and roller maker of Balloon Street, Manchester, emigrated to the USA between summer 1811 and the beginning of 1813 and settled in Christiana Hundred (Fig. 2). All were aged between 17 and 29 and arrived without family — as was typical of most textile immigrants from Britain at this time.[41] Family tradition has it that they evaded Britain's prohibitions on artisan emigration by posing as farmers and sending their tools separately, marked as fruit trees.[42] Be that as it may, they set up on the Brandywine in the vicinity of flour, gunpowder and woollen mills and at least one cotton spinning operation. We know little about their capital resources except that in August 1813, George, Thomas and Isaac Hodgson, soon after the fifth brother Henry arrived, were secured in business by several Brandywine manufacturers and Philadelphia merchants, at a time when the demand for textile machinery was rapidly outstripping supply. Several local capitalists (including Caleb Kirk, who had an older cotton and wool mill, and E. I. du Pont who operated gunpowder works, a woollen mill and was just venturing into cotton with Duplanty, McCall & Co.) formed the Brandywine Mill Seats Company to speculate in mill property. The Hodgsons contracted to purchase from this group a property on the Brandywine at Rockford previously belonging to Job Harvey, 'below the old dam', for the sum of $5000, repayable in nine annual instalments of $555.55 commencing on 1 May 1814.[43] They remained at Rockford until 1818; George Hodgson moved to another site near Rokeby also on the Brandywine before 1822, by which date the brothers had split up, one going West.[44] George stayed on the Brandywine until his death in 1849 though his machine works continued under his name until 1854 at least.[45]

As a kin-knit group of machine makers, they were in a much stronger position than Daniel Large, or any of the other eight or nine machine shops on or near the Brandywine,[46] to exploit opportunities for the specialization of labour presented in 1813–1814 by the fast-expanding regional capital goods market. Their work for Duplanty, McCall & Co. reveals the situation in which they found themselves and the ways in which they contributed to the mechanization of the region's cotton industry: primarily by adding to the supply of machine makers and lowering capital goods prices; by allowing their machine making skills to pass to native Americans; and by introducing the latest British equipment to the region.

At first the Duplanty, McCall & Co. partners approached other manufacturers for recommendations to machine makers: E. I. du Pont contacted the McKims, leading cotton manufacturers in Baltimore, while Archibald McCall made enquiries as we have seen, in Philadelphia.[47] The McKims' machine makers (John Duff and W. H. Richardson) had orders that would keep them busy until the end of 1813, so the Brandywine partners looked nearer home[48] — but now armed with a schedule of Baltimore machinery prices.[49] Within ten days of hearing from the Baltimore machine makers, Duplanty had three sets of quotations in front of him (Table 3). The Baltimore one could be ruled out because Richardson & Duff could not start work on a new order until the end of the year, nine months hence, and in any case their equipment imposed heavier transportation costs. The quotation from Alrichs & Dixon, a firm established in Wilmington in 1808,[50] specified prices which were fractionally higher than Richardson & Duff's; like the Baltimore machine makers', theirs did not include accessories (change gears, bobbins, cans, leather covering on rollers, chiefly); and once the makers 'deliver them to the waggons at the shop door, we consider [them] out of our charge.'[51] The Hodgson brothers, in contrast, offered

TABLE 3. MACHINE MAKERS' PRICES QUOTED TO DUPLANTY, MCCALL & CO., BRANDYWINE VALLEY, NR. WILMINGTON, DELAWARE, 1813–1814 (PRICES IN DOLLARS)

Machine	Richardson & Duff Baltimore, 12 March 1813[1]	Alrichs & Dixon Wilmington, 22 March 1813[2]	George Hodgson & Brothers			McFee & Newton, Wilmington, 1814[6]
			Brandywine		Contract, 22 March 1813[5]	
			Undated but March 1813[3]	Undated but (March?) 1813[4]		
Blowing machine	300	300–400	150	150	150 (returnable if not approved)	400
Carding machine (without card clothing)	200	200	190 (white pine covering)* 195 (walnut)* 205 (mahogany)*	200	200 (mahogany covering)	
Drawing frame (per head)	35	25–40	38		38	
Roving frame (per head)	30	25–40	31	31	31	
Stretcher (variant of mule for roving stage) (per spindle)				(108 sp.)		6.25 (114 spindle stretcher)
Mule (per spindle)	2.50 (240 sp. mule)	2.60 (240 sp. mule)	3 (216 sp. mule)	3 (216 or 240 sp. mule)		
Throstle (per spindle)	5.00 (120 sp. throstle)	5.00–5.50 (without change gears, creel bobbins or covering on top rollers)	5.50 (with all accessories) (120 sp. throstle)	5.50	5.50	

* Each with iron frame and spur gears.

Sources: [1] EMHL, Longwood MSS, Group 6, Box 1, Richardson & Duff to E. I. du Pont, 12 March 1813. [2] Ibid., Group 6, Box 2, Alrichs & Dixon to R. Duplanty, 22 March 1813; ibid., Alrichs & Dixon quote, 1813. [3] Ibid., undated quote from Hodgsons: 'List of Machinery for Messrs Du Pond & Co.' [illegible] prices and written by one of the Duplanty, McCall & Co. partners. [5] Ibid., agreement dated 22 March 1813.

prices the same or marginally less than their competitors', except for the blowing machine — half the price of their rivals' products — and mule spindles — twice the going price. In addition the Hodgsons offered their drawing, roving and spinning equipment 'compleate in Every thing',[52] except for the card clothing on their iron-framed carding engines. On 22 March 1813, the very day he received Alrichs & Dixon's quotation, Duplanty signed a contract with the Hodgsons in which he extracted even more favourable terms from them. The immigrant machine makers agreed:

> The whole of the said machinery to be put up by ourselves, and in case some alteration is wanted, the same to be made at our own expense, it being well understood that the machinery herein stated shall be considered as delivered To & received by R. Duplanty & Co. at the time only when it shall have been Tried and have proved to work To satisfaction.'[53]

Duplanty was clearly suspicious of the low priced blowing machine and stipulated in the contract 'one Blowing machine, to be returned after Trial if R. Duplanty & Co. do not like its performance.'[54] The Hodgsons matched their keenly competitive prices and installation arrangements with the promise of speedy delivery dates: the blowing machine, one breaker and two finisher cards and the drawing frame would be set up by 15 October 1813; two breaker and three finisher cards, one roving frame and one throstle spinning frame a month later; and the rest, one breaker, one finisher, one roving frame and one throstle, by 15 December 1813, equipment totalling $4360. If the Hodgsons failed to keep to these dates, delivery was overdue by only a few months: the ten cards were recorded in the company's ledger as completed on 30 April 1814.[55] Perhaps to test the quality of the Hodgsons' work and certainly to avoid their high mule spindle prices, Duplanty ordered at least one more throstle frame of 120 spindles and a pair of mules (totalling 480 spindles) from Alrichs & Dixon.[56]

From their viewpoint the Hodgsons were not wholly exploited. Some of the prices they quoted were comparable with (i.e. slightly higher than) those prevailing in Manchester whence they had just come.[57] Above all the contract helped to secure their foothold in the Philadelphia region market. It no doubt led to the purchase of a millseat and workshop, as already noted, five months later.

The Duplanty, McCall & Co. factory ledger, opened in April 1814, shows that over the working life of the factory, until it closed in 1819, the Hodgsons supplied 19.4% of the factory's machinery, as measured in dollar costs.[58] They supplied much more than this of the original production line. A total estimate of the Hodgsons' contribution to this particular Brandywine mill is impossible to make because of the appearance in the 1814–1819 ledgers of items (like the 10 cards) which really belonged to the 1813 accounts.

What is especially interesting is that when machinery prices shot up in 1814, as Robert McCall remembered,[59] the firm bought few complete machines. Their purchases, as their account with the Hodgsons shows, were increasingly of spare or key parts. This account (Table 4) contains a number of features which throw further light on the Hodgsons' role as machine makers. In most cases their prices were standardized according to weight, except for gears, priced according to the number of teeth; and throstle spindles, flyers and bobbins, individually priced. The basis of the cast-iron price presumably was the cost of the iron plus labour costs for making the wooden patterns and pouring castings, plus profit. For steel spindles, flyers and rollers, requiring machining, the price reflected the wages of machinists as well as the cost of steel which was not easy to get in wartime.

TABLE 4. STATEMENT OF ITEMS SUPPLIED BY THE HODGSONS TO DUPLANTY, MCCALL & CO., 1814–1815

	Date	Item	Amount ($)
1814	27 July	Castings, 139 lb @ 12½ cts	17.37½
	1 Aug	Castings, 29 lb @ 12½ cts	3.62½
	6 Aug	Castings, 126 lb @ 12½ cts	15.75
	6 Aug	Castings, 39 lb @ 12½ cts	4.87½
	9 Aug	Lignumvitae, 62 lb @ 6 cts	3.72
	9 Aug	14 roving cans @ $1.75	24.50
	9 Aug	4 brass guides, 1½ lb @ 75 cts	1.12½
	9 Aug	1 brass wheel, 37 teeth @ 4 cts	1.48
	15 Aug	Castings, 12 lb @ 12½ cts	1.50
	24 Aug	17 front and 34 back mule rollers, 136½ lb @ 5 cts	68.25
	3 Sept	Castings, 7 lb @ 12½ cts	0.87½
	11 Sept	17 boxes of mule rollers, 137½ lb @ 5 cts	68.75
	21 Sept	A Casting, 73 lb @ 12½ cts	9.12½
	31 Oct	Castings, 3 lb @ 12½ cts	0.37½
	31 Oct	42 lb of cast wheels @ 15 cts	6.30
	31 Oct	2 weights, 16 lb @ 10 cts	1.60
	12 Nov	1 set of mule rollers, 139 lb @ 5 cts	69.50
	12 Nov	10 dozen throstle spindles and flyers @ $9.50	95.00
	17 Nov	Mule weights, 619 lb @ 10 cts	61.90
	21 Nov	Castings, 20½ lb @ 12½ cts	2.56
	21 Nov	Castings, 4½ lb @ 12½ cts	0.56
	28 Nov	Castings, 51½ lb @ 12½ cts	6.43½
	28 Nov	Castings, 7½ lb @ 12½ cts	0.90
	31 Dec	10 carding engines, $200 each	2000.00
	31 Dec	1 set of mule rollers, 138 lb @ 50 cts	69.00
	31 Dec	2 sets of mule rollers, 276 lb @ 50 cts	138.00
	31 Dec	Weights, 14 lb @ 10 cts	1.40
1815	17 Feb	10 dozen throstle bobbins	4.16
	24 March	17 boxes of mule rollers, 145 lb @ 50 cts	72.50
	3 May	Worm wheel, 30 teeth @ 5 cts	1.50
	5 May	2 pair of card rollers, 73½ lb @ 40 cts	29.40
	12 May	1 set of mule rollers, 142 lb @ 40 cts	56.80
	16 May	Castings, 65 lb @ 12½ cts	8.12½
	8 June	6 carriage wheels, 36 lb @ 12½ cts	4.50
	8 June	Turning wheels	3.00
	8 June	2 pinions, 19 teeth: 38 teeth @ 2 cts	0.76
	17 July	2 large end rollers, 10½ lb @ 40 cts	4.20
	23 July	12 large weights, 184 lb @ 10 cts	18.40
	24 July	Quantity of plates, 159 lb @ 10 cts	15.90
	26 July	5 cast supports, 13½ lb @ 12 cts	1.68½
	9 Aug	1 cast wheel, 5½ lb @ 12½ cts	0.68½
	11 Aug	Castings, 7½ lb @ 12½ cts	0.94
	11 Sept	Castings, 5½ lb @ 12½ cts	0.69
	27 Sept	2 hearts [heart wheels], 4 lb @ 12½ cts	0.50
	27 Sept	Brass worm wheel, 16 teeth @ 5 cts	0.80
	12 Oct	8 twisting rollers, 38 lb @ 40 cts	15.20
	17 Oct	18 boxes of throstle rollers, 173 lb @ 40 cts	69.20

Date	Item		Amount ($)
17 Oct	Blower arms, cast, 37 lb @ 12½ cts		3.62½
26 Oct	8 twisting rollers, 35 lb @ 40 cts		14.00
26 Oct	2 drawing heads @ $35		70.00
11 Dec	5 rollers, 21 lb @ 40 cts		8.00
		Total	3079.45

Sources

Duplanty, McCall & Co. Papers, In-file 1816 (128). Co. to George Hodgson Brothers. Eleutherian Mills Historical Library, Wilmington, Del.
Another version of this account is Longwood MSS, Group 6, Box 2. Also in the EMHL.

Some of the crucial elements in the spinning equipment, namely rollers and spindles, seemed moderately priced compared to the cost of a complete machine. Mule and throstle rollers, sold by weight, were four to five times the price of dead weight cast iron and averaged $1.34 each. Spindles and flyers were 79 cents a set (Table 4). A complete 120-spindle spinning frame (throstle) sold at $660; spindles and flyers at $94.80, were therefore 14.4% of the price of the machine and the rollers (assuming one set of three rollers served six spindles as in the throstle illustrated in Rees' *Cyclopaedia*), at about $80.40, 12.2% of the price of the machine — a little more than the proportion of 20% for rollers and spindles claimed for British machines a decade later.[60]

The relatively low prices of these castings and forgings apparently became an incentive to manufacturers to obtain only metal components from specialized machine makers, especially, perhaps, after the war when prices began to move downwards with rollers falling from 50 to 40 cents a pound. Having learned how the new machines were assembled and operated, the Duplanty, McCall & Co. managers used their millwrights or carpenters to build additions to their production line. William Boyd was one man employed in this way. Using wood, he made 'the end, roller beams and braces below' for one throstle and the ends for another. Other hands in the mill evidently completed assembly. William Murphy, a millwright, had one of his men make some of the reels.[61] In short, the new technology imported with immigrants like the Hodgsons was partially transferred to American manufacturers as two of the steps in machine making — machine design and the establishment of static and dynamic interrelationships between a machine's crucial parts — passed into the hands of American millwrights and carpenters.

The other vital stage, the manufacture of crucial metallic components like spindles and rollers, was also readily transferred to native Americans by the Hodgsons. Among their early pupils were Franklin and Titian Peale, sons of Charles Willson Peale (1741–1827), the Philadelphia portrait painter and naturalist. George Hodgson took the boys, under an initial arrangement for a six years' apprenticeship, in November 1813. Their father reported that the Hodgsons then had a foundry, 14 blacksmiths and 15 water-powered turning lathes at work.[62] Within a year, the two Peales had apparently almost finished learning how to make machinery, and, by spring 1815, were building equipment on their own account.[63] Whether they had in fact mastered all the skills of machine making in less than two years is open to question. The important point is that the Hodgsons displayed none of the secretiveness frequently found among manufacturers and machine makers, immigrant and native-born alike, in this early industrial period.[64]

Besides helping to lower Philadelphia region textile machinery prices and sharing their new industrial skills with native Americans, the Hodgsons made a major contribution to mechanization in regional textile manufacturing by building a proportion of the region's stock of best practice equipment. Since they came from Manchester in 1811–1812 at the end of a period of innovation in cotton spinning,[65] theirs was knowledge of the latest British machine designs and techniques. What little is known about the design and the metal castings of the equipment they built on the Brandywine confirms their innovatory role. A blower, completing the separation of cotton fibres (a process begun with picking apart the impacted masses of cotton from the bale), was one of the opening machines reported in Rees' *Cyclopaedia* in November 1812 as 'used in some of the most improved mills' in England.[66] The Hodgsons' blower, one of these, apparently experienced some teething troubles for it took 12 months or more to come into operation.[67] Similarly progressive was their introduction of breaker and finisher cards of metal construction. Specified in their preliminary quotation, but not in the quotations of their rivals,[68] the two types of carding engine were well known in Britain and seem to have been introduced to America by Slater in the 1790s. The Hodgsons' design followed a radical improvement on the first generation of American cards because it used cast-iron framing and metal arches instead of wooden frames which, subject to warping, were liable to cause difficulty in operation and increase maintenance costs.[69] Details about the other equipment supplied by the Hodgsons are not known. Daniel Large, we have seen, was making throstles with cast-iron frames, again because this recent British practice minimized distortions in tolerances between key moving parts.[70] Presumably, the Hodgsons also supplied metal frames, if required, for their drawing, roving and throstle frames. This best practice equipment made by the Hodgsons was supplied to mills not only on the Brandywine but also to firms like Charles Willson Peale's cotton mill at Belfield in Philadelphia.[71]

Besides Large and the Hodgson brothers there were a number of other immigrant machine makers from Britain in the Philadelphia area. Among those used by Duplanty, McCall & Co. was William McKenzie who arrived in 1806 and by 1813 was a settled millwright in Wilmington, Delaware. In 1815 he installed a drive shaft for what proved to be inefficient waterlooms. The job took him 39 days and he charged $1.60–1.70 a day.[72] Another, James Wagstaff, a spindle maker who arrived in Wilmington in 1813, turned more than 800 mule spindles for the company in 1815, at an average of 20 cents each and later the same year, another 204 spindles at 30.5 cents each — thereby lowering the machinery prices charged by the Hodgsons.[73] William Hartley, reedmaker, who emigrated in 1811 and settled in Philadelphia, supplied the company with reeds between 1815 and 1817.[74]

Thomas Siddall

If British immigrants represented a substantial section of the Philadelphia region's pool of metal machine making skills, and if they also helped to introduce the latest technology from Britain, were they likewise at the frontier in developing new technologies in their new American situations? Duplanty, McCall & Co.'s efforts to acquire a powerloom reveal the prominence of British immigrants among the inventors, builders and developers of this the 'sunrise' technology of the day. The one the firm dealt with most extensively was Thomas Siddall who introduced, and developed, an early powerloom from Britain.

Before 1815, several looms in the region were the subject of patents: those of Robert Lloyd (1809), Thomas Mussey (1811), Silas Shepard (1812), Seth Craige (1814), all of Philadelphia; and Thomas Siddall (1814) of Philadelphia County, Pennsylvania.[75] Some of these came to the attention of Alexander McCall. Two, belonging to individuals named Jarvis and Miller, he rejected on the advice of Daniel Large, as no better than existing unimproved waterlooms.[76] About the same time, early 1814, Thomas Cooper (1759–1839), the respected English-born chemistry professor and Manchester radical, told Duplanty, McCall & Co. about one in Germantown that worked 'well, quickly and neatly, and in that respect was like the looms in use in England and in Boston.'[77] News came too of Mr Levering's looms in Baltimore. McCall went to see the Germantown loom built by Thomas Siddall, a Lancashire immigrant who, with other Siddalls, arrived in Philadelphia in the War of 1812 period. The loom's salient improvement, claimed by Siddall as his own, was a warp protector stop motion — to halt the loom before the batten smashed a jammed shuttle through the warp shed.[78] Siddall planned to patent the device, and presumably it was in his patent of 9 July 1814. But he could not claim to be the first to design a warp protector stop motion. Large, the English millwright, accurately corrected Cooper's view that Siddall's warp protector stop motion was unique, by pointing out that it was already in use in Boston and had been employed for a long time in England — in fact since Cartwright's 1787 patent.[79] He could have added that it was in Thorp's American patent of 1812. Siddall in fact was pursuing an English type of powerloom technology, which as perfected by Horrocks was more complex than that currently being developed by Moody and Lowell.[80] Siddall's contrivance comprised an iron catch which held back the lathe or batten until the shuttle returned to one of its boxes. For his waterlooms, 34 inches wide and built by a machine maker under his direction, Siddall charged $50. He claimed they could each weave one yard in 20 minutes and four could be run by only one weaver.[81] If we suppose that there were 60 picks per inch, a not unusual weft spacing, then Siddall's loom ran at 108 picks per minute, about twice the rate of a handloom weaver. Since one weaver ran four of Siddall's looms, the productivity increase looked like a ten-fold one at least, and in summer 1815 a ten-fold production increase was indeed anticipated by McCall.[82] But there were problems. McCall found that Siddall's waterwheel was 'lop-sided' and conveyed irregular movements to the loom, and its inventor 'seemed shy of working it long at a time.'[83] Siddall, not unnaturally, rated his loom above others available, like that of Seth Craige in the Globe Mill, Philadelphia.[84] McCall wanted to examine powerlooms in Baltimore before deciding to purchase the Siddall loom.[85] But Siddall's machine in April–May 1814 looked good. Siddall had applied for a patent to guard and confine its use; he was installing 18 looms in a new weaving shop adjoining his factory; he made the daring bet that four of his looms could, under his supervision, weave 100 yards of No 30 yarn in 12 hours;[86] and, he claimed 'the finest Muslin's wch are made in Scotland & [are] wove in these looms.'[87]

What diverted Duplanty, McCall & Co. into powerloom weaving — most likely expectations of higher unit profits on fine cotton goods made from their mule-spun yarn — was never made wholly explicit in the correspondence from the company's Philadelphia agent to its Brandywine manager. What is certain is that, at the end of March 1814, nearly months after the search for a powerloom began, McCall was warning that the American trade embargo and even the war would soon come to an end, the consequence of which would be 'that both yarns and cotton manufactures of every kind will fall considerably in price.'[88] Faced with this prospect, McCall suggested that, as a hedge against falling prices, the company take on a government contract. Duplanty's fear, that the government would continue the war and be unable to honour its contracts, was brushed aside by McCall who argued that in this case the contract could be cancelled (and the cotton goods sold to private customers).[89] Predicting peace, and the government's ability to pay contractors, McCall signed a contract with the Commissary General of Purchases on 13 April 1814 to supply.

> Fifty Thousand yards of white twill'd cotton Drilling, 27 inches wide, weighing from 6½ to 7 ounces per yard, free from broken seeds and streams, fit for uniform clothing for the army with a regular & sufficient proportion of filling to chain for good & handsome work & substance, of a good & regular color, and well driven in the loom.[90]

Based on a standard sample and subject to inspection, the drilling would be bought by the US government at 40 cents a yard when delivered over the succeeding 12 months. McCall expected the company to spend up to $2000 on new powerlooms, a sum, he reckoned, amply covered by profit from the government contract, since the weaving could be done with powerlooms at 6 cents a yard. And, if the contract were cancelled, any unwoven yarn could be sold just prior to peace at 72 cents a pound.[91] With these calculations in mind, the company bought a powerloom from Siddall. Glimpses of its teething troubles are caught in McCall's letters.

Siddall would not let fellow Englishman Daniel Large alter the loom so that it would weave the twill agreed in the government contract.[92] Whether Siddall made the change in the shedding mechanism himself is unknown. The loom at last arrived, late in June 1814.[93] In transit from Philadelphia it was damaged — a shuttle box broke off, as did a cam and part of the warp protector stop motion. Without Siddall's warping and dressing machines, the company had to wait on Siddall for warps. And, to add to these frustrations, Siddall's bill increased to $101 — $75 for the loom (he claimed the one he sent was broader than his models), $25 for the patent rights and $1 for the packing box.[94] In March 1815 the Brandywine factory was still trying to get the Siddall loom into production, and was searching for a weaver who would make relatively fine goods on it. McCall located a man named Patterson who was running several of Miller's waterlooms and he gave advice from time to time until spring 1816.[95] In summer 1815 Siddall, now in partnership with James Phelps, was building new powerlooms at Frankford just north of Philadelphia. At his request the Brandywine mill sent some of its fine yarn warps up to Frankford so that Siddall could make adjustments on his machines that might be made in the powerloom in the Duplanty, McCall & Co. factory.[96]

At this point the company's experiment in stocking loom weaving was failing because its French hosiers were moonlighting.[97] It was therefore imperative that the company find a way to manufacture its fine yarns, which Philadelphia weavers would not handle, and so survive against importations without altering its fixed capital investment in fine spinning equipment. A Philadelphia weaver, William Huston, was deputed by McCall to try Siddall's loom.[98] Meantime McCall urged Duplanty, the mill manager, to expand his weaving operations and to order eight or ten more powerlooms. Duplanty could not see the increasing importations but McCall's words had a note of urgency: 'you can never expect to do any good in the small way in which you now move. You must go upon a larger scale or you will be in a worse situation every day.'[99] Mechanization and economies of scale (i.e. higher turnover on lower profit margins) were the solution to falling cotton goods prices and the shortage of fine handloom weavers. If only a viable powerloom could be found and replicated. By early December 1815 the race to find such a powerloom, before the bottom dropped out of the cotton goods market, had been lost.[100] By spring 1816 one of the Miller looms turned up in the store of a Wilmington grocer. Patterson, one of the factory's weavers, insisted that it was the best available, but by now the company was relying on handloom weavers working coarser yarns for western and government customers.[101] In 1820, after the factory closed, 'one power loom on Siddall's plan' was inventoried without comment among the spinning equipment on the fourth storey.[102] Duplanty, McCall & Co. evidently did not buy one of Siddall's warping machines (patented 27 March 1815), possibly because it did not appear in time, possibly because it too could not handle fine warps.[103]

The failure of Siddall's loom on the Brandywine was obviously not wholly due to Siddall himself. Poor timing; a small investment in the new technology; separation, in ownership and distance, of research and development from the factory plant, were reasons implied in McCall's letters to Duplanty.[104] But the prime reason why the Siddall powerloom had eventually to be consigned to oblivion was the one which the Brandywine managers failed to realise until importations flooded the market. Committed to fine spinning, the Brandywine partners wanted their loom 'to work rather finer than common.'[105] Of course, the first powerlooms, with harsh, jerky mechanical movements, would strain and break the more fragile fine cotton warps. That the Brandywine firm inappropriately applied the available technology was suggested by E. I. du Pont's strong approval of the weaving in Siddall's factory in May 1815 and by Siddall's own successful production of 20 yards of bed tick per day per powerloom in 1816, using No 10 yarn, a very coarse and therefore strong yarn able to withstand the beating and shedding actions of a loom without Horrocks' variable batten speed motion (patented in England in 1813).[106] Thus it seems that if Siddall was trying to copy British examples, to adapt his powerloom to finer yarns, he singularly failed. His powerloom was no more suited to this purpose than the one developed at Waltham. Even so, Siddall's activities confirm that individual British immigrants were to the forefront in early attempts to mechanize weaving in the Philadelphia region.

Conclusion

On the Brandywine, one of the two centres of machine making in the Philadelphia region during the War of 1812 period, it is clear that immigrant machine makers played several roles. They were indeed introducers of a few specific pieces of new British equipment, like the four cone roving frame and the blowing machine. At the prototype stage, they were clearly active in setting up powerloom pilot plants in the Philadelphia region. But since this was done at a point when British manufacturers were also pioneering powerloom technology, they had the opportunity to pursue new directions. This, if Siddall was typical, they failed to do. Instead, immigrant machine makers in the Philadelphia region followed the lines indicated by British models: a more complex technology suited to the production of higher quality cotton goods. But in missing Horrocks's variable batten speed motion, Siddall fell short of this goal. The concern for higher quality production was no doubt in part dictated by the established patterns of handloom manufacture in and around Philadelphia. Whether a Boston-type investment in a thoroughly American technology in the Philadelphia region would have yielded a different technology we cannot tell. It does seem, however, that the prominence of British machine makers as equipment suppliers for the new Brandywine factories helped to tie the first mills to the pursuit of a British style of powerloom weaving technology.

Equally importantly, the immigrant machine makers, by their numbers and competition for contracts, lowered the prices of machines and of machine parts. They also spread the use of cast iron in machine frames, like cards and throstles. They further accelerated the internal diffusion, within the region, of their Lancashire machine-making methods by training young Americans and by selling key metal components (rollers and their stands, spindles and flyers, gears, guides, plates, wheels, levers — all mostly cast iron but some steel and brass) which local millwrights and carpenters, copying immigrants' models, could assemble into woodframed machines. The complex process of industrialization in America, Professor Thomas Cochran has recently reminded us, was confined neither to New England nor to textiles. This paper certainly affirms the spread of advanced industrial technology through the activity of immigrant machine makers in the Philadelphia region, albeit in textile manufacturing.[107]

Acknowledgements

The preparation of this article would not have been possible without the use of collections in the Eleutherian Mills Historical Library at Greenville, near Wilmington, Delaware, USA. For this access I am most grateful to the Eleutherian Mills–Hagley Foundation Research Committee who awarded me a grant-in-aid to work there in summer 1969 and to the Academic Advisory Board of the Regional Economic History Research Centre (at the EMHL) who awarded me a Fellowship to spend a further month there in August 1979. This second visit was supported by a travel grant from the British Academy, to whom I am also grateful. I am further grateful to Ken Ponting for the invitation to present the first version of this article to the SSRC/Pasold Conference on Textile History held at St John's College, Oxford, on 26 September 1981. Finally, my thanks to Helen Jeffries and Alexandra Kidner, secretaries with the *Dictionary of Business Biography* project at the Business History Unit, who patiently and obligingly reduced my scrawl to legible typescript.

References and Notes

[1] David J. Jeremy, *Transatlantic Industrial Revolution: the Diffusion of Textile Technologies between Britain and America, 1790–1830s* (Cambridge, Mass: MIT Press, 1981). This is hereafter cited as *TIR*.

[2] Diane Lindstrom, *Economic Development in the Philadelphia Region, 1810–1850* (New York: Columbia University Press, 1978), 66, 95, 100–101.

[3] Peter C. Welsh, 'The Brandywine Mills: a Chronicle of an Industry, 1762–1816', *Delaware History*, 7 (1956), 17–36.

[4] For the 1810 spindleages *see* Tench Coxe, *A Statement of the Arts and Manufactures of the United States of America in the Year 1810* (Philadelphia: A. Cornman, Jr., 1814) and for 1820 spindles, *TIR*, 277.
[5] The 1820 figures are from ibid. while the 1831 figures are from Friends of Domestic Industry, New York Convention, *Report on the Production and Manufacture of Cotton* (Boston: J. T. and E. Buckingham, 1832), 16.
[6] *TIR* chapters 1 and 4 for background and detailed references.
[7] *New York Daily Gazette*, 29 April 1791 cited in Rita S. Gottesman, *The Arts and Crafts in New York, 1777–1799: Advertisements and News Items from New York City Newspapers* (NY: New York Historical Society, 1954), item 369.
[8] *TIR*, 158.
[9] National Archives, Washington DC, RG 59, War of 1812 Papers, US Marshals' Returns of Enemy Aliens. Hereafter cited as DNA RG 59 Returns of Enemy Aliens.
[10] Carroll W. Pursell, Jr., *Early Stationary Steam Engines in America. A Study in the Migration of a Technology* (Washington, DC: Smithsonian Institution Press, 1969), 35, 43–44. For Smallman *see* ibid., 29–30, 32, 34, 35, 42. John P. Crozer, later to become a leading cotton manufacturer in Delaware County, years later recalled that Large was the leading steam engine manufacturer in Philadelphia. *See* John Price Crozer, *Biographical Sketch of John P. Crozer Written by Himself* (Philadelphia: Grant, Faires and Rodgers, *ca* 1861), 11–12.
[11] DNA RG 59, Returns of Enemy Aliens; *Paxton's Philadelphia Directory* 1813.
[12] *TIR*, 58–59, 90.
[13] For Duplanty, McCall & Co.'s history *see* Roy M. Boatman, 'The Brandywine Cotton Industry, 1795–1865' (Hagley Research Report, 1957), 20–49. The partners were two pairs of brothers, Archibald and Robert McCall and Victor and Eleuthère Irénée du Pont and Raphael Duplanty. Duplanty, who, like the du Ponts, fled from the French revolution, had worked in the USA for E. I. du Pont as an accountant and then was involved in the design of the du Pont woollen factory.
[14] Eleutherian Mills Historical Library, Wilmington, Delaware, (hereafter EMHL), Accession 500, Duplanty, McCall & Co. Papers (hereafter DMC), Archibald McCall to Raphael Duplanty, 31 March 1813.
[15] See Samuel H. Needles, 'The Governor's Mill and the Globe Mills, Philadelphia', *Pennsylvania Magazine of History and Biography*, 8 (1884), 379–382.
[16] DMC, McCall to Duplanty, 31 March 1813. Coxen was another steam engine maker and English immigrant; *see* Pursell, *Early Stationary Steam Engines*, 33.
[17] For Wetherill *see DAB* and for his steam engine, Pursell, *Early Stationary Steam Engines*, 43.
[18] All quotes from DMC, McCall to Duplanty, 31 March 1813.
[19] Ibid., same to same, 5 April 1813.
[20] *See TIR*, 91 and sources.
[21] DMC, McCall to Duplanty, 5 April 1813.
[22] Ibid., same to same, 14 April 1813.
[23] Ibid.
[24] Ibid., same to same, 14, 24 April 1813.
[25] Ibid., same to same, 26 January 1814, 18, 28 October 1814, 16 January 1815.
[26] Ibid., same to same, 21 January 1815.
[27] For the Belper water wheels *see* Robert S. Fitton and Alfred P. Wadsworth, *The Strutts and the Arkwrights, 1758–1830: a Study of the Early Factory System* (Manchester: Manchester University Press, 1958), 205–217.
[28] DMC, Daniel Large to Duplanty, 3 August, 4 September, 1 November 1813.
[29] Boatman, 'Brandywine Cotton Industry', 26. Dimensions and weights given by Robertson Buchanan: *Essays on the Shafts of Mills*, (London, 1914), 33 ff, show that Large's specification was well within the limits of the wooden water wheel (a wooden wheel of diameter 16 ft and a width of 9 ft was expected to weigh 10½ tons). This not only confirms that Large was confronted with a materials problem, but also suggests that his mill wrighting was competent to a degree. I am very grateful to Professor Louis C. Hunter for supplying this reference as well as comments and other sources on weights of water wheels.
[30] DMC, deposition of James Blackie in Horace Binney (the firm's lawyer) to Duplanty, 16 November 1819; ibid., Duplanty to Binney, 18 November 1819; ibid., joint deposition of Thomas Oakes and William Murphy, undated, in general file. For Oakes's English origins see *PMHB*, 37 (1913), 471–479.

31 DMC, McCall to Duplanty, 17 March 1814. As early as August 1814 Duplanty was refusing to pay Large. *See* ibid., McCall to Duplanty, 29 August, 5, 9 September 1814.

32 Ibid., Large to Duplanty, 23 April 1813; McCall to Duplanty, 14 June 1813.

33 Fitton and Wadsworth, *Strutts and the Arkwrights*, 198–203 and *passim*.

34 DMC, McCall to Duplanty, 1 September 1813. The external staircase at Waltham has been hailed as a novel feature but perhaps only the tower element was novel. *See* Bryant F. Tolles, Jr., 'Textile Mill Architecture in East Central New England: an Analysis of Pre-Civil War Design', *Essex Institute Historical Collection*, 107 (1971), 236.

35 DMC, William B. Leonard (later manager of the famous Matteawan works) to Duplanty, 15, 27 September, 8 November 1813.

36 Ibid., McCall to Duplanty, 5, 27 May, 14 August 1814.

37 Ibid., same to same, 2 June 1814.

38 Ibid., same to same, 21, 25 April 1818. Joseph Bamford emigrated to the USA in spring 1805. By August 1812, when he was aged 39 and living in Filbert Street, Philadelphia, with a wife and 5 children, he was working as a machine maker. *See* DNA RG 59, Returns of Enemy Aliens; also *Archives of Useful Knowledge* 3 (1813), 191.

39 Crozer, *Biographical Sketch*, 10–12.

40 US Federal Record Centre, Waltham, Massachusetts, First (Massachusetts) Circuit Court, Case files of *Paul Moody* v. *Medway Cotton Factory*, October 1821, affidavit of Samuel Stevens. My thanks to Ms. Betty-Bright P. Low of the EMHL for checking the Philadelphia directories.

41 DNA RG 45, Returns of Enemy Aliens. John Hodgson was listed as a turner of 24 Balloon Street, Manchester, in 1804, and as a roller maker in 1808–1809. *See Dean's & Co Directory*. I am very grateful to Dr R. S. Fitton for checking the early 19th century Manchester directories for the immigrant machine makers in the Philadelphia region. See *TIR*, 144–159 for profiles of textile immigration during the War of 1812 period.

42 Norman B. Wilkinson, 'Brandywine Borrowings from European Technology', *Technology and Culture*, 4 (1963), 3.

43 EMHL, Longwood MSS, Group 6, Box 3, Articles of Agreement between Caleb Kirk (of Christiana Hundred), James Jefferies (of Chester Co, Pennsylvania), E. I. du Pont, John Warner (of Christiana Hundred), William Warner (Philadelphia merchant) and John Torbert of one part, and George, Thomas and Isaac Hodgson of the other part, 23 August 1813. For Caleb Kirk and the Brandywine Mill Seats Company, *see* Boatman, 'Brandywine Cotton Industry', 12–13, 79–80. For Kirk *see also* George H. Gibson, 'Fullers, Carders and Manufacturers of Woollen Goods in Delaware', *Delaware History*, 12 (1966), 34.

44 Boatman, 'Brandywine Cotton Industry', 94–96. *TIR*, 231.

45 Boatman, 'Brandywine Cotton Industry', 94–96.

46 Ibid.

47 EMHL, Longwood MSS, Group 6, Box 1. Alexander McKim to E. I. du Pont, 22, 24 February 1813; Robert McKim to E. I. du Pont, 24 February, 12 March 1813.

48 Ibid.

49 Ibid., Richardson and Duff to E. I. du Pont, 12 March 1813.

50 Boatman, 'Brandywine Cotton Industry', 96.

51 EMHL Longwood MSS, Group 6, Box 2. Alrichs & Dixon to R. Duplanty, 22 March 1813.

52 Ibid., 'List of Machinery for Messrs Du Pont & Co', undated.

53 Ibid., agreement for spinning machinery made by G. Hodgson & Brothers, 22 March 1813.

54 Ibid.

55 EMHL Acc. 500, Ledger 195, Duplanty, McCall & Co, Factory Ledger 1813–1825, f. 81.

56 EMHL, Longwood MSS, Group 6, Box 2, Alrichs & Dixon to Duplanty, 29 March 1813.

57 *TIR*, 188.

58 EMHL, Acc 500, Ledger 195, 73, 81–82, 132–135, 164. The total sum expended in the machinery account 20 May 1814–30 April 1819, was $16,566.95 of which $3,212.74 was paid to the Hodgsons.

59 DMC, Robert McCall to Archibald McCall, 18 January 1825. In 1814 'not one piece of machinery stated in the account [William Boyd's?] was done in that year, when the price of work was nearly double of what it was afterward'.

60 *TIR*, 26.

[61] DMC, Robert McCall to Archibald McCall, 18 January 1825. *See also* ibid., 'Copy of Memo by R. Duplanty of work done by William Boyd'.
[62] American Philosophical Society, Charles Willson Peale Letterbooks, Peale to John de Peyster, 12 November 1813, (information from Roy M. Boatman's research notes in the EMHL).
[63] Ibid., Peale to Rembrandt Peale, 31 July 1814; Peale to John de Peyster, 17 November 1814; Peale to Angelica Peale, 14 February 1815; Peale to Rembrandt Peale, 19 February, 12 March 1815; Peale to Angelica Peale, 24 March 1815.
[64] *TIR*, 36–40, 322 n. 19 and *passim* (*see* index).
[65] Ibid.
[66] Rees, *Cyclopaedia* sv. 'Manufacture of Cotton'.
[67] DMC, Robert McCall to Archibald McCall, 18 January 1825. A blowing machine was known also to Richardson & Duff of Baltimore (EMHL, Longwood MSS, Group 6, Box 1, Richardson & Duff to E. I. du Pont, 12 March 1813) and to Alrichs & Dixon (EMHL, Longwood MSS, Group 6, Box 2, undated sheet headed 'Alrichs and Dixon').
[68] Ibid., and EMHL Longwood MSS, Group 6, Box 2, George Hodgson & Brothers 'list of Machinery for Messrs Du Pont & Co'.
[69] *TIR* 80, for illustration of Slater's card built of wood. The Hodgsons' preliminary quote specified 'Cards, with Iron frames' (source: previous note).
[70] *See* above, note 32 and for English practice which started around 1805–1810, *TIR*, 306 n. 41.
[71] Boatman, 'Brandywine Cotton Industry', 95.
[72] DNA, RG 45, Returns of Enemy Aliens; DMC, William McKenzie's bill, 8 December 1815.
[73] DNA, RG 45, Returns of Enemy Aliens; EMHL, Acc 500, Ledger 195, 144.
[74] DNA RG 59, Returns of Enemy Aliens; EMHL, Longwood MSS, Group 6, Box 20, William Hartley to Duplanty, 3 March, 25 May 1815.
[75] US Patent Office, *A List of Patents Granted by the United States from April 10, 1790, to December 31, 1836* (Washington, DC: Commissioner of Patents, 1872), *passim*.
[76] DMC, McCall to Duplanty, 17, 23 February, 13 April 1814.
[77] Ibid. *See DAB* for Cooper.
[78] DMC, McCall to Duplanty, 19 April 1814.
[79] Ibid., same to same, 26 April 1814; *Emporium of Arts and Sciences*, new series, 3 (1814), 459.
[80] *TIR*, 92–103.
[81] DMC, McCall to Duplanty, 19 April 1814. *See also* EMHL, Longwood MSS, Group 6, Box 2, Thomas Siddall to Duplanty, 29 November 1814 and 6 August 1815.
[82] DMC, McCall to Duplanty, 19 April 1814 and 11 July 1815.
[83] Ibid., same to same, 26 April 1814.
[84] Ibid., same to same, 26 April 1814. Craige was one of the few Philadelphia County manufacturers with powerlooms in May 1814. *See* ibid. McCall to Duplanty, 30 May 1814.
[85] Ibid., same to same, 26 April 1814, McCall wanted to run a Baltimore loom alongside Siddall's.
[86] Ibid., same to same, 20 May 1814.
[87] Ibid., same to same, 20 June 1814.
[88] Ibid., same to same, 31 March 1814.
[89] Ibid., same to same, 4 April 1814.
[90] National Archives, Washington DC, RG 92, Textual records of the Office of Quartermaster General, Special Files: Consolidated Correspondence, 1794–1890, Box 273, Duplanty, McCall & Co. contract with Callender Irvine, Commissary General of Purchases, 13 April 1814.
[91] DMC, McCall to Duplanty, 4 April 1814.
[92] Ibid., same to same, 27 May 1814.
[93] Ibid., same to same, 20 June 1814.
[94] Ibid., Duplanty to Siddall, 19 November 1814, draft on verso of Siddall to Duplanty, McCall & Co., 10 November 1814.
[95] Ibid., McCall to Duplanty, 31 March 1815. Also ibid., same to same, 23, 25 May 1816.
[96] EMHL, Longwood MSS, Group 6, Box 2, Siddall to Duplanty, McCall & Co., 13 August 1815.
[97] DMC, McCall to Duplanty, 18 June 1815.
[98] Ibid., same to same, 3 August 1815.

[99] Ibid., same to same, 6 October 1815.

[100] Ibid., same to same, 7 December 1815.

[101] Ibid., same to same, 6 April, 7, 21, 23, 25 May, 1816; Boatman, 'Brandywine Cotton Industry', 34–35, 40.

[102] DMC, Inventories, 1820.

[103] *TIR*, 99.

[104] For example, McCall expressed strong hopes that Thomas Siddall would move to the Brandywine, DMC, McCall to Duplanty, 16 May 1815.

[105] EMHL, Longwood MSS, Group 6, Box 2, Siddall to Duplanty, McCall & Co., 6 August 1815.

[106] DMC, McCall to Duplanty, 24 May 1815; ibid., same to same, 14 February 1816.

[107] Thomas C. Cochran, *Frontiers of Change: Early Industrialization in America* (New York: Oxford University Press, 1981). *See* especially chapter 4, where he emphasizes the relative importance of industries associated with the cotton gin, power saw, flour mill, steamboat and iron rolling mill. The 'backbone of production', he asserts, ran from Albany and Troy down the Hudson valley to New York then went from Paterson to Trenton, New Jersey, and then down the Delaware River through Philadelphia and beyond Wilmington into Maryland; in the west, heavy industry for local needs followed the Ohio from Pittsburgh to Cincinnati (ibid., 90).

VII

Transatlantic Industrial Espionage in the Early Nineteenth Century: Barriers and Penetrations[1]

As advances in computer technology in the capitalist West helped to trigger a final round in the Cold War contest for military-industrial knowledge in the 1980s, prior to the collapse of Soviet Communism, industrial espionage reached a new pitch:

> A Soviet scientist is caught red-handed slipping lasers into his pocket in a Marseilles research lab, a diplomat is arrested rooting in a French rubbish dump, a military attaché rooting in a Maryland tree trunk. Electronics experts in Holland have been caught at it. So have airline employees in Paris and trade mission officials in Cologne. Academics collect intelligence for the Academy of Science, students snoop for the GKNT (Soviet State Committee on Science and Technology): Tass covers for the KGB, Aeroflot for the GRU (military intelligence service). Polish machine companies were involved in Los Angeles, German trade fairs were involved in Leipzig. The list of players and passive participants could be expanded almost indefintely. The plain fact is that no Eastern institution is untouched by the global Soviet drive for Western technology.[2]

This may be compared to the much more dignified and culture-laden, but none the less siphonic, approach of one American industrialist approaching Britain's arsenal of industrial secrets in the 1820s:

> The knowledge of the useful arts, now cultivated as a branch of public instruction, is not infrequently to be gained either by entering apartments filled with the smoke of furnaces, and resounding with the deafening noise of machinery, or by conversing with men devoted to the common handicraft labors of life. European travellers have commonly preferred to breathe the pure air of a beautiful country, where the works of a munificent Creator are offered for contemplation and enjoyment; or to indulge exclusively in the pleasures afforded to taste and intellect, by the examination of splendid buildings, paintings, statues and libraries. Although the principal design of the following pages is to record useful facts, yet, in accommodation to the taste of general readers, sketches of scenery, habits and manners, have been introduced to vary and enliven the subjects of remark. The castle and the abbey, the theatre and the gaming house, the cottage and the palace, as well as the work shops of Europe, have all, in turn been visited by 'The Practical Tourist' in his circuitous rambles over some of the most interesting countries of Europe.[3]

Despite the distance in time between these two comments on industrial espionage, it is the contention of this paper that not much has changed, either in the objectives or the techniques of that publicly deplored but privately fostered activity. Knowledge, whether economic or technical, is not free, and industrial espionage has been practised as one of the least costly, though morally questionable, ways of wresting commercially valuable knowledge from a competitor. The topic is a large one and this article looks at only two aspects, barriers and penetrations, in the early transatlantic context, from the viewpoint of American competitors. Espionage on the part of European rivals has been closely studied by J. R. Harris, Eric Robinson and W. O. Henderson. Their work sustains the picture of enormous foreign interest in developments in Britain and of a heavy traffic in people, knowledge, and to a lesser extent equipment, in this early industrial period.[4]

VII

Transatlantic Industrial Espionage

Barriers to Industrial Espionage

War

Americans with the incentive to pillage European industrial techology faced daunting barriers. Between 1800 and 1840, the period of most rapid technical change in Britain and the USA — certainly for the machine tools and textile industries — widely-felt European wars disrupted the normal flows of trade and travel for no less than fifteen of the forty years. In the years 1807–14 a trade embargo and then war with Britain reduced flows of skilled artisans to a trickle. In the five years 1809–14 I could trace only 806 textile immigrants as having arrived in the USA (though some may have abandoned their trades after arriving; also the evidence is not perfect). By comparison, in the eight years 1824–31, some 4,999 textile workers arrived in the USA. That is, in wartime the flows of skilled immigrants were reduced to a quarter of peacetime levels (160 v. 625 per annum), though war was but one factor in explaining pushes and pulls on migration to America.[5]

Not only did war reduce numbers of skilled immigrants arriving in the USA, it also poisoned this free fountain of new industrial knowledge. During the War of 1812 all British immigrants who had not yet become American citizens had to register as enemy aliens and to live 40 miles away from tidewater, in case they tried to betray their new homeland. From high acclaim as a national asset, industrial immigrants (like the rest) fell under suspicion of being a national liability. One American citizen on the Maryland–Delaware border reported to the government,

> A particular friend of mine who has purchased a mill seat and is erecting a large cotton works at Ellicot's Mills, having observed three or four Englishmen in that neighbourhood of a suspicious character, gives me the following information . . . (one) is a manufacturer . . . These men deal in cattle and young McCoy [the innkeeper's son] is ready to prove that he saw [two] drive a drove of cattle past his house to Annapolis or its neighbourhood since the British fleet lay off that place. It is believed that the renting of the farm and the pretext of being manufacturers is a mere cover to their designs as British emissaries.[6]

The violence of war affected the carriers of new industrial knowledge. An anonymous informant told of a 'respectable Englishman at a Porter house the upper end of the Bowery [New York City]' who 'became intoxicated' and late in the evening 'inadvertently said Washington was a British Tory'. A Major Horn was listening and thereupon 'seized the tongs & with a side stroke cut his ear in two, then gave him three contusions in the upper part of the skull & left him as he supposed in a dying state. But fortunately for the man who has a large family of young children he is likely to recover'. The correspondent went on, 'I need not point out to you the benefit foreigners have been to the United States. They deserve to be protected'. And added, 'Since writing the above I have learned the man is a Yorkshireman of good habits & is a woolen cloth Manufacturer, has resided in America four years & is at present manufacturing blankets'.[7]

In Britain European war led to the tightening up of the restraints on the emigration of skilled workers (like textile manufacturers) and the export of machinery. The Board of Trade was informed in December 1811 that the Commissioners of Customs had received 'a letter from a Person at Bristol, representing that a Mechanic had left that Place by Coach with intent, as is supposed, to embark on board one of the Packets at Falmouth, taking with him a Trunk suspected to contain Tools, Brasses and Iron Work used in the construction of

Machinery employed in the Cotton Manufacture of this Kingdom'. The Board was told that the Collector of Customs at Falmouth had been alerted and directed to intercept and detain the aspiring conveyor of the high tech of the day.[8]

Language and Geography

Besides war, industrial spies setting out from the USA faced language barriers if they contemplated visiting France, Austria, Russia or the German states. Fortunately for Americans, Britain was the first industrial nation, not overhauled until after 1860. There were also barriers of distance. The 3,000 miles of ocean between the European and American continents took up to 35 days to cross in 1794 by the fastest means available, the Post Office packets, and 21 days to return to Europe. By the 1820s the picture was not much different, and remained unchanged until the advent of steamship services in 1840. Faster scheduled sailing services were organized in 1819 by the Black Ball Line: once a month from New York to Liverpool and back, reducing average times to about 35 days westwards and 21 days eastwards.[9] Time was not the only factor in transatlantic travel. Even in calm weather, sea sickness frequently confined wealthier cabin passengers to bed or bunk for some of their voyage and increased the discomfort and misery of those in steerage. Storms upset even those with the strongest stomachs and nerves: 'the tossing of the ship stirs up the effluvia of the bilge water and other odours, peculiar to a cabin, which are of themselves almost sufficient to create nausea'.[10] Expense — which was about £30 for cabin passengers and about £3 10*s*. for steerage passengers (one way) by 1800 and not much different in the 1830s — was hardly a consideration for the industrial spy.[11] In short the American who undertook industrial espionage in Europe in the early nineteenth century faced very considerable physical and psychological tests before he reached his quarry. No doubt conditions on the Atlantic passage in the early nineteenth century helped to sift out the more suitable candidates for the task.

Industrial secretiveness

Assuming Britain was the main target for American industrial spies in the early nineteenth century, it is clear that a very mixed reception would have greeted predatory visitors. This can be seen by distinguishing between private and public protection, between new and old technology, between consumer goods manufacture and capital goods manufacture.

In this period information on the new, mostly mechanical, technology was to be found primarily in the country's leading industrial centres: London, Manchester, and Glasgow, and to a lesser extent Newcastle, Birmingham, Leeds, and Bristol. Demand for the new technology came primarily from the expanding textile industries and later from the much newer railway system. In these high tech centres metal working facilities and skills depended on supplies of iron from South Wales, the West Midlands, Sheffield, and the North East. Within the high tech centres of the early industrial period — that is, chiefly in London, Manchester, and Glasgow and their hinterlands — machine shops and mills were the central repositories of the new technical information.[12] How easy were they to penetrate?

Transatlantic Industrial Espionage

In general, secretiveness was the prevailing attitude awaiting American visitors. Since the Middle Ages novitiate craftsmen had been sworn to keep their masters' trade secrets and this attitude persisted, even intensified, when there were technical improvements to be guarded, machine-breaking mobs to be feared, patent application procedures to be observed, and labour piracies to be checked. Sharpest watch against intruders was kept where the technology was most recent. In the 1780s and 1790s Robert Owen recalled that cotton mills regularly kept their outer doors locked against strangers. And many a mill resembled a medieval fortification with perimeter walls and gatehouse.[13] (See Fig. 1.) One American visitor to a mill near Stroud in the 1820s noted 'It is surrounded by a brick wall ten feet high which incloses nearly three acres of ground including the dye-houses, store rooms and other buildings connected with the manufacture of cloth. The mill is situated in a beautiful vale, which would be pleasant for a country seat. At the gateway is a porter's lodge, who admits only the workmen and those who have permision to enter'.[14] In the mid-1830s Manchester manufacturers and merchants were so secretive that they moved goods in and out of their warehouses by back doors and at night. Curious visitors admitted to Manchester factories had to be careful what questions they asked for fear of arousing suspicions.[15]

As late as the 1820s the Manchester manufacturers tried to suppress the publication of technological information which might convey their new knowledge to foreign rivals. The first thorough description of the new cotton manufacturing technology was not published until 1812 when the article on 'Manufacture of Cotton' appeared in the latest installment of Abraham Rees' monumental *The Cyclopaedia: Or Universal Dictionary of Arts, Sciences and Literature* (45 vols, London: Longman *et al.* 1802–20); and a set of cotton processing rules, specifying sizes of gears, relative speeds of rollers &c, did not become available until James Montgomery published his *The Carding and Spinning Master's Assistant* (Glasgow: John Niven Jr, 1832).[16] However, as new 'best practice' technology spread and became representative technology, familiar to foreign rivals, there was less foreign inquisitiveness and domestic protective measures relaxed somewhat. The reported openness of the Glasgow cotton manufacturers in the 1820s stemmed from their use of what had become a widely known technology for spinning coarse yarns. Another reason why some manufacturers seemed open to foreign rivals was because the locus of their technology lay not in the line of machines in their mills but in the workers who worked with or in conjunction with those machines. Thus production methods still requiring heavy inputs of manual skill, like those in the fine cloth industry in the West of England, could be displayed to visitors without fear of easy espionage. Workers would have to be seduced, and not just individuals but whole teams whose skills had been acquired by lengthy apprenticeships (three to seven years).[17]

British machine makers were far more open to foreign customers than were the manufacturers. They had capital equipment for sale and foreign manufacturers were as much potential customers as domestic ones. James Nasmyth, one of the larger machine makers in Manchester in the late 1830s, gladly exhibited his works and designs to visitors, confident that the advantages of secretiveness were far outweighed by the augmented trade gained by openness.[18] So earlier had Boulton and Watt in Birmingham and Matthew Murray in Leeds.[19] This openness on the part of machine makers was widespread throughout the capital goods industries, as MacLeod has recently shown.[20]

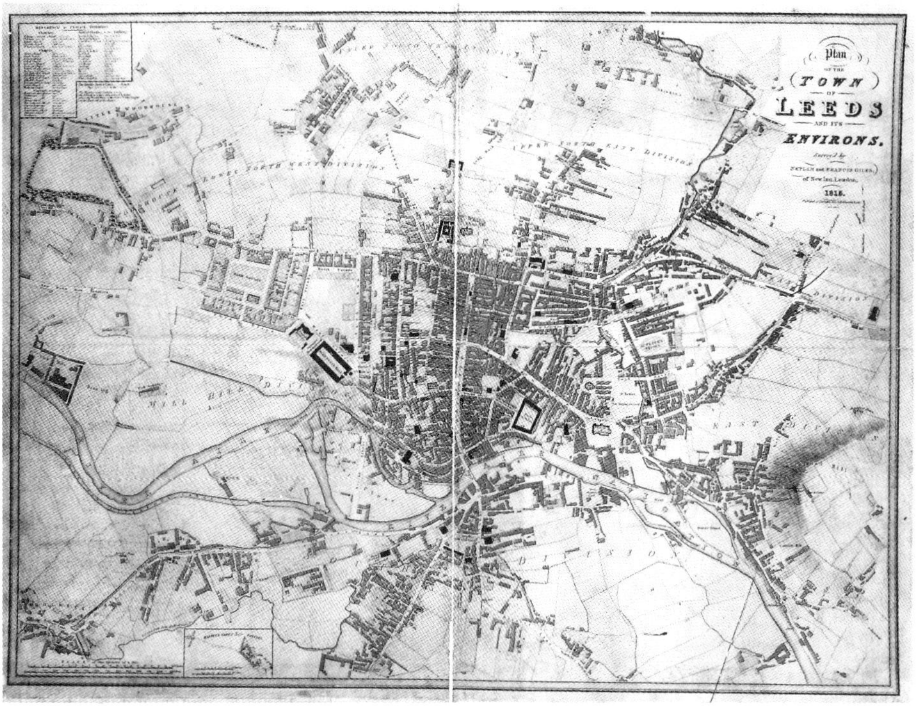

FIG. 1. Plan of Leeds in 1815, showing (on far left) Benjamin Gott's mill, the Park Mills, at Bean Ing (first quadrangle of buildings built 1792–93; rebuilt after fire in 1799; second quadrangle of buildings added in 1815). The quadrangular plan typified many early mill complexes and expressed a defensive, secretive posture on the part of the manufacturers.

Source: Netlan and Francis Giles, 'Plan of the town of Leeds and Its Environs, 1815'. I am grateful to the staff of the Local and Family History Library, Leeds City Libraries, for supplying the copy of this map and for permission to reproduce it here.

Industrial secretiveness was also imposed by the state. Whether embodied in the 'dead' (implements or machines) or 'living' instruments of trade, the new technology and much of the old, would have been theoretically halted at Britain's ports by severe prohibitory laws. That at least was the legal position. In reality these laws, which were enacted between the 1690s and the 1780s in response to manufacturers' pressure groups, were far from effective. They threatened severe consequences to violators: agents faced a £500 fine for each worker enticed abroad, and twelve months' imprisonment; the export, or attempted export, of textile, metalworking, clockmaking, leatherworking, papermaking and glass manufacturing equipment incurred a £200 fine (£500 for printing equipment), forfeiture of equipment seized and twelve months' imprisonment. During the French Wars procedures were tightened up. An Order in Council of 1795 obliged captains of foreign vessels to submit a list of passengers' names, ages, occupations, and nationalities before clearing. The Passenger

Act of 1803 limited the number of passengers British vessels could carry. And packet captains were instructed to offer free passages to artisans willing to return to England. However, with the advent of a free trade ideology after 1820 the prohibitory laws were drastically modified. All restrictions on emigration were lifted in 1824. Machinery exports came under a licensing system which continued to check exports of new manufacturing technology until it too was abandoned in 1842. Since the machine tool makers — the capital goods manufacturers — expressed relatively little concern about machinery exports, their equipment (lathes, planes and the like) was more often licensed for export than was consumer goods manufacturing equipment.[21]

Yet not all machine makers were as open as Nasmyth. An illustrative case comes from the experiences of John Smith (1796–1886), a machinist who in 1816 at the age of twenty-one had emigrated from his native Brechin in Scotland to the USA with a shilling, a Bible and his machinist's tools. He settled at Andover, Massachusetts, and became partner in a machine shop specializing in cotton machinery. A decade after leaving Scotland he returned on a winter visit (1826–27), not least to survey improvements in Glasgow's textile machine shops. With his Scottish background and connections we might expect him to have gained regular and easy access to the workshops he previously frequented. He did not in every instance. Early on his stay in Glasgow (12 January 1827) he noted in his diary 'This morning I proceeded as yesterday calling upon some of the public works for admission, but there was no admittance to be had for me. I found there was no way I could make out to get into the works unless I found some person or friend to introduce me. I called at Cook's where I formerly worked, but there was no admittance. I saw some of my old shop mates, Thomas Gillespie, and several others who still remained with Cook'.[22] Yet a few days later he recorded, 'I visited a number of public works this week, spinning mills, etc., etc., also Girdwood's machine shop with John Scott, taking me through all the apartments where they put up the small machinery, throstle frames, etc. Some parts of the work I liked much, the cast iron roller beams polished up bright, the make of the stands and the rollers very nice, as the fluting engine leaves them without filing or polishing, the flutes being perfectly smooth and bright in the hollow'.[23] At Cook's machine shops Smith was obviously regarded as a skilled but outside machinist who would quickly spot the different, the new, the innovative modification, and thus might steal away improvements to another business, another country.

Technical traditions

There were yet other barriers in the path of foreign spies, prying into the structure and nature of Britain's 'high tech' industries of the early nineteenth century. Interlacing both machine building and manufacturing processes were a maze of technical traditions. In machine building, for example, the cutting of screws conformed to no single standard until the 1860s, with the widespread adoption of Joseph Whitworth's average dimensions of pitch, depth and form which he had formulated in 1841. Before the 1840–60 period each machine shop had its own screw and bolt contours and these were not easy to copy, being the product of craft and labour-intensive conditions and skills. Even the Whitworth compromise standard screw thread needed three kinds of cutters and two lathes in its manufacture.[24]

In manufacturing, the localized structure of the pre-industrial economy had likewise created an undergrowth of technical practices whose standards and nomenclatures can only have bewildered the intruding foreigner. In textiles, grades used in wool sorting, systems for stating the fineness of card clothing, methods of measuring yarn fineness, and standards for expressing the fineness of cloth all varied from district to district.[25] In other industries the thicket of craft practices on which industrial technology was imposed presumably also existed, not least because of local dialects (or languages in the case of Scotland, Wales and Ireland) and regional or local systems of weights and measures.[26]

The pace of technical change

Because in the eighteenth century technical improvement was increasingly seen as an engine of economic growth, because invention was becoming more related to economic change, foreign industrial spies needed a rising level of knowledge before they could even understand the new technologies they found in early industrial Britain. The well-known measure of technical change is patent registration and a brief reminder of British rates and conditions provides a rough idea of the size of this barrier to industrial espionage.

In the 1750s about a dozen patents, for all kinds of invention, were registered each year in England. By the 1790s the level stood at about 90 a year. Between 1800 and 1830 it moved up from 100 to 200 and in the 1830s the rate doubled to over 400 a year.[27] In textile processing and machine making, the high tech area of the day, some 303 patents for mechanical devices were taken out between 1790 and 1830.[28] This may not sound like many patents. However, it should be noted that some patents covered several devices; that the rate of registration was rising; that (as Dutton confirmed) patenting was becoming more economic (i.e. more closely related to the market); that there was a shift towards more, rather than less, complex capital equipment (as well as a more capital-intensive technology), designed to produce high quality goods; and that patents, being so expensive in Britain, were but the public manifestation of localized technical modifications which might eventually be recorded in patents.

Certainly not all major textile inventions were patented but most were by the 1790s.[29] (Crompton's mule is one of the oft-cited exceptions.) This assertion is based on a reading of the 303 textile patents of 1790–1830 and on the comparative cost of British patents of invention in the early nineteenth century. At £70 in 1815 and at least £150 in 1829, they were around three times the annual wage of a British cotton factory overseer. In comparison, American patents cost only $30 to register, about a fifth of a skilled American artisan's annual wage. British patentees therefore needed more confidence and stronger expectations of commercial returns from their patented inventions than American counterparts.[30] It may be that British patents were longer in moving from the workshop to the preliminaries of patent registration, so that winners more frequently emerged in Britain than in the USA. At any rate, foreign acquisitors (themselves or their agents) needed to consult British patent lists and the patents in their area of interest.

But there are other points to be made when considering the patent situation from the viewpoint of American borrowers. First, they, like modern historians, had no means of knowing exactly which inventions were covered by a patent and which were not. They were therefore obliged both to search the London patent offices and patent journals and also to ferret around the machine makers and manufacturers in the industrial districts. Only about a quarter of British textile patents were reported in the *Repertory of Arts*, the main patent journal, from its foundation in 1794 until 1830, though the journal did print a list of titles, identifying the subject of each patent. To obtain copies or make summaries of patents, enquirers had to make costly searches in London. The rolls of vellum on which the patents were recorded were scattered between three offices (the Petty Bag, the Rolls Chapel, and the Enrolment Offices) and were neither adequately nor reliably indexed before the 1830s (indeed, no proper indexes were published until Bennet Woodcroft produced his in the 1850s). Searches demanded time and money. By 1829 it was costing between two and forty guineas to secure a patent copy. A final point: since the struggle over Arkwright's patents in the 1770s, patent case law required that the specifications be comprehensible only to mechanics familar with existing machines. This meant that as the knowledge frontier advanced so it would become increasingly difficult for the uninitiated to understand patents.[31] All this had two implications for industrial espionage. First, it suggested that the optimum industrial spy would himself be a manufacturer or a machine maker, someone with a strong technical background, who would understand exactly what he was reading or observing. Second, it suggested that it could well be more effective for the spy to spend much more time in the provincial manufacturing districts than in the patent search rooms in London.

What then were the best methods for overcoming these barriers to the sharing of new industrial knowledge? Recruitment of immigrants? Smuggling machinery across the Atlantic? American visits to Britain's industrial districts? Reliance on published descriptions of the new technology, such as patent reports? Which routes Americans perceived as the most efficient can be seen by looking at a few cases.

Penetrations of Britain's Reservoir of Advanced Industrial Technology

Espionage strategies, networks and limitations

In the 1980s East–West struggle for the possession of the latest high tech, in which Eastern Bloc countries were invariably the predators, and United States corporations the 'victims', a battery of techniques, from highly refined electronic devices to crude blackmail, were deployed by the KGB and similar state instruments in the old Soviet satellites on one side and the CIA and US Customs officials on the other.[32] In some respects their world seems vastly different from that of early industrial spies. Americans were then the predators and British industrialists and Customs officials the victims. Yet there are similarities. To capture America's space technology the Soviets relied on an espionage strategy which encompassed a wide spectrum of technical knowledge. This was embodied in equipment ranging from Vermont ball bearing grinders (for missile or space vehicle gyroscopes) to Sperry Univac mainframe computers. Secondly, the Soviets and their agents focused not only upon individuals who could release new knowledge but also upon distribution networks with freight forwarding and storage facilities organized through a number of shell companies and bogus destinations, to conceal true ownership and the final destination of smuggled equipment. Thirdly, the Soviets still had to struggle with the critical disjunction between equipment (whether hardware or software) and complementary human skills. One instance illustrates some of these points. The Bulgarians in 1982–83 illegally bought from the US Digital Equipment Corporation of Maynard, Massachusetts, three VAX–11–780 mainframe computers, highly sophisticated machines having 'substantial weaponry and nuclear as well as naval and commercial applications'.[33] The first of these mainframes was trucked from Massachusetts to Minnesota by an American freight firm; then it was sent by Lufthansa cargo plane to Cologne, under a valid US export licence for export to West Germany. The West German shipper, Datagon, had a sister firm in Dorset to which it sent the computer. From Dorset Datagon despatched the computer to a scientific institute in Bulgaria. The British managers of Datagon pleaded ignorance of the embargo on the equipment and the two other machines followed the same route a few weeks later. And then western distributors should have heard no more of the transaction. But this was not the end of the story. The Bulgarian importers paid for the first machine outright but sent only a fractional deposit for the other two. Soon after, the Bulgarians complained that the computers were faulty. In fact there was nothing wrong with them. The truth was that the Bulgarians lacked operational expertise — service. Ivor Edwards one of the British managers confessed 'there is no worse mess than a computer mess'.[34]

Almost exactly the same components can be traced in American attempts to secure illegally Britain's early industrial technology. While their attempts, unlike the Soviets', were not centrally controlled, they were spread across a wide spectrum of techniques. For example, when in 1835 Daniel Treadwell, newly appointed Rumford Professor on the Application of Science to the Useful Arts at Harvard, made a visit to Britain, he took with him five assignments from the Boston mercantile-manufacturing community. The treasurer of a Pennsylvania coal company wanted workers able to smelt iron with coke (by the Darby process, well known in England since the 1750s); another merchant wanted a detailed account of methods of making iron sheets and hoops; another wanted globes and compasses from London; Amos and Abbott Lawrence, then about to form the Boott Manufacturing Company, a cotton company at Lowell, wanted particulars of English machinery for making cotton twist for export and commercial information about manufacturing costs and yarn sizes suitable for markets in northern Europe and China. Lastly, a cotton mill manager from Maine wanted a variety of commercial and technical information, not least to see whether it would be cheaper for him to import key machine parts from England. The mill manager also had some inventions which he wanted Treadwell to patent and market in England. While these espionage strategies were not orchestrated by the government, clearly the visible hand of management was creating a broad knowledge-acquisition strategy when market opportunities arose.[35]

Knowledge acquisiton and distribution networks were similarly of crucial importance in American industrial espionage ventures during the early nineteenth century. The net flow of immigrants from Europe to the USA had led some Americans to believe they could rely on immigration to bring over new technology. In 1787 Tench Coxe, the Philadelphia merchant, had proposed that the newly-formed Pennsylvania Society for the Encouragement of Manufactures and the Useful Arts should send a Committee 'to visit every ship arriving with passengers from any foreign country, in order to enquire what persons they may have on board, capable of constructing useful machines, qualified to carry on manufactures, or coming away as with a view to that kind of employment'.[36] This was far too haphazard a method to be solely relied upon, though it might have served the envisaged purpose as an adjunct to more aggressive techniques. Even by the 1820s, fresh skilled immigration to the USA from Britain could not keep pace with the labour requirements of America's expanding textile industries. How then did American rivals acquire knowledge of Britain's early industrial technology? How was that knowledge. whether 'hardware' or 'software' or embodied in skilled technicians, relayed back to the USA? Since it is impossible to assemble a representative picture of unco-ordinated covert activity in the distant past, anecdotal evidence has to suffice. Several episodes have been taken in what follows to illustrate some recurring elements in the evidence on industrial spying. Most attention has been given to Zachariah Allen.

Networks of access

Zachariah Allen (1795–1882) was one of at least three sons of Zachariah Allen Sr, a cotton goods importer of Providence, Rhode Island, and his wife Anne née Crawford. He was orphaned at the age of nine when, presumably, his brothers took over the firm. One of them, Philip Allen (later a US Senator) was ten years older than Zachariah. From a school at Medford, Massachusetts, Zachariah was sent to the recently-founded Phillips' Exeter Academy, New Hampshire. Modelled on the English dissenting academies, Phillips emphasized training for professional skills at the expense of the classics. From Phillips Exeter he entered Brown University, Providence. Though fascinated by chemistry and physics (then partially understood), he was articled to a Providence lawyer after graduating from Brown in 1813. He was admitted to the Bar in 1815 and also took a course at Brown Medical School. Meantime he engaged in mercantile operations and took a growing interest in the infant American cotton industry since his brother Philip had established a cotton mill at Smithfield, Rhode Island, in 1813 and his other brother Crawford also engaged in cotton manufacturing. On marrying Eliza Harriet, daughter of Welcome Arnold, an eminent merchant in Providence, Rhode Island, he took his new wife on a bridal tour across the Allegheny Mountains, not least to survey western markets. By 1822 Zachariah Allen was a member of the Providence town council and, no doubt with the support of his inheritances, both paternal and marital, that year he set up a mill village named Allendale near Providence where he started woollen manufacturing. To the city of Providence he introduced an improved fire engine and a central furnace system for heating houses.[37]

Within a decade Allen's woollen mill, one of 22 in the state, represented an investment of $94,000 (this included money annually paid for wages and other materials), ran 600 spindles and 21 broadcloth looms and employed 27 men, 30 women and 10 children. Principally it made cloth for the New Orleans market.[38] Clearly here was a well-educated and inquisitive man, flourishing in a milieu of Yankee improvement, and with the capital to pursue the quest for new knowledge. More specifically, he was a patriot engaged in practical woollen manufacturing. His prime rivals in Southern markets were British, and to a lesser extent French, woollen manufacturers.

In 1825 Allen decided to visit Europe. He was in fact one of 83 American manufacturers, engineers and tradesmen known to have visited Britain and to have returned to the USA between 1824 and 1831.[39] Several years later he stated publicly that his purpose in visiting Europe was 'to examine the effects of the important improvements in machinery upon the state of society at the present day'.[40] That clearly emerges from one of his unpublished pocket diaries. The first entry reads 'Wednesday April 13th 1825. I left Manchester at half past 2 o clock on my route to survey the manufactories of woollens in Yorkshire'.[41] He had in fact arrived in England very early in 1825, as another larger diary which he kept indicates. Disembarking in Liverpool, he went by way of Stockport and Macclesfield to Birmingham, Blenheim, Windsor, and London. From London he made three sweeping tours, the first to Manchester, Leeds, and Sheffield; the second to Portsmouth, Salisbury, Bath, Bristol, Cheltenham, and Stratford; the third down to Dover and across the Channel to Calais and thence Paris, Lille, Brussels, Leyden, Amsterdam, and back to London. On his final traverse across the British Isles he travelled north by way of Boulton and Watt's Soho works and the Stockton and Darlington Railway to Edinburgh, Glasgow and Paisley. From Greenock he went over to Ireland, travelling from Belfast to Dublin. Before sailing back to America from Liverpool he toured North Wales and ascended Snowdon. Keeping Allen in Europe for about a year, his tours are recorded in the account he published in 1832, *The Practical Tourist, or Sketches of the State of the Useful Arts, and of Society, Scenery &c. &c. in Great Britain, France and Holland* (2 vols, Boston: Richardson, Lord and Holbrook; Providence: A. S. Beckwith, 1832). But this cloaks much of the industrial espionage in which he engaged, activities which clearly emerge in his original pocket diaries, one of which has survived. This small volume spans his itinerary from Manchester to Saddleworth, Huddersfield, Leeds, France, Holland, and the West of England. Opening on 13 April 1825, the pocket diary peters out by the following August.

What kinds of network facilitated Allen's movements in England and the Continent? Networks were crucial, as John Smith's experience illustrated, in opening doors. They were also important in pointing foreign visitors to those doors. Doubtless Allen carried letters of introduction to American contacts in Britain. This was a customary procedure. Johann George Bodmer, the Swiss engineer and inventor, who toured British industrial districts in 1816 and 1817, relied on a chain letter system, collecting letters of introduction in one district to open doors in the next on his itinerary.[42] John Smith, returning to his native Scotland in 1827, carried letters to their relatives from emigrants in New England, which eased Smith's movements.[43] Allen certainly made contact with the family of one of his immigrant employees: 'In the afternoon', he wrote, 'carried the letter of Peter Buckley (the head workman at finishing who has worked for me more than two years) to his father in Saddleworth. The old man and his family seemed very much rejoiced at hearing from him, as they had received no intelligence of his welfare for three years. Was compelled, after repeatedly declining, to drink some sour wine, offered me, that I might not affront them by refusing their well meant civility, although it was a severe tax on my politeness'.[44] Occasionally Allen tried to gain admittance to an English mill without letters of introduction. This was not always successful. Allen and his American friend turned up at one mill, a mile and a half down the Avon from Bath (which suggests it was a mill at Twerton) without a letter of introduction. Allen noted, 'the proprietor stated that he had so many applications of visitants that he had made it a positive rule to admit no person without introduction, and we returned disappointed to our lodgings at the White Heart [*sic*] inn'.[45] Obviously this manufacturer was not averse to visitors: he simply wanted to limit numbers and restrain if not vet his intruders.

Allen's British networks had an American as well as a local dimension. Early on his English stay in 1825, while in London, he met Abraham H. Schenck (1775–1831), an

American textile manufacturer, with whom he appears to have travelled for much of his European tour. Schenck was one of the family proprietors of a cotton mill and machine shop, the Matteawan Factory, set up at Fishkill in Dutchess County in the Hudson Valley about fifty miles north of New York City. In 1820 the Matteawan Factory represented an investment of $100,000 (buildings and machinery); ran 1,600 spindles and 44 powerlooms and some handlooms; produced 300,000 yards of cotton shirting, sheeting, stripes, plaids, bed ticks &c. a year; and employed 52 men (including 16 machine makers, 10 foremen, dyers &c.), eight women and 92 children. In 1820 the Schencks reported to the US federal government that their factory had started operations in 1815 and, despite the post-war depression, had been expanded with the addition of the 'most approved machinery'.[46] While neither Allen nor Schenck owned factories as large as those found in the northern New England region after 1815 (the so-called Waltham System), clearly the two men together represented some of the most experienced, the most innovative American textile manufacturers of their day, well aware of American developments and keenly alert to the possibilities of European inventions for American use. Together they could appraise either cotton or woollen manufacturing technologies with the eye of practical men.

Their American contacts were of various kinds. In Leeds they received much co-operation from William Davy, the American Consul. He had been moved from Hull to Leeds in 1818 in an attempt to monitor at source those English woollen manufacturers who were suspected of involvement in arrangements for smuggling Yorkshire woollens into the USA via Canada. (The Yorkshire manufacturers were implicated by their willingness to so label and package their cloths that their true origins were concealed and they might be passed off as American goods, thus surmounting the US tariff.) Davy took Allen to see Hirst's woollen mill in Leeds, one of the largest in the country, employing 500 persons. Another day Davy's son, Albert, escorted Allen and Schenck to a works for printing cassimere shawls, then to a dyehouse and lastly to the great flax mills of John Marshall (1765–1845) then running 10,000 spindles. Possibly the American consul's contacts facilitated an introduction to Benjamin Gott (1762–1840) who personally escorted the two Americans through every department in his large factory (750 employees) — which Allen believed produced as much broadcloth as all the woollen mills in New England (See Fig. 1). Gott took Allen and Schenck out to his country house where he dined his American guests and proudly displayed his gentrified comforts.[47]

Besides the consular service there were other American contacts in England, not least because of the existence of a small expatriate community. In London there were American merchants and bankers, like Joshua Bates (partner in Baring Brothers), Timothy Wiggin and Henry Higginson, and one or two engineers like Thomas Robinson Williams who also acted as a patent agent. In the manufacturing districts one or two Americans had set up as machine makers, the most successful being Joseph Chesseborough Dyer (1780–1861), a Rhode Islander who settled in England in 1811. He established a card clothing manufactory first in Birmingham and then in Manchester where he was located in 1825 when Allen and Schenck arrived. In his large diary Allen noted, 'May 10th we visited a manufactory of machine cards in which the wire is set in the leather by the machinery invented by Whitmore [Amos Whittemore] and first used in the United States. Mr Dyer, to whom this establishment belongs, is also an American'.[48] Allen met Dyer and had a lengthy conversation with him. In the impersonalized, published version of his tour Allen described both

Dyer's works and those of another American, Jeptha Avery Wilkinson. A mechanic from Otsego in upstate New York, Wilkinson had settled at Manchester in 1817 and set up his steam-driven reed making machine, another marvel of self-acting labour-saving invention.[49]

Information goals

If networks, heavily American based, gave Allen and Schenck admittance to the early-nineteenth-century equivalents of late-twentieth-century Silicon Valley, what sort of information did the visitors seek and how did they collect it? One kind was economic information. The size of English factories, numbers employed and factor prices, especially machinery prices and wage rates, are sprinkled through Allen's pocket diary. He prefaced his notes on the West of England section of his tour record with a list of wage rates he had found cited in a London newspaper. When he came to publish his *Practical Tourist* he printed comparative wage rates (of textile operatives, machine makers, carpenters, masons, servants) to illustrate that wage rates in England were higher than in France but lower than in the USA (where the cost of living was reckoned about half that in England). This all fuelled the contemporary debate about comparative manufacturing costs and whether the USA was catching up with Britain in the race to industrialize.[50] Since much of this kind of information was in the public domain it did not require much espionage to secure. Neither did the geographical and social information with which Allen heavily packed the published version of his journal. Yet, monarchical principles and class differences, social customs such as English drinking habits, or social values such as attitudes to the introduction of labour saving machinery were not unrelated to the American interest in acquiring Britain's new technical knowledge and in rivalling her spectacular industrial performance.[51] So Allen recorded as much information as he could about the European cultures he visited.

The industrial knowledge which he wanted especially was of a technical nature, whether related to design or processing, whether embodied in factory layout, machinery or plans, processing techniques or chemical 'recipes'. Significantly he was not solely interested in recently-patented machinery. Equally, he recorded comments on commonly deployed machinery and carefully noted when that was out-dated by the American models he knew so well. For example, on emerging from the largest mill at Bradford-on-Avon, Allen noted 'I did not perceive anything new or interesting in this place'.[52] In contrast he rated the millwork at Harris and Stevens mill (the Stanley Mills at Kings Stanley) near Stroud as 'in a style superior to any I have seen elsewhere' and in the same mill he saw 'the most beautiful blue cloths here, a parcel of which he [Stephens] was preparing to send to the United States. Several women were engaged in picking out the names of the manufacturers marked in silk letters upon the end of the cloth, as he observed the purchaser did not wish it should be known in the United States from whence he obtained his supply of cloths. The colours of the blue cloths are remarkably brilliant & lively. He uses the finest sorts of Saxony wool. The filling appeared to be remarkably slack & large for the quality of the wool'.[53]

Allen's highest priority was to see newly-invented equipment or older machines or processes especially different from those found in the USA. Most rewarding were the machine shops and woollen mills of Yorkshire, to which he returned after visiting France and Holland. In Leeds John Jones, a brush manufacturer, showed Allen one of his recently

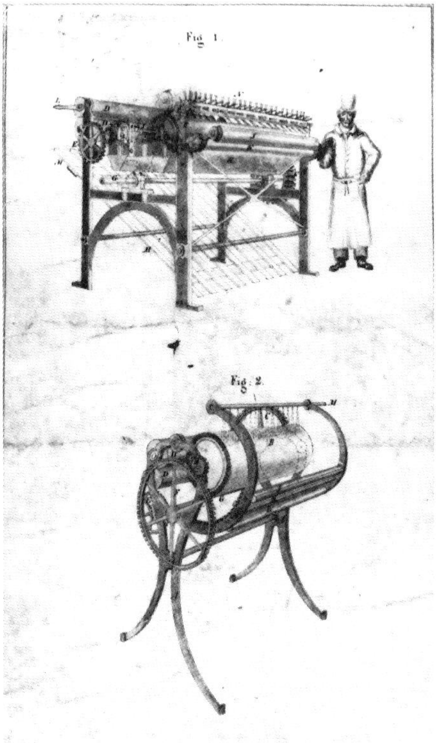

FIG. 2. Drawings accompanying the original patent for John Jones's steam brushing machine or dresser, for napping cloth (dated 27 January, 1824): the machine Zachariah Allen examined and sketched when visiting Leeds in 1825. Compare with illustration 3.

Source: Public Record Office, C66/4266, entry 1. I am grateful to the staff of the PRO for locating this drawing and for permission to use it.

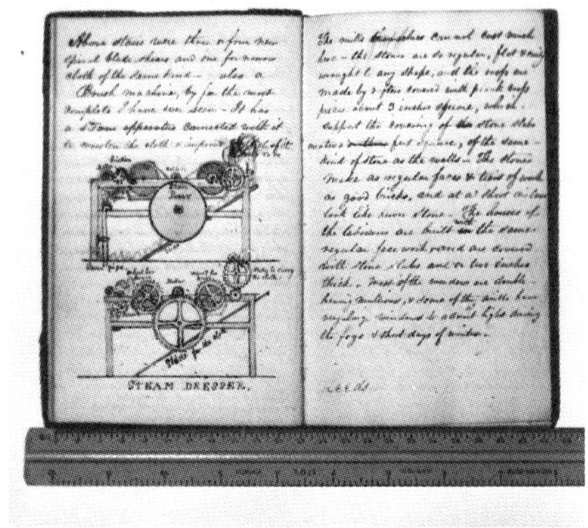

FIG. 3. Pages from Zachariah Allen's 'Journal of European Trip, 1825' showing his elevations of Jones's cloth dresser which Allen inspected at Leeds. Compare with illustration 2.

Source: Rhode Island Historical Society, Zachariah Allen Papers. Photograph by the author.

patented steam brushing machines priced at £118. Allen estimated that the brushes alone, formed of wire twisted with bristles, would cost £35. (The machine was used to raise the nap on flannels, the steam apparatus fixing the nap after it was brushed.)[54] (See Fig. 2.) Near Gomersal Allen visited machine maker Samuel Stead who was building 200–250-spindle woollen mules (like cotton mules in many respects) a great improvement on the old jenny. Stead's mules cost £65 (or $230) for all the ironwork which, completed with the addition of woodwork, in the USA Allen reckoned would make a total cost (including freight) of $600.[55] (Mentally he must have compared this price with the $2 a spindle which was the going rate for a woollen jack in New England in the mid-1820s.)[56] Near Kirkstall Abbey Allen found an extensive woollen mill where the 'carding engines were very beautiful machines',[57] and where there were three kinds of mechanical cloth shears: the old hand shears driven by power; a newer machine in which the spiral cutting blade traversed the cloth longitudinally (and which Allen was told 'did not work well and often damaged cloth'); and 'the new kind of spiral blades running across the cloth from list to list'.[58] The old hand shears under power were used to make the first cut and the newest machine to finish the face of the cloth. In the same place half a dozen new fulling stocks had just been installed and Allen noted that, except for the cloth box which was lined with wood, they were entirely made of iron.[59]

It is evident from his pocket diary that Allen was most excited by the technical improvements he found in the machine shops and largest factories in Yorkshire. More capital intensive than that found in the West of England, this new technology was more appropriate to America's natural resources and factor prices. Ironically there was quite as much interest in innovation in the West of England as in Yorkshire at this time: indeed in the period 1790–1830 patentees in the West of England registered 70 patents related to woollen manufacturing, as compared to 41 from Yorkshire patentees.[60] Yorkshire woollen men learned much from the West of England: the novel shearing frame Allen saw, which cut from list to list, was originally patented by John and William Lewis, Stroud manufacturers, and their engineer William Davis, in 1818.[61] Paramount in the manufacture of fine cloths for so long, West of England clothiers regarded the new technology with caution since it was not initially suited for quality production. Even so, there were progressive machine makers in Gloucestershire, Wiltshire, and Somerset. Could it be that Allen's network of access to the West was less efficient than the network he employed in Yorkshire? This might explain why he more frequently approached West of England woollen manufacturers rather than their machine makers. On the other hand, he might have been less interested in machines than in the manufacturing techniques of the West of England, which he knew lay behind the high quality of the region's best cloths.

Information collection

Whatever the relative regional strengths of Allen's access networks it is clear that, once inside a British factory, he was swiftly accumulating information related to the goals of his visit. He gathered that information in a variety of ways, most of which seem obvious. It is worth reiterating that Allen collected his information personally wherever he could. An intelligent well-informed woollen manufacturer himself and the end user of much that he learned, he could best decide what information to select, what to record, and how to record it. His qualifications made him an ideal industrial spy.

Inside a machine shop or factory, visual observation allowed Allen to survey British structures, designs, processes, materials, and then to focus upon any features different from those in America. At one place he was most interested to inspect a mill devastated by fire, for it was incontrovertible proof of the efficacy of a fire proof construction. The mill belonged to Abraham Rhodes, a bombazette (worsted) manufacturer near Leeds, whose products were well known in the American market. Walking around the blackened structure Allen reported, 'The fire appeared to have extended itself through three storeys of the main wing of the building. The floors were uninjured except in a few places where the heat had broken the flat stones that lay upon the iron beams and joists. The roof, formed of iron rafters and stone slabs, was as entire as before the fire took place. Several of the panes of glass were broken by the heat, but the sashes being of cast iron were entire. The principal damage appeared to have been done to the stock and the machinery, which had become much rusted by the steam and heat. The walls, and every other part of the building, were merely blackened by the smoke'.[62]

Where he was permitted Allen even took measurements of facilities. Near Kirkstall Abbey he noted

I went over a large new mill the walls of which the masons have just topped off. The dimensions are width 11yds, length 55yds, 5 stories high, walls of lower story 2 feet thick, leaving the mill 29 feet wide in the clear. The foundation for the steam engine is very permanently and strongly constructed of large blocks of hewn stone. The pillars and beams of this building are all of cast iron . . . . One of the beams which I measured was ten feet long, ¾ of an inch thick with a bottom about 3 inches wide at right angles with the perpendicular post. They crown or arch a little in the centre where they measure about 12 inches deep.[63]

Allen's pocket diary is evidence of the industrial spy's reliance on careful notetaking. Allen was observant, methodical, exact. In Gott's mill he noted dimensions, layout of machinery, sequence of processes, and singled out fulling for more detailed description (particularly the roller rinsing machine, which washed the cloth).[64] Where he deemed it necessary he added explanatory sketches. Some bear the marks of being done on the spot – in pen, with lines wavering.[65] Two sketches, however were drafted with an accuracy needing ruler and compasses. In a Huddersfield cloth finishing establishment he found 'machinery . . . in the very best order and upon the most improved plan' including a 'Brush machine, by far the most complete I have ever seen'. And of this he made two elevations.[66] (See Fig. 3.) A week later in Leeds, he recorded something of his method of making a machine drawing. 'Saturday April 23rd 1825. Went to a machine shop to see some Gig mills and carding engines. The gig machines are built after the plan upon which most of those I have seen in operation in the best mills are constructed. From several visits made to these machines we have been able to get a pretty accurate draft — which is copied on the next page'. (See Fig. 4.)[67] Whether Allen and Schenck made their sketches covertly or openly is unclear.

Allen made some progress in cutting into the undergrowth of Britain's regionalized technical language and traditions. In Gott's mill he noted of the handloom weavers, 'Their work is calculated by the *string* of about 3⅓ yds or ten feet in length, and they average about one string per day of the fine cloth such as is made by this mill'.[68] The previous day (26 April 1825) he recorded the success he had had in obtaining a manufacturer's trade secret from a machine maker: 'A Reed maker in Leeds procured for me the following table from William Hirst's head weaver'. The table showed the relative fineness of warp and filling yarns for cloths of varying reeds (which controlled the closeness of warp yarns).[69]

On his way through Britain's woollen manufacturing districts Allen collected recipes which he thought might be useful back in Allendale. One was for 'Cement for covering Cylinders for cards and emery cylinders'.[70] Five recipes (for milling or fulling cloth; for using the 'washer machine'; for scouring cloth in the grease; for raising the nap; and for finishing blue cloth) came from the Stroud area.[71]

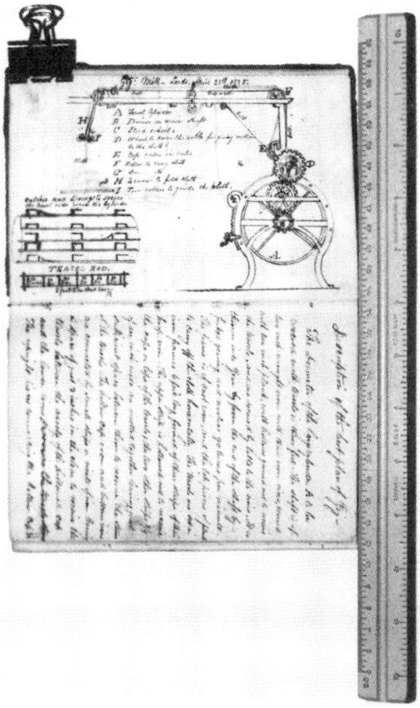

FIG. 4. Pages from Zachariah Allen's 'Journal of European Trip, 1825', showing his elevation of a gig mill which he examined at Huddersfield. Compare with illustration 5.

Source: Rhode Island Historical Society, Zachariah Allen Papers. Photograph by the author.

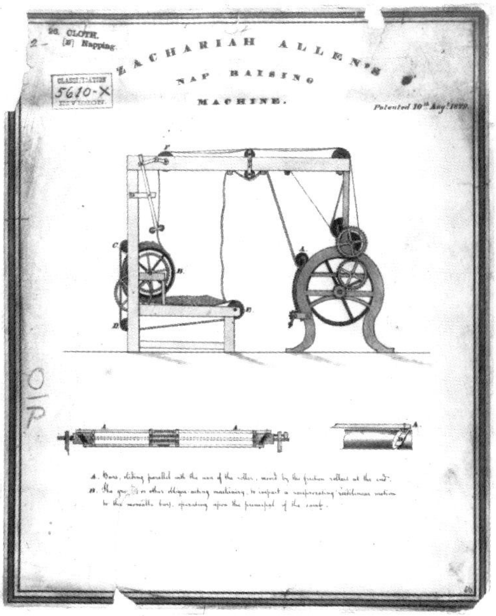

FIG. 5. Zachariah Allen's United States' patent for an improvement in the gig mill: a grooved cam imparting lateral movement to the bars holding the teasels. Compare with illustration 4, the Huddersfield gig, to which Allen added the lateral movement. Compare with illustration 4.

Source: US National Archives, RG 241, Patent no 5610, drawings, 10 August 1829.

Given a chance, Allen added complementary and portable objects and items to his notes and drawings. 'The shape of the French teasels [set in the gig mill they brushed the cloth nap] is uncommonly fine. We procured some seed both at Paris and at Louviers', he recorded.[72] Into his baggage too went the occasional small machine component. One belonged to the gig mill he had sketched in Leeds: 'Each workman has a key of different shapes to take off and put on the iron teasel frames upon the cylinder. I have exact iron patterns of the frames, keys &c to take to America'.[73]

Lastly, in this survey of the techniques of one American industrial spy, it is evident that Allen could rely on indirect sources of information. Three kinds were mentioned in his pocket diary. One comprised public sources like newspapers, which spasmodically reported wage rates and, more regularly, cloth prices.[74] Another was a manufacturer's supplier, particularly his machine maker or implement maker. The Leeds reed maker who extracted information from Hirst's head weaver for Allen's use, was one example. There were others. While at Leeds, Allen 'went with Mr Schenck to see some carding machines, which the machine maker who made them informed me I might examine. Calling there the day after to look at them more closely the owner, a rough looking Yorkshire man became very jealous, and accused us of wishing to get away his trade and at last vented his choler on the machine maker, declaring he should never come into his premises again'.[75] Allen proceeded to note diameters, breadths, and speeds of the card rollers and the prices of breaker and finisher cards ($450 and $250 respectively). Allen's tactic of exploiting the divergent market interests of manufacturers and machine makers was of course only possible when an industry had developed specialist sectors as a result of growing divisions of labour. Then and since, approaches to the suppliers of a target have been a classic weapon in the armoury of the industrial spy.[76]

The third indirect source of information came to Allen from employees, with some prompting from the American manufacturer, as one entry in his diary disclosed:

> Stroud. Cains Cross inn. July 26th 1825. In the evening two clothiers called to see us from Lewis mill to whom we had spoken in their mill upon the subject of manufacturing. Having promised them a pound or about $5 if they would give us the best information in their power upon the different branches with which they were acquainted, they commenced, after drinking a pretty good stock of beer and brandy as a preface.'[77]

There followed the five recipes, the value of which could only be assessed by a practical manufacturer. Allen was sufficiently convinced of his informers' competence, integrity, and degree of sobriety to record their information.

Networks of distribution

For Zachariah Allen the problems of knowledge distribution were minimal because the information he chose to collect was for his own use. He was not acting as an agent for an American manufacturer. He was not trying to export machinery to the USA. He was not apparently searching for skilled workers to join his manufacturing village at Allendale, Rhode Island. He was his own industrial spy and the knowledge he selected was stored in his notebooks, his memory and in a few items in his baggage. Once he was back home his knowledge was able to be safely distributed chiefly to the end user, himself. In the event he made much of it public in the USA by patenting two devices (the gig mill in 1829, see Fig. 5; and a steam press in 1830)[78] and by publishing a more literary and less technical version of his diaries.

Far different were the problems of those American manfacturers trying to arrange the recruitment of British workers or the importation of British machinery. Only the slightest traces of their acquisition networks under Britain's prohibitory laws have emerged in surviving papers but even in the period of legal emigration (after 1824) the problems of finding well motivated, appropriately skilled workers were very considerable. During the 1820s American textile industries faced growing demand, fuelled by population growth and western settlement. To meet it, production was considerably expanded. Over 350 new American cotton mills or firms and 270 woollen mills or firms were started between 1820 and 1831. The workforce in the US cotton industry rose from 12,000 to 62,000 over this period and the US woollen industry from 3,500 to 10,500.[79] A proportion of these new positions was filled by new immigrants from Britain. Some firms actively sought key workers in England or Scotland.

This was the practice of the larger New England cotton companies, with treasurers and directors drawn from Boston's mercantile and financial community. The Dover Manufacturing Company of Dover, New Hampshire, for instance, had the services of an agent in England named Swan who proved valuable in the mid-1820s when the Dover management competed against managements at Lowell and Taunton, Massachusetts, in the race to capture the latest British calico printing technology.[80] In the correspondence between the Dover agent (factory manager) in New Hampshire and his treasurer (executive director) in Boston there are hints of considerable reliance on Swan. On 31 March 1825, the agent told the treasurer 'am glad we can possess ourselves of the English mode of sizing yarns. Hope it can be easily applied to our mode of Dressing. Shall shortly have your copies of letters from Mr Swan which will tell us all'.[81] A couple of months later there is allusion to 'Mr Swan's contract with the Englishmen'.[82] Swan's recruits included handloom weavers and a mule spinner whose habits and lack of managerial skills drove the agent to dismiss him within six months, the spinner's fellow immigrants clubbing together to pay his fare back to England. The agent paid the mule spinner's wages and 'gave him a present of $10 with which he appears well satisified with the Company, which I considered of some importance too, as we may want to get more workpeople from England which his spleen and report might possibly have affected'.[83] To secure the company's reputation and the credibility of their English recruiter, the company agent obtained a signed statement from three of the spinner's fellow immigrants stating that 'the said company have in all respects complied with the engagements of Mr Swan so far as we are concerned & that Mr Leard [the spinner] was a man who from habits of intemperance & ignorance of manufacturing except mule spinning was not calculated to promote the interest of any manufacturing company'.[84]

Looking back in 1827 over the problems of two years of trying to operate vertically integrated cotton manufacturing at Dover, the agent analysed the cause of impeded expansion in a long and informative letter to his Boston treasurer. One cause was the defects of the skilled workers sent out from England:

> Of the six sent out by Mr Swann, the spinner was a worthless wretch, who had neither character nor knowledge of his profession except on mule spinning; the carder was well versed in part of his business but knew nothing of double speeders or stretchers, when not in such a state of intoxication as not to know the difference between day and night, which was not often the case, and was altogether so useless that his mysterious disappearance was a relief to us. The dresser or sizer had been accustomed to work on a frame as one of our females do here, of course could 'handle yarn' well, and with the experience he has gained here, I think him qualified to take charge of the dressing room in No 4 [mill] when it goes into operation. Of the three weavers who had been overlookers in England and who were qualified to become immediately overseers with us, it may be said they understood individually on handlooms and their overlooking amounted simply to their having had charge of three of four looms in their houses or cellars. They had no knowledge of the discipline of a room, keeping the accounts or anything connected with it and not one of them would have been competent to take charge.[85]

(Two eventually became second overseers).

The point the Dover agent was making was that the English workers were largely inappropriate for his American needs. If their intemperance did not conflict with his industrial discipline, the puny scale of their workshop overlooking experience made them wholly unfitted for the factory management positions they had been recruited to fill. Plainly had the Dover agent been able to tour Lancashire as Zachariah Allen did, he would have diminished the blunders of Swan the recruiter. That the agent was not sent was a reflection of perceived priorities between organizing the expansion of production at Dover and the acquisition of new technology and skilled workers in distant England. When the firm at last decided to send one of their own people to Britain for these purposes, they chose a local * academic. Dr Arthur C. Porter, a professor of chemistry at the University of Dartmouth, was also an Edinburgh University graduate, so he promised well. However, Porter proved little more successful than Swan in distinguishing men with both managerial skills and respect for the work ethic. Peter Bogle, from Chorley, Lancashire, who was hired as printery superintendent at a salary of £300 a year and was enthusiastically greeted on reaching Boston in March 1827, soon dashed Yankee hopes. The problem lay in his excessive secretiveness, which surprised even fellow workers. His refusal to explain the best method of purifying zinc sulphate led to several thousand yards of calico being more or less damaged. In autumn 1827 Porter told the Dover agent, 'Mr Bogles services are I consider not only of no use in the concern but absolutely highly pernicious to your interest'.[86] Bogle was dismissed but a short time later died of cholera before he could move on.[87]

The regionalized technical traditions found in Britain also caused some confusion in the USA. William Partridge, a Gloucestershire dyer who emigrated to the USA in 1808, discovered that biers (reed units) formed the basis of one New Jersey factory's yarn numbering system: 'so preposterous a mode of doing it', he thought.[88] The mix up he attributed partly to the managers' and workers' ignorance, partly to the admixture of workers from Britain's three main woollen regions, the West of England, Yorkshire and Ireland. If the living instruments of trade and information transfer were so unpredictable and unsuited to American circumstances in the 1820s and 1830s, were networks of distribution any more successful when it came to the dead instruments?

In evaluating machinery smuggling to the USA in the early nineteenth century, British efforts to check these illegal exports, reveal something of the strengths, objects and networks of American endeavours. British controls over the machinery export business had very mixed results. Between 1825 and 1843 the Board of Trade received 2,098 applications for licences to export machinery. Of these 218 (or 10 per cent) were refused. Of all applications, 1,409 (or 67 per cent) related to textile machinery. Of applications refused, 92 per cent covered textile equipment. In other words foreign customers were most interested in Britain's new textile technology and heavily interested in spinning, weaving and finishing machines nearly all of which, by the evolving definitions made on the advice of British manufacturers, could not be legally exported. (Prohibited textile equipment came to be defined as any machinery used in carding or in any subsequent processing stages, though after 1830 calico printing machines were licenced and in 1842 the law was massively relaxed). Ironically, machine tools were exported in quantity: between 1830 and 1840 over 1,300 metal working rollers, 41 lathes, and 20 cutting engines went abroad.[89]

The American interest in legally exported machinery was much slighter than that evinced in Europe: less than 8 per cent of the 2,098 applications to export machinery made in the years 1825–43 cited the USA as a destination. It would be possible to export equipment to Europe and thence the USA, but this would only increase costs and risks of loss through interception, without obvious advantage: British prohibitions did not vary between destinations apart from colonial ones (for which some few exceptions were made). Of the

158 applications to export machinery to the USA some 99 (or 63 per cent) cited textile machinery, slightly less than the proportion for all destinations. A lower percentage of American refusals involved textile machinery (80 per cent) than did all refusals (92 per cent). However, a slightly higher proportion of total applications for the USA were refused: nearly 13 per cent, as opposed to 10 per cent refusals of all applications. The general picture is that Americans showed much less overall interest in the machinery legally allowed out of Britain, and less interest in 'high tech' textile machinery, than European states. Possibly Americans more quickly learned how the system worked and therefore only expended time and money on legal paths which they knew would succeed, otherwise choosing smuggling. Alternatively, Americans were simply more self-sufficient in their needs for textile technology by the mid-1820s. This is suggested by the strength of their machine-making resources, a substantial slice of these coming from the immigrant machine makers who had established themselves in the USA before 1820.[90] So talented were some that their inventions were taken back to Britain and adopted here.[91] A measure of American self-sufficiency is also suggested by the American focus on what amounted to the latest British technology, in textile production and in machine tools. Judging from the register of applications after 1825 Americans most desired calico printing equipment, especially multi-colour printing machines, engraving machines, and engraved or plain shells (rollers) for use in printing machines. This was hardly surprising since the period of licensed machinery exported coincided with American cotton manufacturers' move into calico printing.[92]

Within the framework of the law, foreign customers and their agents, apparently, rapidly ascended a learning curve, as Table 1 shows. Foreign customers for textile machinery shifted their distribution networks from London to the provinces. Between 1826 and 1843

TABLE 1. APPLICANTS AND APPLICATIONS TO EXPORT TEXTILE MACHINERY, 1826–42

Applicants						Applications			Average Applications Per Applicant		
	1 Total No.	2 London Men	3 London Middle-men	Unknown	T/LM	4 Total No.	5 Via London Men	6 Via London Middle-men	C4/C1	C2/C5	C3/C6
1826	9	7	4	–	77.8%	12	10	7	1.33	1.43	1.75
1830	20	12	9	4	60.0%	34	23	20	1.70	1.92	2.22
1834	43	24	17	12	55.8%	79	42	38	1.84	1.75	2.24
1838	70	27	23	29	38.6%	148	79	73	2.11	2.93	3.17
1842–(1)	55	21	17	22	38.2%	73	33	28	1.33	1.57	1.65
1842–(2)	64	21	18	30	32.8%	117	37	33	1.83	1.76	1.83

SOURCES
PRO, BT 6/151, 152 (Machinery Books, 1825–43).
London directories.
NOTE. The licencing regime was relaxed on 25 August 1842. Hence the figures for that year are divided at that date (1842–(1) and 1842–(2)).

the London share of all applicants for export licences fell from 78 to 33 per cent. By 1838 over half the applicants came from outside London. Among the London men the proportion of middlemen rose from nearly 60 per cent to over 70 per cent within eight years. This suggests that the trade rapidly fell into the hands of specialist merchants, insurance brokers, ship brokers, ship agents, ship chandlers, shipowners, commercial agents, warehouse operators, wharfingers, wholesale and packing ironmongers, but mostly merchants — the occupations given by applicants for export licences (and identified from the London directories).[93] See Table 1. However, machine makers and especially those in the provinces were by the late 1830s — early 1840s responsible for a growing number of applications. Of 25 applications from machine makers in 1842 only two came from London men. Three came from Glasgow, three from Sheffield, one from Birmingham and 12 from Lancashire (one from John Hetherington & Co. of Manchester; one from Hibbert & Platt of Oldham; three from Higgins & Son of Manchester; and seven from Curtis & Parr & Madeley, also of Manchester). Various dodges were evidently tried. John Jones, the Leeds brush machine maker whom Zachariah Allen visited in 1825, busied himself with one. He made both direct and indirect applications for machinery export licences. In 1827, 1828 and 1834 he himself applied to the Board of Trade for licences. In 1829, 1830 and 1834 he made application through W. Sanders, merchant of Mincing Lane, London. At any rate the export licensing system was moving into the provinces and to some extent into the hands of machine makers.[94]

While applicants for export licences increasingly came from the provinces, evidence shows that 40 per cent of machinery destined for illegal export (well over half of it textile machinery) was seized in the port of London. In the ten years, 1830–39, Customs made 289 seizures of machinery, as can be seen from Table 2. Of these 289, some 206 involved textile machinery and another 18 seizures (10 at Liverpool, 8 at London) involved equipment unidentifiable because it had been dismantled. Distribution networks for American industrial spies trying to get machinery across the Atlantic illegally plainly existed, for equipment was moving from machine shops to dockside warehouses or yards in order to merit Customs' seizure. Such furtive networks meshed together in the ports of Dover, Hull and nearby Goole, Liverpool and London. A glimpse of the twisting routes taken by machinery destined for illegal export comes in two letters from Kirk Boott, agent of the

TABLE 2. CUSTOMS' SEIZURES OF MACHINERY DESTINED FOR ILLEGAL EXPORT, 1830–39

Port	1830	1831	1832	1833	1834	1835	1836	1837	1838	1839	TOTAL
Dover	5	–	7	2	3	–	–	1	–	5	23
Hull and Goole	4	2	3	7	28	5	3	2	8	9	71
Liverpool	2	7	13	5	2	8	2	5	6	1	51
London	6	3	4	5	25	17	19	16	14	12	121
18 Other Ports	1	1	3	4	1	2	–	1	8	2	23
TOTALS	18	13	30	23	59	32	24	25	36	29	289

Source: Customs House Archives, Customs 37/56 'Seizures, 1825–56'

Merrimack Manufacturing Co. of Lowell, to his directors, while he was on a machinery shopping expedition to Britain in the winter of 1826–27. Accompanied by John Dynley Prince (1779–1860) a highly skilled printworks manager who had been recruited to the company in 1825 at a salary of £1,000 a year, Boott purchased an engraving machine made by John Worswick of Manchester and costing £500. In January 1827 Boott described his plan to smuggle it out of England. It was 'to get the cases containing the machine, secretly conveyed from the maker's to a friend's warehouse in Manchester, who will immediately send them to another friend's in Birmingham, where I will have them repacked and then forwarded to Liverpool as "hardware"'. In February he wrote, 'I have heard of a Jew-broker in Liverpool, who will ship and guarantee for fifteen or twenty per cent, and I am not sure I shall not conclude to pay this insurance'. By mid-February the engraving machine, dismantled and cased, was in Birmingham. Presumably it reached Lowell, Massachusetts, as safely as Porter's machine purchases arrived at Dover, New Hampshire.[95]

Costs and benefits

The impression left by contemporary evidence is that Britain's protectionist laws against the outflow of men and machines failed to deter the efforts of industrial spies. On the other hand these laws raised the costs of industrial espionage. The risk of Customs seizure inflated the cost of freight insurance. In 1825 smugglers' insurance premiums were reportedly 30 to 45 per cent of the value of the equipment; by 1841 they were 10 to 25 per cent for small machine parts.[96] From the perspective of the foreign customer this was not much of a deterrent. In the long run it could well be worth paying this excess if the equipment were entirely new, for it could then be replicated ('reverse engineered' to use modern jargon) to introduce some new product or processing technique, or adapted to reduce American factor costs.

Illegal emigrants incurred the personal costs of anxiety about breaking the law and, when caught, imprisonment. Writing from Delaware in 1821 an English handloom weaver (who had emigrated alone in 1817) asked his brother living near Shrewsbury to 'write to my son John [in Manchester] to make a bag and put his shuttles and bobbins in, and put them in his bed till he gets out to sea. If he has a side engine [drop box, presumably] he may take it and conceal it under his berth. Tell him to keep on good terms with the Customs House Officer. But if you are in any way afraid to risk it, do not attempt to do it'.[97]

That such fears were not imaginary was illustrated by the experience of two Manchester cotton spinners in 1811. Early in March that year a Liverpool Customs officer, James Kipping, took the Customs cutter *Viper* across the Mersey to the American ship *Union* then waiting to be cleared for New York. He mustered all the passengers and examined their baggage. In a chest belonging to David Wooding he found a marriage certificate identifying Wooding as a spinner from Manchester. The next day the Liverpool magistrates issued a warrant for Wooding's arrest. Wooding admitted that his father-in-law, Dennis Manion, another cotton spinner, was also on the *Union*. The pair were detained in Liverpool Bridewell and in subsequent hearings it emerged that they had been employed for the past five years by McConnel and Kennedy, one of the largest fine spinning firms in Manchester.[98] Exactly how long the pair languished in prison or how much they were fined is unknown. That they paid some penalty for their offence is almost certain. That the laws leaked is even more evident from the subsequent progress of the two spinners. Manion next turns up as a spinner working in the Globe Mill, Philadelphia, in September 1814, at which date he recorded that he had been in the USA for three years and three months, i.e. he had reached the USA in June 1811, three months after being arrested by the Liverpool magistrates.[99]

Against the risks of fines and imprisonment for illegal emigrants before 1824 and the costs of substantial insurance machinery premiums on smuggled machinery before 1843, there were overwhelming benefits awaiting industrial espionage. In the case of the USA, immigration (some of it illegal) from early industrial Britain could have provided between 25 and 50 per cent of the rising demand for experienced textile workers in the 1820s. It would seem that managers (or self-proclaimed managers) from Britain's textile trades could have supplied 46 per cent of the new cotton factories that opened in the USA in the 1820s, and 70 per cent of those in the new American woollen mills of the period. On the machine-making side, experienced immigrant machine makers from Britain could have met 41 per cent of the 1820s expansion in the American workforce of machine makers.[100] These are probably optimistic estimates. They are still remarkable.

When tracing individual achievements, it is clear that industrial espionage conducted by Americans in Britain secured major triumphs. Francis Cabot Lowell, visiting Britain in 1810–11, obtained designs for the first American powerloom.[101] Timothy Wiggin, another Boston merchant, in 1822–25, recruited John Dynley Prince, a leading Manchester printworks manager, who became responsible for building up one of the largest and most advanced printworks in the USA, the Merrimack Printworks at Lowell, Massachusetts.[102] In the 1830s American recruiters made one of their biggest catches when they attracted James Montgomery from Glasgow to Saco in Maine. Through his publications, through his practice as a cotton mill manager, and through his migrations from New England to New York and then South Carolina, Montgomery contributed to and was part of the diffusion of textile technology across the Atlantic and then from New England to the American South.[103] Zachariah Allen's diaries tell us in fullest detail how some of this industrial espionage was done.

Conclusion

Despite the formidable obstacles facing American manufacturers, they patently succeeded in tapping Britain's expanding stock of industrial technology. Once peacetime conditions returned in 1815 they broke the defences raised by industrial firms and the state in British factories, ports and patent offices, aided in their efforts by the net outflow of industrial workers. The most efficient way of securing Britain's latest industrial secrets was for an American manufacturer to visit Britain personally. Securing new knowledge (however embodied) and distributing it back to the USA required the support of networks of contacts. These might be American consuls, expatriate Americans in Britain, the British-located relatives of immigrants in America, or middlemen operating in a web of commercial relations in the major ports and industrial centres. The devious routes by which equipment went illegally to the USA are reminiscent of the high tech industrial espionage of the 1980s, though this had a military context. The difficulties experienced by Bulgarians trying to operate their illicitly-acquired computers echoed the trials of American managers when their immigrant recruits proved non-communicative or accustomed to a different work discipline or technical tradition. The lesson of this article is that industrial espionage, if ever morally or commercially justifiable, is not a simple matter of sending a machine, a formula, or even a skilled worker from country A to country B. Technology transfer, for that is what industrial espionage seeks to achieve, is a much larger and more complex process demanding a range of technical, economic and cultural preconditions in the recipient state as well as carriers appropriate to the technology concerned.

References

[1] This article was originally drafted for presentation at a workshop on 'Industrial Espionage' organised by Professor John R. Harris and myself at the Ninth International Economic History Congress in Berne, Switzerland, 26 August 1986. My thanks are due to him and, most recently, to Dr Gillian Cookson for commenting on the manuscript.

[2] Jay Tuck, *High Tech Espionage: How the KGB Smuggles NATO's Strategic Secrets to Moscow* (London: Sidgwick & Jackson, 1986), pp. 176–77.

[3] Zachariah Allen, *The Practical Tourist*, 2 vols (Providence RI, 1832), I, pp. 5–6.

[4] John R. Harris, *Essays in Industry and Technology in the Eighteenth Century: England and France* (Aldershot: Variorum, 1992); A. E. Musson and Eric Robinson, *Science and Technology in the Industrial Revolution* (Manchester: Manchester University Press, 1969); W. O. Henderson, *Britain and Industrial Europe, 1750–1870*, 3rd edition (Leicester: Leicester University Press, 1972).

[5] David J. Jeremy, *Transatlantic Industrial Revolutions: The Diffusion of Textile Technologies between Britain and America, 1790–1830s* (Cambridge, MA: MIT Press, 1981), p. 149. Hereafter cited as *TIR*. See also Charlotte Erickson, *Leaving England, Essays on British Emigration in the Nineteenth Century* (Ithaca NY: Cornell University Press, 1994).

[6] US National Archives, RG 59, War of 1812 Papers, microfilm M–588, roll 1, W. Jones to General John Mason, Commissioner General of Prisoners, 31 August 1813.

[7] Ibid., anonymous letter from New York to James Madison, 13 January 1813.

[8] Public Record Office, BT 5/21, p. 70, minutes of 4 December 1811.

[9] David J. Jeremy (ed.), *Henry Wansey and His American Journal, 1794* (Philadelphia: American Philosophical Society, 1970), p. 146; Howard Robinson, *Carrying British Mails Overseas* (London: Allen & Unwin, 1964), p. 115. For the Black Ball Line: Robert G. Albion, *The Rise of New York Port (1815–1860)* (New York, 1939; reprinted Newton Abbot: David & Charles, 1970), pp. 38–43.

[10] Allen, *Practical Tourist*, II, p. 421.

[11] R. J. Dickson, *Ulster Emigration to Colonial America, 1718–1775* (London: Routledge, 1966), p. 87; Allen, *Practical Tourist*, II, pp. 409–10.

[12] For descriptions of these districts see the accounts of travellers seeking industrial information e.g. W. O. Henderson (ed.), *Industrial Britain under the Regency: The Diaries of Escher, Bodmer, May and de Gallois, 1814–1818* (London: Cass, 1968), and Allen, *Practical Tourist.*

[13] Robert Owen, *The Life of Robert Owen, Written by Himself*, 2 vols (London: Wilson, 1857), I, p. 31; *TIR*, pp. 36–39.

[14] Rhode Island Historical Society, Providence, RI, Zachariah Allen Papers, Allen's 'Journal of European Trip, 1825', p. 126. This notebook is unpaginated but I have added page numbers to my xerox copy (pp. 1–155) and these are used to locate quotations from this source in the rest of this article. I am grateful to the RIHS for permission to quote from this source.

[15] David J. Jeremy, *Technology and Power in the Early American Cotton Industry: James Montgomery, the Second Edition of His 'Cotton Manufacture' (1840), and the 'Justitia' Controversy about Relative Power Costs* (Philadelphia: American Philosophical Society, 1990), p. 18. In the highly competitive world of early

industrial textile manufacturing it was not only knowledge of hardware that was sought, nor was theft of intellectual property confined to foreigners. Before the Design Act of 1842, British calico printers cheerfully pillaged one another's designs. See David Greysmith, 'Patterns, Piracy and Protection in the Textile Printing Industry, 1787–1850' *Textile History*, 14, no. 2 (Autumn 1983).

[16] *TIR*, 68–69; Negley B. Harte, 'On Rees's *Cyclopaedia* as a Source for the History of the Textile Industries in the Early Nineteenth Century', *Textile History*, 5 (1974); Jeremy, *Technology and Power*, pp. 6–12.

[17] *TIR*, pp. 36–39.

[18] John A. Cantrell, *James Nasmyth and the Bridgewater Foundry. A Study in Entrepreneurship in the Early Engineering Industry* (Manchester: Chetham Society, 1984), pp. 44, 136–39.

[19] E. Kilburn Scott (ed.), *Matthew Murray: Pioneer Engineer. Records from 1765 to 1826* (Leeds: Edwin Jowett, 1928), pp. 11–12, 37–42, 49, 80; Jennifer Tann, 'Marketing Methods in the International Steam Engine Market . . .' *Journal of Economic History*, 38 (1978), p. 378. I am obliged to Gillian Gookson for these references.

[20] Christine MacLeod, 'Strategies for Innovation: The Diffusion of New Technology in Nineteenth-Century British Industry', *Economic History Review*, 45 (1992).

[21] David J. Jeremy, 'Damming the Flood: British Government Efforts to Check the Outflow of Technicians and Machinery, 1780–1843', *Business History Review*, 51 (1977). For the origins of the laws see John R. Harris, 'The First British Measures against Industrial Espionage' in Ian Blanchard, Anthony Goodman and Jennifer Newman (eds), *Industry and Finance in Early Modern History. Essays Presented to George Hammersley on the Occasion of His 74th Birthday* (Stuttgart: Franz Steiner Verlag, 1992).

[22] 'Journal of John Smith (1796–1886). From November 28, 1826 to March 9, 1827', *Essex Institute Historical Collections*, 106 (1970), p. 101.

[23] Ibid., p. 102.

[24] Bruce Sinclair, 'At the Turn of a Screw: William Sellers, the Franklin Institute, and a Standard American Thread', *Technology and Culture*, 10 (1969).

[25] *TIR*, pp. 50–51; William Partridge, *A Practical Treatise on Dy[e]ing* (New York: H. Walker, 1823; reprinted Edington, Wilts: Pasold Research Fund, 1973), pp. 19–20, 46–47; David J. Jeremy, 'British and American Yarn Count Systems: An Historical Analysis', *Business History Review*, 45 (1971).

[26] G. B., *Parliamentary Papers (House of Commons)*, 1820 (314), VII, 'Second Report of the Commissioners . . . [on] Weights and Measures', Appendix A.

[27] H. I. Dutton, *The Patent System and Inventive Activity during the Industrial Revolution, 1750–1852* (Manchester: Manchester University Press, 1984), p. 2.

[28] *TIR*, p. 55.

[29] Dr Cookson's view, that 'the patent system was hardly used for textile innovations' flies in the face of the evidence if *major* modifications only are considered. See her 'Co-operation or Competition? Innovation in the Yorkshire Textile Industry, 1780–1850', paper given at the Economic History Society Conference, April 1994, p. 4; and her 'Innovation, Diffusion, and Mechanical Engineers in Britain, 1780–1850', *Economic History Review*, 47, no. 4 (November 1994), p. 751. I have not seen her recently-completed dissertation, 'The West Yorkshire Textile Engineering Industry, 1780–1850' (York DPhil 1994).

[30] GB, *Parliamentary Papers (Commons)*, 1829 (332), 3, 'Reports from the Select Committee on the Law Relative to Patents for Inventions'.

[31] *TIR*, pp. 45–49.

[32] This paragraph is based on a reading of Tuck, *High Tech Espionage*.

[33] Ibid., p. 66.

[34] Ibid.

[35] Jeremy, *Technology and Power*, pp. 13–15.

[36] *American Museum*, 2 (1787), p. 253. For the background of intense transatlantic industrial espionage in the 1790s see David J. Jeremy, 'British Textile Technology Transmission to the United States: The Philadelphia Region Experience, 1770–1820', *Business History Review*, 47 (1973); and Anthony F. C. Wallace and David J. Jeremy, 'William Pollard and the Arkwright Patents', *William and Mary Quarterly*, 3rd ser., 34 (1977).

[37] For Allen's biography see the *Dictionary of American Biography* and J. D. Van Slyck, *Representatives of New England Manufacturers* (2 vols, Boston: Van Slyck & Co., 1879), 1, pp. 1–15.

[38] US Congress, *Documents Relative to the Manufactures of the United States* Executive Document No. 308, 1st Session, 2nd Congress (2 vols., Washington, D.C.; Duff Green, 1833), 1, pp. 951–52, 976. This is known as the *McLane Report*.
[39] This was the number I have found in US Passenger Lists and in miscellaneous sources.
[40] Allen, *Practical Tourist*, 1, p. 5..
[41] Allen, 'Journal', p. 1 (see note 14).
[42] Henderson, *Industrial Britain under the Regency*, pp. 7–16, 75–111.
[43] 'Journal of John Smith', p. 96.
[44] Allen, 'Journal', pp. 9–10.
[45] Ibid., p. 114.
[46] US National Archives, RG 29, MS Returns of the 1820 Census of Manufactures, microfilm M–279, roll 10, entry 1, 425; *TIR*, pp. 238, 378.
[47] Allen, 'Journal', pp. 41–67. For Gott see William B. Crump (ed.), *The Leeds Woollen Industry* (Leeds: Thoresby Society, 1931). For Marshall see W. G. Rimmer, *Marshall's of Leeds, Flax-Spinners, 1788–1886* (Cambridge: Cambridge University Press, 1960); for the Davys see National Archives, RG 59, Consular Despatches, microfilm T–474, roll 1.
[48] Rhode Island Historical Society, Zachariah Allen papers, 'Journal, European Trip, February – July 1825' (4 vols., photocopy), 1, p. 96. For Whittemore and his card clothing machine see *TIR*, passim., and Henry G. Kittridge and A. C. Gould, *History of the American Card-Clothing Industry* (Worcester, MA: T. K. Earle Manufacturing Co., 1886).
[49] For Allen's visits to Americans in Manchester see Allen, *Practical Tourist*, pp. 170–72. For Americans in England see *TIR*, pp. 240–51.
[50] Allen, *Practical Tourist*, 1, p. 49. For the debate see Jeremy, *Technology and Power*, pp. 29–35.
[51] Allen, *Practical Tourist*, pp. 159, 177, 214, 348. For the diverging business cultures of Britain and USA at this time see David J. Jeremy, 'British and American Entrepreneurial Values in the Early Nineteenth Century: A Parting of the Ways?', in Robert A. Burchell (ed.), *The End of Anglo-America. Historical Essays in the Study of Cultural Divergence* (Manchester: Manchester University Press, 1991).
[52] Allen, 'Journal', p. 116.
[53] Ibid., pp. 127–28. For Stanley Mills see Jennifer Tann, *Gloucestershire Woollen Mills* (Newton Abbot: David & Charles, 1967), pp. 149–51. The cloth mill was built on the site of an old grist mill in 1813 by George Daniel Harris and others; one of his later partners was Charles Stephens.
[54] Allen, 'Journal', p. 27; GB Patent No. 4897 (27 January 1824).
[55] Allen, 'Journal', pp. 28–29.
[56] *TIR*, p. 229.
[57] Allen, 'Journal', p. 29.
[58] Ibid., p. 30.
[59] Ibid., p. 31.
[60] *TIR*, p. 57.
[61] Julia de L. Mann, *The Cloth Industry in the West of England from 1660 to 1880* (Oxford: Clarendon Press, 1971), pp. 303–4.
[62] Allen, 'Journal', pp. 71–72.
[63] Ibid., p. 31.
[64] Ibid., pp. 52–63.
[65] Ibid., pp. 13, 19.
[66] Ibid., pp. 18–20.
[67] Ibid., pp. 33–34.
[68] Ibid., p. 53.
[69] Ibid., p. 51.
[70] Ibid., p. 154.
[71] Ibid., pp. 134–40.
[72] Ibid., p. 103.
[73] Ibid., p. 40.
[74] Ibid., p. 113.
[75] Ibid., p. 72.

[76] See, for example, Peter Hamilton, *Espionage, Terrorism and Subversion: An Examination and a Philosophy of Defence for Management* (Leatherhead: Heims, 1979), pp. 81, 271.
[77] Allen, 'Journal', p. 134.
[78] US Patents dated 10 August, 1829 and 2 February, 1830.
[79] *TIR*, pp. 161–63.
[80] Ibid., pp. 104–15.
[81] New Hampshire Historical Society, Concord, NH, Dover Manufacturing Company, Agent's Letterbooks (three spanning January 1825–September 1828), Matthew Bridge (assistant agent) to William Shimmin (treasurer), 31 March, 1825. I am grateful to the NHHS for permission to read and quote from these letterbooks.
[82] Ibid., same to same, 27 June, 1825.
[83] Ibid., John Williams (agent) to Samuel Torrey, 24 September, 1825.
[84] Ibid., Bridge to Torrey, 14 October, 1825.
[85] Ibid., Williams to Shimmin, 26 January, 1827.
[86] Ibid., Porter to Shimmin, 19 September, 1827.
[87] *TIR*, pp. 112, 170.
[88] Partridge, *Practical Treatise*, pp. 46–47.
[89] *TIR*, pp. 43–45; Jeremy, 'Damming the Flood'.
[90] See, for example, David J. Jeremy, 'Immigrant Textile Machine Makers along the Brandywine, 1810–1820', *Textile History*, 13, no. 2 (Autumn 1982).
[91] *TIR*, pp. 240–51; David J. Jeremy, 'Technological Diffusion: The Case of the Differential Gear', *Industrial Archaeology Review*, 5, No. 3 (Autumn 1981) for the remarkable coincidence of the simultaneous American and British re-invention of the differential gear for use in the roving frame and dynamometer respectively.
[92] *TIR*, pp. 104–17.
[93] It is possible that some London men were not listed in the Post Office directories, so this head count is not definitive.
[94] Public Record Office, BT6, vols 151, 152, Machinery Books (Register of applications to the Board of Trade to export machinery under licence, 1825–43).
[95] *TIR*, p. 112.
[96] *TIR*, p. 43.
[97] Charlotte Erickson, *Invisible Immigrants: The Adaptation of English and Scottish Immigrants in 19th Century America* (Coral Gables, Florida: University of Miami Press, 1972), p. 269.
[98] Public Record Office, BT 1/76, f. 36, James Kipping, Liverpool Customs Surveyor on Special Service, to the Collector of Liverpool, 8, 9, 12 March, 1811.
[99] *TIR*, p. 138.
[100] *TIR*, pp. 144–64.
[101] See Robert F. Dalzell, Jr., *Enterprising Elite: The Boston Associates and the World They Made* (Cambridge, MA: Harvard University Press, 1987), pp. 5–25.
[102] *TIR*, pp. 109–10.
[103] Jeremy, *Technology and Power*.

Errata

p. 114, line 6, Dr Arthur C. Porter *should read* Dr Arthur L. Porter

VIII

Invention in American textile technology during the early nineteenth century, 1790–1830

This is an attempt to measure inventive activity in one of the key industries of the early industrial period, textile manufacturing, in the U.S.A. If inventive activity represents efforts to meet economic demand with technical solutions on the supply side of the market, we should like to know more about its extent and the conditions under which it flourished. How did American activity compare with that of Britain, the first industrial nation? Did it change over time? How did it vary between the cotton and woolen branches of the textile industry and between regions in the U.S.A.? Who were the inventors in these periods and which regions attracted most effort?

The answers to these questions, insofar as they can be given, have been largely derived from analyses of British and American patent registrations, in which a number of limitations are inherent. Firstly, in looking at patents quantitatively, we are overlooking the qualitative commercial differences between one invention and another. However, numbers of patents are a measure of expectations of profit, and to that extent, in aggregate, reflect commercial reality, rather than mere eccentricity.[1] Secondly, not all inventions were patented; indeed patents are the tip of the iceberg of inventive creativity. Both in Britain and the U.S.A. in the early nineteenth century lots of examples of unpatented inventions can be cited;[2] good examples are found in the recently-discovered collection of the drawings of Isaac Markham

[1] The point is made by Jacob Schmookler. See, for example, his shorter work, 'Economic Sources of Inventive Activity,' in Nathan Rosenberg, ed., *The Economics of Technological Change: Selected Readings* (Harmondsworth, Eng.: Penguin Books, 1971), 117–36.

[2] Great Britain, *Parliamentary Papers (Commons)* 1829 (332), III, 'Report from the Select Committee on the Law Relative to Patents for Inventions,' 134–38, evidence of John Farey who reported that motivations for failing to patent included complacency, profitability (in case of medical cures) and fear of lawsuits. The chances of failing to keep an invention secret and the promise of profitability outweighed these motivations (ibid., 138–43). In the U.S.A. the dead spindle appears to have been one of the few important textile inventions not patented.

(1795–1825) of Middlebury, Vermont.[3] We assume that in most cases the inventions covered by a patent represented the improvement which the inventor believed held the greatest chance of technical and commercial success; conversely we assume that few major technical improvements after 1800 were not patented. On these premises rests the case for regarding patents as an index of inventive activity. Thirdly, the evidence for American patents before 1836 is hard to use to any extent because in that year the U.S. Patent Office burned down and, for the most part, only those patents still in force (i.e., less than 14 years old) were reconstructed. This leaves only the annual patent lists for the early years of the American federal patent system, from 1790 onwards; and these lists give merely the date, name, and address of the patentee, and the short title of the patent.[4] Frequently a title conceals the exact nature of the improvement, e.g., 'cloth dressing, sizing' could be regarded as a woolen finishing patent, but one with that title awarded to Paul Moody, January 17, 1818, covered sizing cotton warps. In grouping patents by title I have tried to reduce the chances of wrongly categorizing a patent by grouping 'non-specific' titles with those definitely belonging to cotton manufacturing, on the grounds that, in general, innovations with no fiber-identity in their titles (and applicable to processes common to wool and cotton) were more likely to apply to cotton than to woolen manufacturing. As a result of ambiguity in the titles, it becomes impossible to make the fundamental distinction between household and factory technology. Given the transition from domestic to factory technology in this period, this flaw in the evidence can only be regretted and certainly never amended. I have tried to reduce the incidence among the American patents of inventions intended for household use (which cannot appropriately be compared to the factory inventions that almost wholly filled the British textile manufacturing patents of the period) by confining American patents to those awarded to patentees in the eastern seaboard states between New Hampshire and Maryland. These states accounted for at least 90 percent of American cotton and woolen factory production before 1830. But the result of this discrimination is only an approximation of patented factory inventions because factory inventions by patentees in the western states are omitted and household inventions by patentees in the eastern states, chiefly New York, are included.

---

[3] Papers recently brought to my notice by the late Michael B. Folsom, Executive Secretary of Charles River Historic Industries, Waltham, Mass., and discovered by Polly C. Darnell of the Sheldon Art Museum, Middlebury, Vermont.

[4] United States, Patent Office, *List of Patents for Inventions and Designs, Issued by the United States, from 1790 to 1847*, comp. Edmund Burke (Washington, D.C.: J. & G. S. Gideon, 1847), 67–96. Patent specifications were frequently published in the *Journal of the Franklin Institute* but this did not appear until 1826.

Table 1

Investment in cotton manufacturing per head of population in Britain and the U.S.A.., 1790–1830/31

	1790	1830/31	Differential $\frac{(1830/31)}{(1790)}$
Britain			
$\frac{I}{P}$	$\$\ \frac{3,360,000}{14,531,000} = \$0.2312297$	$\$\ \frac{72,000,000}{23,875,000} = \$3.0233046$	13.074897
U.S.A.			
$\frac{I}{P}$	$\$\ \frac{72,000}{3,929,000} = \$\ 0.0183252$	$\$\ \frac{45,000,000}{12,901,000} = \$3.4881016$	190.34453
£ = \$4.80	I = Investment, P = Population		

Sources

Population

Britain (1790 and 1830) (1) Brian R. Mitchell and Phyllis Deane, *Abstract of British Historical Statistics* (Cambridge, Eng.: Cambridge University Press, 1962), 5–8. The 1790 figure is an underestimate, using Rickman's estimates for England and Wales

U.S.A. (1790 and 1830) (2) United States, Department of Commerce, Bureau of the Census, *Historical Statistics of the United States, Colonial Times to 1970* (2 vols., Washington, D.C.: Government Printing Office, 1976), 8.

Investment

Britain 1790: An estimate of £700,000 in fixed capital based on Colquhoun's estimate of £500,000 for 1787, quoted in S. D. Chapman, *The Cotton Industry in the Industrial Revolution* (London: Macmillan, 1972), 30.

1830: An estimate of £15 million fixed capital based on £14.8 million for 1834 by McCulloch, ibid., 31, but including fixed capital investment in finishing.

U.S.A. 1790: David J; Jeremy, *Transatlantic Industrial Revolution: The Diffusion of Textile Technologies between Britain and the U.S.A., 1790–1830* (Cambridge, Mass.: M.I.T. Press, 1981), 19.

1830: Friends of Domestic Industry, New York Convention, *Report on the Production and Manufacture of Cotton* (Boston: J. T. & E. Buckingham, 1832), 16.

A. Anglo-American comparisons

Capital Investment Rates

Between 1790 and 1830 more patents for mechanical improvements in cotton and woolen manufacturing were registered in the U.S.A.'s eastern seaboard states than in Britain: 433 as against 276.[5] Ignoring variations in the commercial value of the patents (which it is impossible to compute in any case), we might assume that inventive effort was 56 percent greater in the U.S.A. than in Britain over the period. This makes the American performance look especially favorable and for three reasons. Americans started from a much smaller manufacturing base in 1790, when only about ten mechanized cotton spinning establishments representing some $72,000 of capital investment were in operation (see Table 1). In contrast, Britain then had an industry nearly 47 times larger. Secondly, much American investment in the period would have been devoted simply to setting up factories in which British technology was copied. Thirdly, by 1790, the scope for making basic mechanical inventions appeared greatest at those stages of production where the preindustrial hand methods seemed most difficult to replicate with mechanical, powered equipment, namely cotton and woolen weaving, and cloth finishing. In other words, British inventors had already, by 1790, reached mechanical solutions for the relatively simpler, and more easily exploitable (commercially), preindustrial methods entailed in preparing and spinning cotton and woolen fibers. The difference can be seen by comparing the mechanical structure and dynamics of a waterframe or jenny with those of a powerloom of the 1820s.[6]

But was American inventive effort as great as it might have been? As Table 1 shows, investment in cotton manufacturing per capita of the population over the period rose 13 times in Britain but 190 times in the U.S.A. (these are very rough figures, of course). How much was invested in research and development is unknown. What is clear is that opportunities for inventors to market their ideas expanded nearly 15 times more rapidly in the U.S.A. than in Britain but produced only 56 per cent more patents. One reason for this is that approximately 20–22 per cent of American investment (i.e., that in New Hampshire and Massachusetts) in 1820 and 40 per cent in 1831 incorporated a technology derived from a handful of American patents. But if 40 per

---

[5] The British figures came from my *Transatlantic Industrial Revolution: The Diffusion of Textile Technologies between Britain and the U.S.A., 1790–1830s* (Cambridge, Mass.: M.I.T. Press, 1981), 55 (hereafter cited as *TIR*).

[6] See *TIR*, illustrations 1.1 and 1.5 compared to 3.8 which shows just one of half a dozen of the powerloom's movements, all of which had to be synchronized.

cent of American investment in cotton manufacturing (in 1831) is deducted from the figure of $45 million in Table 1, the per capita figure of $2.09 for 1831 gives an increase of 114 times or an expansion rate over the period 8 to 9 times greater than that in Britain. There are no comparable figures for the much more diffuse American woolen industry.

It is hard to escape the conclusion that, by a quantitative patent measure, Americans were relatively less energetic than British inventors in the cotton manufacturing industry of this early industrial period, and particularly in view of the respective institutional arrangements under which they operated.

Institutional Arrangements for Patenting

> A country without a patent office and good
> patent laws [is] just a crab, and couldn't
> travel any way but sideways or backwards.
> (Hank Morgan in Mark Twain's
> *Connecticut Yankee in King Arthur's Court*, 1889)

The American patent system, as established by the two acts of 1793 and 1800, in force until 1836, was far more efficient in encouraging the disclosure of inventions than the British one. Three features made the American system superior.[7] Firstly, it was vastly cheaper. Whereas a British patent cost in 1815 at least £70[8] and in 1829 at least £150 (legal expenses of £120; plus drafting fees ranging between 3 guineas and £200 but averaging £20), an American patent cost $30 or £6–15s with an extra charge of £4–10s if a lawyer were needed to help draft the initial petition. American patent costs represented at most a fifth of a skilled artisan's annual wage in the 1790s; British patent costs were three times those of a cotton factory overseer, whose earnings were around £50 per annum.

The democratic intentions of the American law extended also to the speed with which a patent could be secured. The payment of $30 to the Patent Office, a petition to the Secretary of State, and an affidavit swearing that the petitioner was the true inventor, together with the specification of the inven-

---

[7] Except where otherwise stated, this comparison of the British and American patent systems comes from *PP (Commons)*, op cit., which includes an appraisal of the American patent law by the American consul in London, Thomas Aspinwall. A fuller exposition of the case law affecting American patents was published in 1826–27 in the first four volumes of the *Journal of the Franklin Institute*, written in 18 parts by Peter A. Browne.

[8] W. O. Henderson, ed., *Industrial Britain under the Regency: The Diaries of Escher, Bodmer, May and de Gallois, 1814–1818* (London, Frank Cass & Co.: 1968), 164.

tion, could be presented and processed in the space of a half a day; though the law allowed up to 15 days for the Attorney General to assure the Secretary of State that the papers conformed to the law. In Britain it took at least six weeks (and by 1829 two to three months) for a patent application to pass through a tortuous process, with additional expenses at each stage, involving the Secretary of State, the Attorney or Solicitor General, the King, the Signet Office, the Privy Seal, and the Lord Chancellor. When the patent was issued the patentee then had six months to lodge his specification. And during this period of eight or nine months he had two great burdens: to ensure that his specification worked in conformity to the title which he initially submitted; and to maintain the strictest secrecy, for fear of piracy and to guard against publication which would *per se* destroy novelty and so vitiate the patent. Maintaining secrecy was especially difficult; consequently interference petitions from rivals were not uncommon during this procedural period, adding an element of uncertainty to the delay in securing a patent.

Thirdly, American patents, regarded by the law as their inventor's reward and his property, could be assigned without difficulty to heirs or transferred to any other party. In Britain patents could not be readily assigned to anyone else since they were not necessarily the invention of the patentee – but simply the patentee's reward for making the invention public. While a patentee in both countries could sell licenses to manufacture the patented invention, only the U.S.A. offered the added incentive to inventors of selling the patent itself.[9]

Thus on grounds of cost, speed, and security in obtaining a patent and then ease of selling the whole patent monopoly, the American system was far more efficient than the British one. And intentionally so for, guided by democratic principles, the framers of the American patent system deliberately excised the procedural excrescences of centuries of European practice. And yet the American system preserved two elements which may or may not have served to damp down patent registration. Firstly, the 1793 act abandoned the examination of patents for novelty which the 1790 Patent Act, like the French system of 1791, introduced. Since the earlier act left the examination entirely in the hands of an overworked Secretary of State, it proved to be a hindrance to the smooth running of the system. After 1793 the degree of novelty and, therefore, merit for the reward of a patent monopoly, was left for the federal courts to test, either by rule of court following an infringer's affidavit against the patentee or by a patentee's suit against an infringer. This was the same as

---

9 For a thorough examination of the development of the US patent system see Carolyn C. Cooper, *Shaping Invention: Thomas Blanchard's Machinery and Patent Management in Nineteenth Century America* (New York: Columbia University Press, 1991), a study which appeared since this present paper was published.

in England. Such a method for testing novelty could well deter potential patentees by injecting an element of uncertainty and possible additional expenses into a patentee's prospects even after a patent was secured. In the 1820s as much as £600 was reckoned necessary in Britain to defend a patent and at least one patentee in the 1820s lost £750 in unsuccessfully defending a textile manufacturing patent. In the United States it seems that the courts favored inventors pitted against manufacturing corporations.[10] Even so, Aza Arnold, a Rhode Island mechanic who sued the proprietors of the Locks and Canals (machine makers for the Lowell corporations) for stealing his roving frame differential, received only $3,500 of the $30,000 he claimed to have lost.[11] On the other hand, a system of simple registration without examination may well have appeared more inviting to potential patentees than one with examiners.

Secondly, at least one contemporary commentator argued that the American fee of $30 was too high. Inventors, he assumed, were poor men who had drained their resources in developing their inventions, and, therefore, ought to be able to register their inventions free of charge. After all, he (John Isaac Hawkins, an itinerant patent agent) argued, the 'men of capital' rather than the 'men of invention' made the fortune, if one was to be made, by what John Farey called 'the triple purpose of making, using and vending the new invention.'[12] If Hawkins was correct then the American system of confining the right to take out a patent to the inventor placed costs on his shoulders which the British system, awarding a patent to anyone who disclosed an invention, allowed the inventor to transfer to the capitalist. Whether this was in fact the case deserves further investigation which is not carried out in this paper.

At any rate, it does appear that, in general, American inventive activity was encouraged by a much higher rate of capital investment in textile manufacturing, 1790–1830, than in Britain. And yet ingenious Americans produced only 56 per cent more patents in this period for the comparable textile industries. Possibly the differential was due more to patenting against the odds in Britain, chiefly because, as John Farey, the engineer, recognized, Britain had the worst patent system in the world but the most 'powerful means of execution' (in which he included money, materials, tools, buildings, workmen, technical periodicals, a transport network, a technical language among workers, and promotional societies.[13] In other words, there were relatively more British patents taken out than the institutional disincentives should have

---

[10] *TIR*, 185–86.

[11] Ibid., 216.

[12] *PP (Commons)*, op. cit. 159.

[13] Ibid., 132.

Table 2

Inventive activity in cotton and woollen manufacturing (mechanical invention in production equipment, excluding machine-making inventions) in Britain and the U.S.A. (N.H. to MD. only) 1790–1830

	1790–1812		1813–1824		1825–1830		TOTALS	
	G.B.	U.S.A.	G.B.	U.S.A.	G.B.	U.S.A.	G.B.	U.S.A.
Cotton Mfg.*								
Preparatory	10	3	4	11	–	13	14	27
Spinning	28	24	19	87	24	60	71	171
Weaving	22	20	19	58	7	26.5	48	104.5
Finishing	17	1	11	1	7	4	35	6
General	–	2	–	1	–	1	–	4
TOTAL	77	50	53	158	38	104.5	168	312.5
Woolen Mfg.								
Preparatory	1	6	–	2	8	3	9	11
Spinning	2	2	3	7	1	6	6	15
Weaving	–	–	2	–	6	–	8	–
Finishing	16	27	38	34	31	18.5	85	79.5
General	–	1	–	3	–	11	–	15
TOTAL	19	36	43	46	46	38.5	108	120.5
GRAND TOTALS	96	86	96	204	84	143	276	433

Sources

Jeremy, *Transatlantic Industrial Revolution*, 55, for British patent figures. United States Patent Office, *List of Patents and Inventions and Designs, Issued by the United States from 1790 to 1847* (Washington, D.C.: J. & G. S. Gideon, 1847), Edmund Burke's *List.*

• Including U.S. patents with 'non-specific' titles.

allowed because of the belief that inventions in Britain stood a higher chance of reaping commercial success than elsewhere. Moving the Anglo-American comparison to a more specific level may help us to see where Americans were relatively slothful and Britons relatively active and vice versa.

British and American Patent Activity, 1790–1830

The first difference between British and American textile patenting activity to emerge from Table 2 is that cotton manufacturing, more attractive to inventors in both countries than woolen manufacturing, received relatively greater support in the U.S.A. (72 per cent of cotton and woolen patents together) than in Britain (61 per cent). In view of the relative position of the British woolen industry, the dominant industry in the economy from the middle ages to the eighteenth century, this is hardly surprising. From Table 2 it is clear that American inventive activity was both substantial and much greater than British activity in two areas only: cotton (and 'non-specific' in the sense that the patent titles, which alone in most instances survived the 1836 Patent Office fire, specified the stage of manufacture but either did not mention a textile fiber or else cited both cotton and wool) spinning and weaving. American activity was greater in cotton preparatory and general manufacturing technology and in woolen preparatory, spinning, and general manufacturing technology, but it was not substantial in absolute numbers of patents granted. In woolen finishing American activity was very substantial but not quite as many patents (79.5: one patent only out of 433 clearly spanned both two processing stages and two fibers) were recorded as in Britain (85).

The whole period 1790–1830 has been broken into three distinct subperiods for purposes of analysis. The first period, 1790–1812, saw the spread of machine spinning in Britain and its establishment in the U.S.A.; the second, 1813–1824, the achievement of powerloom weaving both in Britain and the U.S.A.; and the third, 1825–1830, the consolidation of powered spinning, weaving, and finishing, again both in Britain and the U.S.A. Inventive activity in cotton spinning and weaving was relatively much greater in the U.S.A. than in Britain over the twelve years, 1813–1824, than at any other time in these four decades. Four times as many spinning patents and nearly three times as many weaving patents were awarded in the U.S.A. than in Britain. In woolen finishing American activity was 60 per cent greater than in Britain before 1812; in the middle period it was about the same; in the late 1820s it fell away. This international comparison, therefore, suggests that attention ought to focus on inventive activity in American cotton spinning and weaving in 1813–1824 and on American woolen finishing 1790–1812 and to a lesser extent 1813–1824. This choice can be checked by looking at the American figures extrapolated into equal time periods.

Table 3

Changes in inventive activity in America's (N.H. to MD. only) cotton and woolen manufacturing technologies, 1790–1830

	1790–1812		1813–1824		1825–1830		
	Actual	Pro-rated to 24-year term	Actual	Pro-rated to 24-year term	Actual	Pro-rated to 24-year term	Total (Actual)
Cotton Mfg.*							
Preparatory	3	3.13	11	22	13	52	27
Spinning	24	25.04	87	174	60	240	171
Weaving	20	20.87	58	116	26.5†	106	104.5
Finishing	1	1.04	1	2	4	16	6
General	2	2.09	1	2	1	4	4
TOTAL	50	52.17	158	316	104.5	418	312.5
Woolen Mfg.							
Preparatory	6	6.26	2	4	3	12	11
Spinning	2	2.09	7	14	6	24	15
Weaving	—	—	—	—	—	—	—
Finishing	27	28.17	34	68	18.5	74	79.5
General	1	1.04	3	6	11	44	15
TOTAL	36	37.57	46	92	38.5	154	120.5
GRAND TOTALS	86	89.74	204	408	143	572	433

* Meaning that the short title of the patent is non-specific, e.g. "spinning machine."

† One patent covered two fibers and two processing stages and was therefore divided in half in counting the patents.

B. Relative secular change in American cotton and woolen manufacturing inventive activity

Table 3 shows the extrapolated figures which confirm some of the Anglo-American comparisons. In particular they demonstrate the preeminence of the 1813–1824 period in cotton spinning and weaving, with increases of over five times beyond the numbers registered, 1790–1812. But in addition they indicate that the rate of patenting was in fact increasing in cotton spinning in the late 1820s and then very slightly slackening in cotton weaving. Woolen finishing inventions also look more important in the middle and late sub-periods, the rate of patenting doubling between 1790–1812 and 1813–1824 and slightly rising in the late 1820s.

Clearly the establishment and growth of the cotton spinning industry after 1812 and the modification of its technology to suit American circumstances help to explain the sustained interest in that area. Likewise, the development of American powerlooms and the importation of European models account in part for the dramatic and sustained interest in cotton weaving improvements after 1812 – American patent law after 1800 allowed foreign inventors domiciled in the U.S.A. for at least two years to register American patents but this could be waived by Congress if necessary, and this very occasionally happened but does not seem to have occurred in the textile industries.[14]

In the search for the centers of inventive activity the analysis can be next to narrowed down by means of regional comparisons.

C. Regional comparisons of inventive activity in American cotton and woolen manufacturing activity, 1790–1830

Professor Thomas C. Cochran has recently suggested that by the late 1820s the industrial spine of the U.S.A. (heavy industry and metalwork, at least) ran down the Hudson Valley from Albany to New York, then went from Paterson to Trenton, New Jersey, and down the Delaware Vallwey beyond Philadelphia and Wilmington into Maryland. In the West another spine followed the Ohio from Pittsburgh to Cincinnati.[15] In contrast, light industry, primarily textile manufacturing, has usually been associated with New England, especially with Massachusetts and Rhode Island. Some corrective to this traditional correspondence of textile industry with New England is evident in the patent evidence, set out in Tables 4 to 8.

---

[14] Ibid., 159.

[15] Thomas C. Cochran, *Frontiers of Change: Early Industrialism in America* (New York: Oxford University Press, 1981), 90.

From Table 4 it is evident that just under half (47 per cent of the cotton and woolen manufacturing patents registered in the seaboard states (New Hampshire to Maryland), 1790–1830, came from outside New England (New York southwards) and just over half from New England. The outstanding feature is that New York accounted for the greatest number of patents both over the full 40 years and in each sub-period. Massachusetts came second overall and in each sub-period but in the late 1820s was far outstripped by New York. After Massachusetts came Rhode Island and Connecticut. Surprisingly, Rhode Island inventors seem not to have been particularly active in the early sub-period but forged ahead in the late 1820s. Relatively much stronger (fourth) in the first period than in the subsequent ones was Pennsylvania.

Some interesting relationships emerge from a comparison of patent activity with capital investment. In a general way an increase in patent activity could be expected where industrial expansion signaled increasing inter-firm competition and consequent attempts to reduce operating costs and to preserve technical leadership. Taking cotton manufacturing patents and textile manufacturing patents not specific to any fiber, it is clear from Table 5 that both in 1820 and 1831, with one exception, the states ranked among the top three according to a capital investment measure also appeared in the top three by the measure of numbers of patent registrations. The exception was New York. Whereas it had the largest share of American capital investment in cotton manufacturing in 1820 it fell to fifth place in 1831, being overtaken by New Hampshire and Pennsylvania. Yet despite its relative fall in investment (in absolute terms investment in New York cotton manufacturing during the 1820s grew, but, unlike New England, New Jersey,and Pennsylvania, failed to double in aggregate size), New York continued to lead as a center of inventive activity.

Conversely, in Massachusetts, where capital investment increased eight times, from $1.57 million to $12.89 million, during the 1820s, raising the state's share of the American cotton industry from 14 percent to 32 per cent, patenting activity, as a share of east coast effort, more than halved from 30 to 13 per cent.

How can the contrary movements of patent activity in relation to capital investment in New York and Massachusetts be explained? One explanation for the Massachusetts position (growth in capital investment and fall in patent activity) must lie with the Waltham system of manufacturing which, resting on a handful of patents, spread throughout northern Massachusetts and New Hampshire, promoted by a small group of Boston capitalists in the 1820s and 1830s.

Just because relatively little is known about textile manufacturing in New York State in the early nineteenth century, it is much less easy to explain the situation in New York: with patent activity rising much faster than growth in

Table 4

Numbers of textile manufacturing (mechanical inventions) patents registered in the U.S.A. (N.H. to MD. only), 1790–1830

	1790–1812		1813–1824		1825–1830		
	Actual	Pro-rated to 24-year term	Actual	Pro-rated to 24-year term	Actual	Pro-rated to 24-year term	Total (Actual)
New Hampshire	2	2.09	6	12	4	16	12
Massachusetts	18	18.78	54	108	29	126	101
Rhode Island	4	4.17	28	56	22	88	54
Connecticutt	18	18.78	25	50	19	76	62
New York	26	27.13	66	132	54	216	146
New Jersey	3	3.13	7	14	6	24	16
Pennsylvania	12	12.52	15	30	9	36	36
Delaware	2	2.09	—	—	—	—	2
Maryland	—	—	3	6	—	—	3
District of Columbia	1	1.04	—	—	—	—	1
Totals	86	89.74	204	408	143	572	433

Source: Burke (see Table 2).

Table 5

Fixed capital investment in cotton manufacturing in 1820 compared to patent activity in cotton manufacturing 1813–1824; and investment in 1831 compared to patent activity, 1825–1830

	Capital Investment 1820 %	Patents 1813–1824 %	Capital Investment 1831 %	Patents 1825–1830 %
Maine	1.64		1.88	
Vermont	0.97		0.73	
New Hampshire	5.46	2.53	13.05	3.83
Massachusetts	13.74	30.38	31.74	13.39
Rhode Island	15.17	13.92	15.42	17.22
Connecticut	12.28	11.39	6.95	11.48
New York	18.18	30.38	9.04	40.67
New Jersey	4.44	2.53	4.99	5.74
Pennsylvania	6.70	8.23	9.25	7.65
Delaware	3.77		0.95	
Maryland	12.07	0.63	5.28	
South	0.55		0.71	
West	5.03			
Total	100	99	99.99	99.98
	N = $11,411,122	N = 158	N = $40,614,984*	N = 104.5†

* Excludes $5m invested in printeries and bleacheries for which there is no state breakdown.

† Only one patent split between fibers and stage according to titles.

Sources: (1) Jeremy, *Transatlantic Industrial Revolution*, 276. (2) Friends of Domestic Industry, *Report*, 16. (3) Burke, *Patent List*.

fixed capital. Some clues emerge from a closer look at patents awarded for the stages where inventive activity was concentrated: cotton spinning and weaving and woolen finishing.

D. The centers of inventive activity in the technologies attracting most inventive effort

Cotton spinning

New York Both in the middle period and in the late 1820s patentees came from all parts of the state, though not surprisingly the counties of the lower Hudson Valley saw most activity. Of the 29 patents taken out in 1813–1824, nine came from the lower Hudson counties; nine from the western counties mostly bordering Pennsylvania; five from the counties between Albany and Buffalo following the Mohawk Valley to the littoral of Lake Ontario; and five from northern counties. In the late 1820s ten patents came from the lower Hudson; six from the upper Hudson (between Poughkeepsie and Albany); five from the western counties; two from counties along the Albany-Buffalo line; and one from the northern counties (locations for six unknown).

Some correspondence between rural counties and domestic spinning technology is evident in the period 1813–1824. Then at least 12 of the 29 patents covered domestic spinning devices and 11 of the 12 were registered by patentees in western, northern, or Erie-Mohawk counties. In the late 1820s only four out of 30 patents are known (from their titles) to have covered domestic devices and none of these four patentees hailed from the lower Hudson Valley.

As for the individual patentees, a small number made an impact on advances in industrial spinning technology. And these mostly lived in the lower Hudson Valley: William B. Leonard, who came from Massachusetts but who was manager of the Matteawan Manufacturing Company at Fishkill in Dutchess County, when he patented a spindle in 1819; Gilbert Brewster (d. ca. 1834), originally from Rhode Island but at Poughkeepsie in the late 1820s, who patented a roving frame in 1827 and the eclipse speeder in 1829; and Charles Danforth (1797–1876), also from southern New England, who patented the radical improvement of the cap spindle in 1828 when he was in Rockland County working for the Ramapo Company.[16] Benjamin Brundred

---

[16] *TIR*, index references to Leonard, Brewster, and Danforth and sources quoted there. Leonard in 1813 was agent for the Plymouth Cotton Mill Co. at Plymouth, Mass. and earlier had

(ca. 1791-1853), an English immigrant and machine maker, patented a throstle while at Mamaroneck, Westchester County, in 1824.[17] In New York City four patentees of cotton spinning improvements seem to have been on the distribution side: William Shotwell, a broker (patent, 1813); Francis Price, a merchant (patent, 1824); George Dickinson, importer (patent, 1828); Henry Ruggles, merchant (Patent. 1829) [18]

Among the obscure upstate patentees, two who have been identified in the 1850 Census were born in New England: Elias Wood (b. 1795), a machinist of Lowville, Lewis County, in 1850, who secured a patent for a domestic spinning machine in 1824;[19] and Joel Farnham (b. 1773), a wheelwright of Tioga in 1850, who patented a cotton and wool spinning improvement in 1825.[20]

Massachusetts The 12 years 1813–1824, inclusive, saw a maximum of patent effort in Massachusetts (see Table 6) and of the 27 spinning patents taken out 13 were associated with the Boston Manufacturing Company, the prototype firm in the Waltham system. Of the 13, seven were patented by Paul Moody (1779–1831), the BMC's machine shop superintendent; the remainder came from disaffected machinists led by Jonathan Fisk who tried to carry off Waltham's technical improvements.[21] One other most significant patent was that of George Danforth (1791–1830) of 1824 for the tube speeder, a product of the Taunton district machine shops.[22]

Rhode Island Here both the middle period and the later 1820s saw a great deal of inventive activity in the spinning industry all centered on Providence and largely the product of two mechanics. In the middle years John Brown took out six of the state's 13 cotton spinning patents and in the late 1820s John Thorp registered eight of the 12 patents (in the subject list), including his revolutionary ring spindle. John Thorp (1784–1838) was a well known machinist.[23] John Brown was a relatively unknown mechanic whose work has hitherto escaped notice.

---

worked at Taunton. See James Danforth to Duplanty, McCall & Co., December 16, 1813, in Longwood MSS, Group 6, Box 2, Eleutherian Mills Historical Library [now the Hagley Museum and Library], Greenville, Delaware.

[17] United States, Federal Population Census, 1850 (1,009 rolls of microfilm, M432) roll 461, p. 396. By 1827 Brundred had settled at Paterson N.J.

[18] *Longworth's New York Directory* for 1812, 1825, and 1829.

[19] U.S. Population Census, 1850, roll 523, p. 192.

[20] Ibid., roll 605, p. 87.

[21] *TIR*, 185–86.

[22] Ibid., index to George Pitts Danforth and sources quoted in the text.

[23] Ibid., for Thorp.

Table 6

Cotton and non-specific spinning manufacturing patents 1790–1830

	1790–1812		1813–1824		1825–1830		
	Actual	Pro-rated to 24-year term	Actual	Pro-rated to 24-year term	Actual	Pro-rated to 24-year term	Total (Actual)
New Hampshire			2	4	4	16	6
Massachusetts	4	4.17	27	54	3	12	34
Rhode Island	1	1.04	13	26	12	48	26
Connecticut	3	3.13	6	12	2	8	11
New York	10	10.43	29	58	30	120	69
New Jersey	1	1.04	3	6	6	24	10
Pennsylvania	4	4.17	7	14	3	12	14
Delaware	1	1.04					1
Maryland							
District of Columbia							
Totals	24	25.04	87	174	60	240	171

Source: Burke (see Table 2).

Cotton weaving

New York In patents covering this stage of production New York was barely ahead of Massachusetts in absolute terms (see Table 7). Two sections of the state displayed most interest both in 1813–1824 and the late 1820s: the lower Hudson and the Erie Mohawk counties with seven and four patents respectively. Only one of the loom patents in the middle period avowedly covered a domestic machine. In the late 1820s the patents were for powerlooms, loom motions (e.g., take-up), and loom parts (e.g., metal heddle eyes or wire harness). William B. Leonard again appears, patenting two powerloom inventions (in 1827), one of which covered a woolen powerloom.[24]

Massachusetts Inventors from several centers were responsible for the 16 Massachusetts weaving patents of 1813-1824. Prominent were Waltham (the Boston Manufacturing Company) with three patents, Taunton (John Thorp and Silas Sheppard) with two patents, and Dedham (Calvin Whiting and John Goulding (1791–1877) with two patents.[25] Goulding took out two of the state's five weaving patents in the late 1820s.

Connecticut Here in the middle period inventors of cotton weaving improvements were busiest in Litchfield and Windham counties, to the northwest and northeast of the state. Neither then nor in the late 1820s were any of the patentees responsible for major inventions.

Rhode Island John Thorp appears again, as patentee of two of the state's 13 cotton weaving patents taken out in 1813–1830. From the subject patent list evidence the only distinguishing characteristic of Rhode Island inventors was that they almost all came from Providence, the heart of the state's cotton industry.

Pennsylvania The handful of cotton weaving patents taken out in the two early periods mostly came from the Philadelphia area. While little is known about any of these nine patents, two in the middle period taken out by Thomas Siddall were of significance in establishing the commercial viability of powerloom weaving. Siddall was a cotton manufacturer in the Philadelphia region who emigrated from England; but his improvements, though new to the United States, were already in use in Lancashire.[26]

---

[24] Ibid., 238.

[25] Ibid., index for Goulding and Whiting.

[26] See my 'Immigrant Textile Machine Makers along the Brandywine, 1810–1820,' *Textile History* (Chapter 6 1982).

Table 7

Cotton and non-specific weaving manufacturing patents 1790–1830

	1790–1812		1813–1824		1825–1830		
	Actual	Pro-rated to 24-year term	Actual	Pro-rated to 24-year term	Actual	Pro-rated to 24-year term	Total (Actual)
New Hampshire	1	1.04	2	4			3
Massachusetts	5	5.22	16	32	5	20	26
Rhode Island	2	2.09	8	16	5	20	15
Connecticut	5	5.22	10	20	4	16	19
New York	2	2.09	15	30	10.5	42	27.5
New Jersey	1	1.04	1	2			2
Pennsylvania	4	4.17	5	10	2	8	11
Delaware							
Maryland			1	2			1
District of Columbia							
Totals	20	20.87	58	116	26.5	106	104.5

Source: Burke (see Table 2).

Woolen finishing

New York The preeminence of New York patentees among improvers of woolen finishing methods, with 43 percent of the eastern seaboard's patents in this area for the whole period (see Table 8) far out-balanced the state's share of national fixed capital investment in woolen manufacturing, standing at 27 percent in 1820.[27] Inventive activity was heavily concentrated in the Hudson Valley with 22½ of the 34½ patents (one patent out of all these listed spanned two fibers and two stages of production so was split in two). All the patentees in 1790–1812 came from this region; in the middle period nine out of 15, in the late 1820s two out of four patentees whose places of origin are easily identified.

In the early sub-period most of the patents were taken out for mechanical shears, designed to crop a cloth surface; in the second sub-period mechanical brushing machines (gigs or nappers), used for raising the cloth surface prior to shearing, also figured along with shearing modifications, suggesting some effort to achieve technical complementarity; and in the late 1820s fulling appeared as an additional subject among the New York cloth finishing patents.

Among-the New York patentees were clothiers like Samuel G. Dorr (d. 1794) of Albany[28] and Bishop Parsons (b. 1788), later of Marcellus, Onondago County;[29] and machinists like Samuel Duncan (b. 1788) of Northampton, Fulton County[30] and Beriah Swift (b. 1785 in Massachusetts) who was especially prolific between 1806 and 1824 when he took out four patents, all for cloth shearing devices.[31] Eleazer Hovey, who took out two shearing patents in 1808 and 1811, started as a millwright in New Hampshire but moved to New Lebanon, New York, in the upper Hudson region in order to build shearing machines for village clothiers in the Berkshires.[32]

Massachusetts With such a small number of patentees here any generalizations are hard to make. All 14 patents for woolen finishing, 1790–1830, came from men scattered across the state. The patentees are, however, better known than those in New York. Ebenezer Stowell of Worcester, who took out a cloth shearing patent in 1808, was a woolen and carpet manufacturer.[33]

---

[27] *TIR*, 280.

[28] Ibid., index and sources quoted for Dorr.

[29] U.S. Population Census, 1850, roll 568, p. 356.

[30] Ibid., roll 506, p. 153.

[31] Ibid., roll 497, p. 397. In 1850 Swift had an estate valued at $15,000.

[32] Barnes Riznik, 'New England Wool-Carding and Finishing Mills, 1790–1840,' Old Sturbridge Village Research Report. 1964. 48.

[33] David J. Jeremy, ed., *Henry Wansey and his American Journal, 1794* (Philadelphia: American Philosophical Society, 1970), 62n.

Table 8

Woolen finishing manufacturing patents 1790–1830

	1790–1812		1813–1824		1825–1830		
	Actual	Pro-rated to 24-year term	Actual	Pro-rated to 24-year term	Actual	Pro-rated to 24-year term	Total (Actual)
New Hampshire			2	4			2
Massachusetts	5	5.22	5	10	5	20	15
Rhode Island	1	1.04	4	8	4	16	9
Connecticut	6	6.26	2	4	1	4	9
New York	12	12.52	15	30	7.5†	30	34.5
New Jersey	1	1.04	3	6			4
Pennsylvania	1	1.04	2	4	1	4	4
Delaware	1	1.04					
Maryland			1	2			1
District of Columbia							
Totals	27	28.17	34	68	18.5	74	79.5

† one patent covered two fibers.

Source: Burke (see Table 2).

William Hovey, holder of a shearing patent (1824), started to build shearing machines at Worcester in 1811, having trained as a millwright, like his brother Eleazer.[34] Thomas Hurd, patentee of a napping improvement (1830), was a Lowell manufacturer.[35] Thomas Blanchard (1788–1864), another shearing frame improver (patent of 1813), became 'one of the most celebrated inventors of the nineteenth century' who achieved prominence when in 1818 he improved lathes for turning gun barrels at the federal armories at Springfield, Massachusetts and Harper's Ferry, Virginia.[36]

The evidence of patent litigation in the middle Atlantic states

Unlike New England, textile patents in the Middle Atlantic states (New Jersey, Pennsylvania, Delaware, and Maryland) evidently excited very few actions, either by patentees against infringers or by invaders against patentees. At least no such cases have survived among those extant in the circuit court case files prior to 1830 for these states; admittedly these are somewhat incomplete.[37] In contrast, the Massachusetts and Rhode Island circuit case files contain numerous patent squabbles over textile inventions between 1810 and 1830.[38]

---

[34] Riznik, 'Wool-Carding and Finishing Mills,' 48.

[35] *TIR*, index and sources quoted for Hurd.

[36] Merritt Roe Smith, *Harpers Ferry Armory and the New Technology: The Challenge of Change* (Ithaca: Cornell University Press, 1977), 124–38. See Cooper, *Shaping Invention*.

[37] Circuit court files checked as follows.

A. Equity records of the U.S. Circuit Court for the Eastern District of Pennsylvania, 1790–1847, National Archives film M-985 (23 rolls, Washington, D.C.: National Archives, 1976).

B. Equity docket books (1790–1791; 1790–1813; no others until 1843) of the Delaware Circuit Court, in the Philadelphia Federal Records Center.

C. Equity case files of the U.S. Circuit Court for the District of Maryland, eight boxes covering 1801–1836 in the Philadelphia Federal Records Center.

D. Efforts made in 1979 to locate the early case files of the U.S. Circuit Court for New Jersey failed. They were not held by the New York Federal Records Center, which did have the New Jersey Circuit Court case files from 1850 onwards.

I must record with gratitude the assistance of my wife, Theresa, who went through the film of the Pennsylvania equity records and who shared in sifting through the Delaware docket books and the Maryland case files. We checked for patentees' names and also for cotton and woolen textile subjects. Only one such textile case appeared, the case of John G. Jackson vs. James Benny, 1815–1816, in the Philadelphia court (M-985, roll 10, frames 694–708) and this was about a failure to deliver machinery, not about a patent.

[37] *TIR*, Part IV, passim.

Conclusion

This is a report of a research project which presently can only point to the need for further work before conclusions can be reached. Four weeks' research aimed at identifying patentees in the Middle Atlantic states produced relatively few new identifications from state and city directories, patent litigation, and the federal census of 1850. Clearly, progress can only come from a pursuit of patentees in highly localized sources such as newspapers, rite of passage, or tax and land records.

Two things are abundantly clear from this paper. Firstly, that expectations of profit from inventive activity as measured by numbers of patents registered did not neatly follow industrial expansion. Was this, as I guess, because in northern New England the cotton manufacturing technology underwent a basic modification between 1813 and 1822 which sufficed for more than a decade because interlocking directorates subdued competition? The point requires substantiation in firm level studies for the 1830s. Secondly, the surprising amount of inventive activity in New York state underscores the need for a thorough investigation of early nineteenth-century manufacturing in the Mowhawk and Hudson valleys. Indications are that numbers of mechanics joined the population drifting out of New England in the early nineteenth century; some settled in the industrializing Hudson Valley or further westwards in the state where household manufactures and carding and finishing mills created a demand for their products and challenges to their ingenuity. It would be interesting to discover the extent of New York's reliance on New England for its sources of inventive energy. For this a collective biography of American patentees and mechanics would be one way to move forward.

Acknowledgements

I am most grateful to the Academic Advisory Board of the Regional Economic History Research Center at the Eleutherian Mills-Hagley Foundation for awarding me a month's Fellowship (taken in August 1979) to work on the topic of invention and innovation in early nineteenth-century American textile manufacturing. Travel support came from the British Academy to whom I am similarly grateful.

While at the REHRC I and my wife, Theresa, received many kindnesses from Glenn Porter, Bill Mulligan, and other friends at the EMHL and also from Dr. Andy John and his wife, Pat, and several other old acquaintances from the Evangelical Presbyterian Church at Newark.

Although this piece of research produced more blank walls than fresh pictures from the historical evidence, what was achieved owed much to the indefatigable work of my wife at the microfilm reader and, when we returned to

England, of my daughters, Joanna, who compiled a card index of patentees, and Rebecca, who checked the subject list of patentees against the chronological list, thereby taking the drudgery out of my task of collating the two indexes. To all three I am immensely grateful. The final result is of course my responsibility. And lastly, thanks are due also to Helen Jeffries, *Dictionary of Business Biography* secretary, for struggling once again with my handwriting.

Appendix I: John Thorp's textile patents

Something of the inventive talents of one of the most prolific early nineteenth century inventors in the American textile industry, John Thorp (1784–1848), is captured in the ten patent drawings shown here. They seem to comprise all those in Thorp's name that were reconstructed after the Patent Office fire of 1836 and that are now extant.

At a technical level they show Thorp's bold imagination at work. In 1812 (drawing no. 1; the attached sequence is the strictly chronological one as given in the list of Thorp's patents) he not only aimed to apply power to the handloom but to a loom that would weave figured goods. This required a complex shedding sequence and Thorp intended to effect this with a compound cam, albeit of primitive design (a wooden drum with pegs projecting from its surface). Since the number of distinct cam movements which could be produced by the 360° of a circle was limited, it is almost certain that Thorp's loom never wove a figured weave, an achievement attained by the powered jacquard loom developed in the 1830s.

Seventeen years later and still intrigued by the technical and commercial possibilities of the powerloom, Thorp devised one for weaving narrow goods employing a picking motion which dispensed with the shuttle (drawing no. 2). Instead, he used a lever: which looks like a bell crank, to take the weft yarn across the warp and then loop it round a pin opposite.

The late 1820s, however, found Thorp preoccupied with two kinds of spinning improvements. Firstly, his attention focused on spindle design where he sought higher speeds by substituting or modifying the traditional flyer which, at high speeds, tended to splay out and create wobble. Thorp was not alone in tackling the problem. The Rogers in Oneida County, New York, on the Erie-Mohawk axis, and William B. Leonard at Matteawan and Charles Danforth at Ramapo, both on the lower Hudson, preceded him (see *TIR*, 212–13). The results of Thorp's ingenuity are seen in drawings nos. 3–5 and 8–10. His most radical improvements were the replacement of the flyer with variants of the ring and traveller (drawing nos. 3 and 4). This, he claimed, gave higher speeds, larger bobbins, and a more even yarn – together yielding higher productivity and improved quality (see *Journal of the Franklin Institute* 4 [1829], 64). The alternative, a cap, had already been patented by Danforth in September 1828, a couple of months before Thorp took out his patent (drawing no. 5) which he called a can. Being easily removable from the top of the spindle, it facilitated piecing and doffing, and the concept led Thorp to two more patents (drawing nos. 8 and 9). The ring principle intrigued him further when (drawing no. 6) he devised a split ring to insert twist; how this could have functioned is not immediately clear. Apparently, the traveler gave some trouble (Dr. John Lozier has some evidence of this) and perhaps this was why Thorp finally returned to the flyer, combining it with a split ring (drawing no. 10).

Secondly, Thorp worked on a winding motion for laying the spun yarn on the bobbin in a spiral form, thus allowing unwinding without rotation when the yarn cop lay in the shuttle (drawing no. 7). The specification in the *JFI* 4 (1829), 126–28, states that Thorp's improvement was his motion which laid the yarn up and down the whole length of the bobbin, unlike previous winding motions, including presumably the one patented in 1821 by Paul Moody of Waltham (*TIR*, 181).

In a most thorough investigation of Thorp's life and work, published in 1928 in the *Transactions of the National Association of Cotton Manufacturers*, Charles H. Clark listed eleven patents taken out in his name but guessed that there might have been others taken out jointly with other inventors. I decided to compile a list using two published patent lists, the *JFI* and the patents surviving in the National Archives The result accompanies this note. It did uncover other patents besides those identified by Clark; quite dismaying however was the discovery of serious discrepancies between the published lists and in particular the suspicion that Edmund Burke's subject *List* of 1847, which has been used as the data base for a large part of the preceding article, is less reliable than the chronological lists printed and published annually from 1805 until the late 1820s. However, even the chronological lists carry dates which do not always agree with those given in the reconstructed patents themselves (see far right hand column of Table).

A list of the textile patents of John Thorp (1784–1848)

Date	Address of Thorp	Patent list (1) (13 Patents pre 1830)	Patent list (2) (14 Patents pre 1830)	Reports of patents in *Journal of the Franklin Institute*	Patents in Nat. Archs. RG. 241 Drawing	Spec.
1. 28 Mar. 1812	Providence, RI	Loom, hand and power	A hand and water loom		✓	✘
2. 14 Oct. 1816	Taunton, Mass. (with Silas Shepard)	Winder, socket, bobbin	Socket bobbin winder			
3. 20 Nov. 1819	Providence, RI	Yarn rolls forming on the shuttle bobbin	See 20 Mar. 1820 (No 12)			
4. 10 Aug. 1821	" "	Braiding	Braiding machine			✘
5. 10 July 1826	" "	Braiding	On his patented braiding machine			✘
6. 20 Nov. 1828	" "	Loom, shuttle, tying knots	Called the knot shuttle	*JFI* 3 (1829), 121, for tying knots (nets & weavers' harnesses)*		✘ (28 Nov)
7. 20 Nov. 1828	" "	Spinning and twisting cotton	In spinning and twisting	Ibid., 122, whirling ring spinner*	✓✓	✘
8. 25 Nov. 1828	" "	Spinning, filling and slack twisted yarn	In spinning or filling slack twisted yarn	Ibid. (gives 20 Nov.) can spinner (easy doffing)	✓	✘ (23 Nov)
9. 25 Nov. 1828	" "	Spinning machine	—			
10. 31 Dec. 1828	" "	Spinning machine	In spinning, called ring groove spinner	Ibid., 202, ring groove spinner*	✓	✘
11. 23 Jan. 1829	" "	Spinning machine	A cap spinner, being an improvement of the ring groove spinner	Ibid., 262, cap spinner (21 Jan. 1829)*		

A list of the textile patents of John Thorp (1784–1848)

Date	Address of Thorp	Patent list (1) (13 Patents pre 1830)	Patent list (2) (14 Patents pre 1830)	Reports of patents in *Journal of the Franklin Institute*	Patents in Nat. *Archs. RG. 241* Drawing	Spec.
12. 20 Mar. 1829	" "	(See 20 Nov. 1819) (No. 3)	Improvement in the manner of forming the spire or roll of yarn on a weavers' or shuttle bobbin	Ibid., 417, *JFI* 4 (1829), 126–28*	✓	✗
13. 3 Apr. 1829	Providence, RI	—	Improvement in the can spinner as pat. on 25 Nov. 1828	*JFI* 4 (1829), 49, can spinner*	✓	✗
14. 13 June 1829	" "	Spinning machine	Called Thorp's running cap spinner	Ibid., 130–32*	✓	✗
15. 22 Dec. 1829	" "	—	In weaving narrow stuffs without the use of shuttles	*JFI* 5 (1830), 164–65*	✓	✗ (& 6 Nov. 1829)
16. 11 Nov. 1830	" "	Spinning cotton yarn	Spinning cotton yarns	*JFI* 6 (1830), 388–90*	✓	✗
17. 31 Dec. 1838	" " (with Wm. G. Angell)	Loom, weavers' harness for loom	Not checked			
18. 26 Feb. 1840	" " (with Wm. G. Angell)	Loom, heddle stuff, making weavers' eye harness	" "			
19. 27 Sept. 1844	North Wrentham, MA	Hook spinner and twister, whirling and rotary	" "			

* Full patents in *JFI* 4 (1829), 62–70.

Source:
Patent list (1): Edmund Burke, *List of Patents for Inventions and Designs, Issued by the United States, from 1790–1847* (Washington: J. & G. S. Gideon, 1847).
Patent List (2): U.S. Patent Office, *A List of Patents Granted by the United States, from April 10, 1790 to December 31, 1836* (Washington: 1872). This is a photolithograph of the annual chronological lists.
Clark: Charles H. Clark, "John Thorp-Inventor of Ring Spinning," *Transactions of the Nat. Assoc. of Cotton Manufacturers*, 124–25 (1928), 72–95.

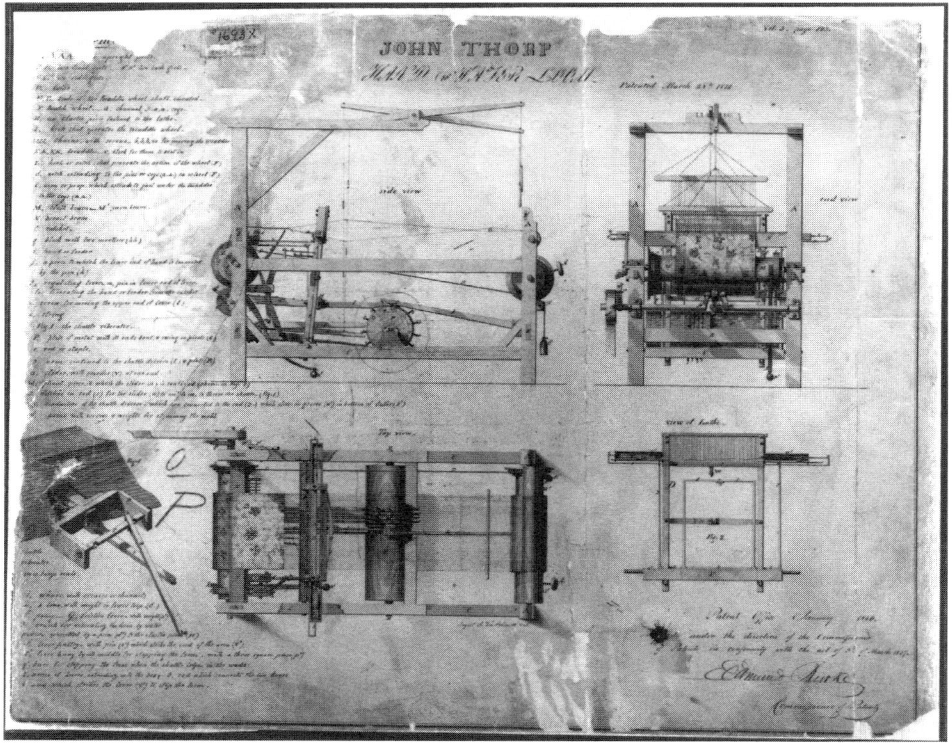

Drawing no. 1 Hand and Water Loom (Courtesy National Archives)

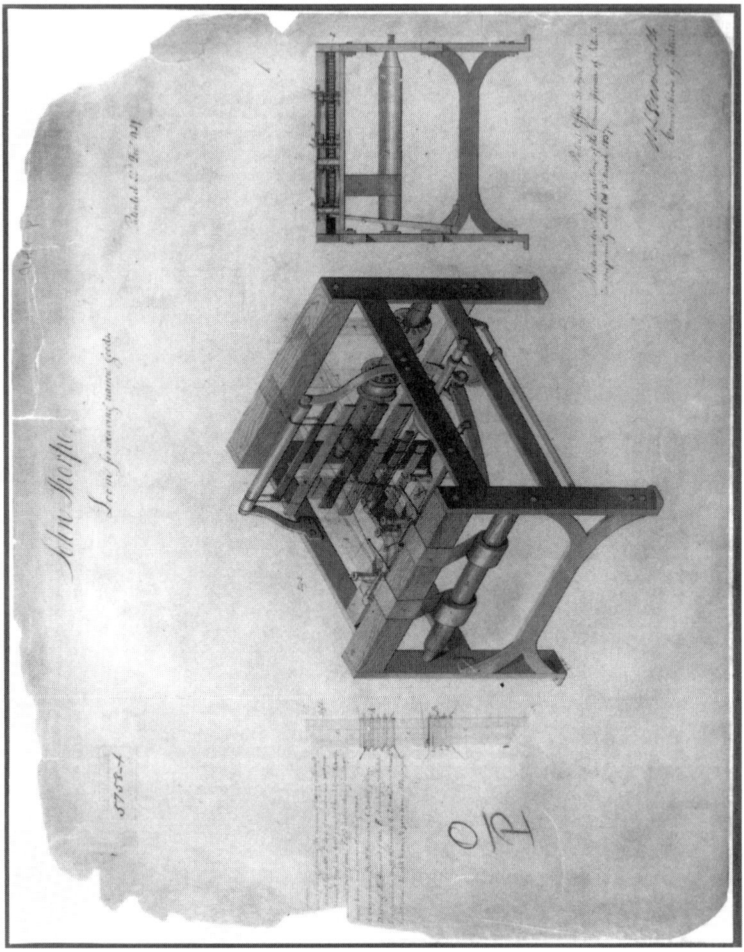

Drawing no. 2 Lcom for Weaving Narrow Goods (Courtesy National Archives)

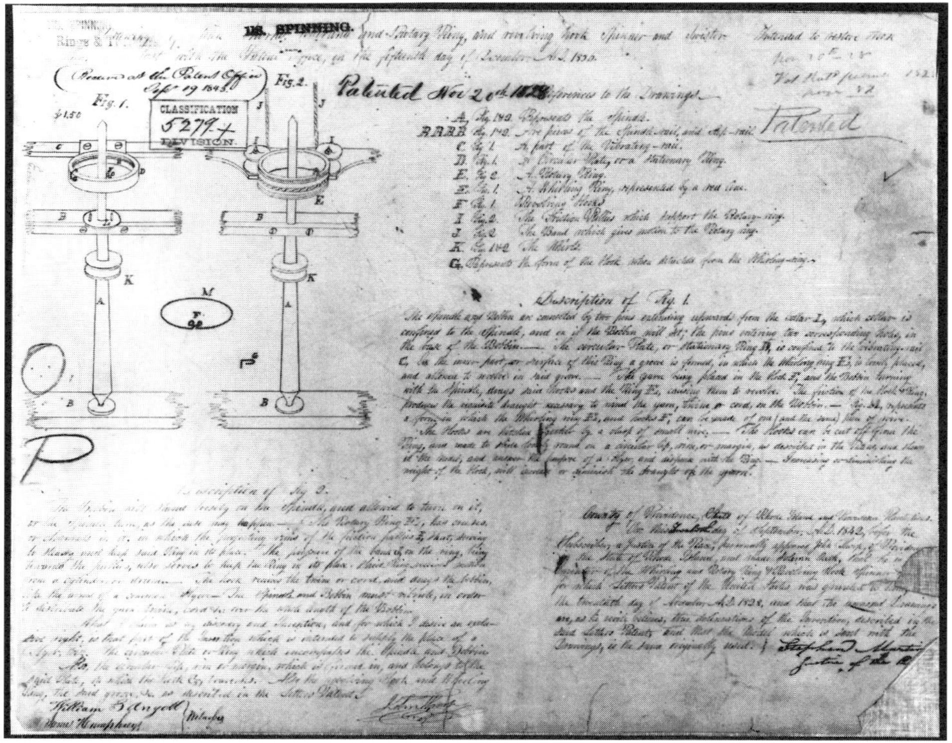

Drawing no. 3 Spinner and Twister (Courtesy National Archives)

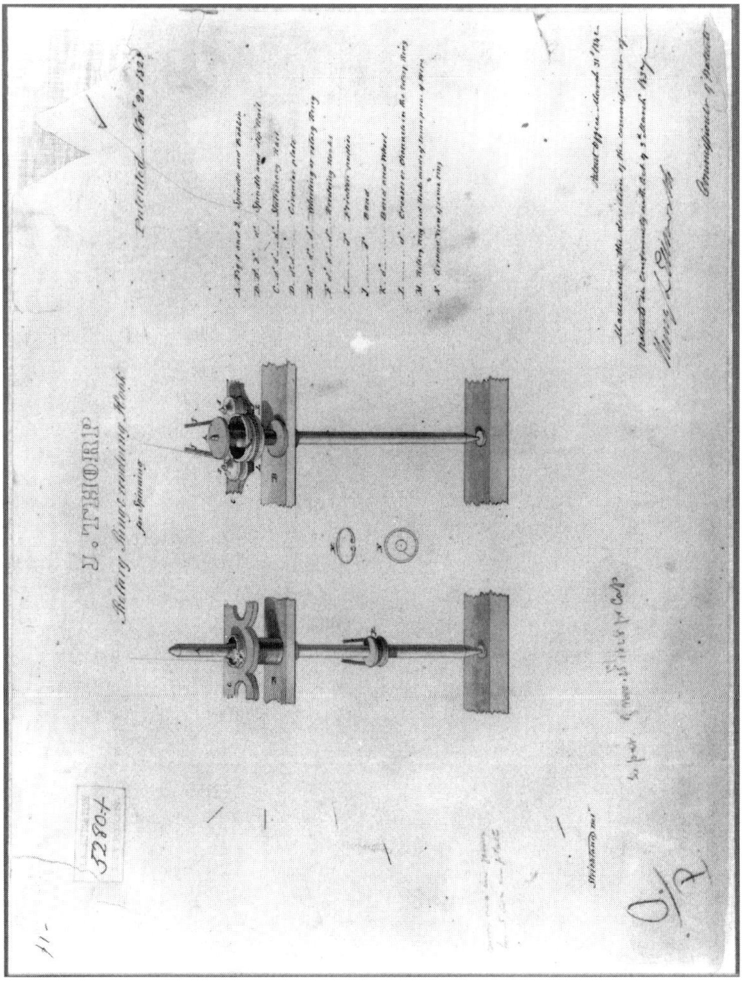

Drawing no. 4 Rotary Ring and Revolving Hook (Courtesy National Archives)

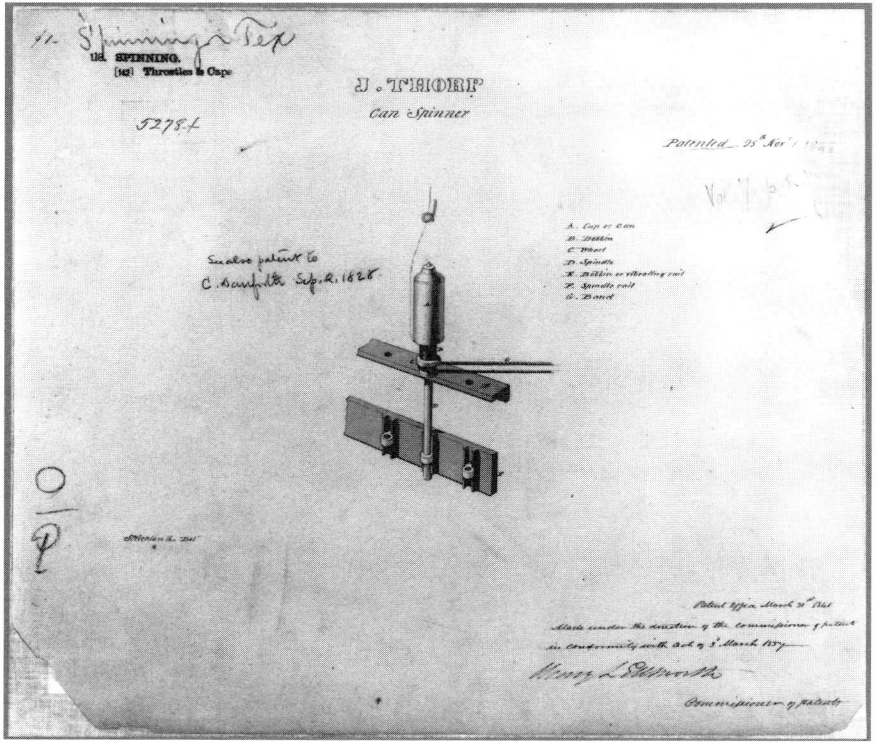

Drawing no. 5 Can Spinner (Courtesy National Archives)

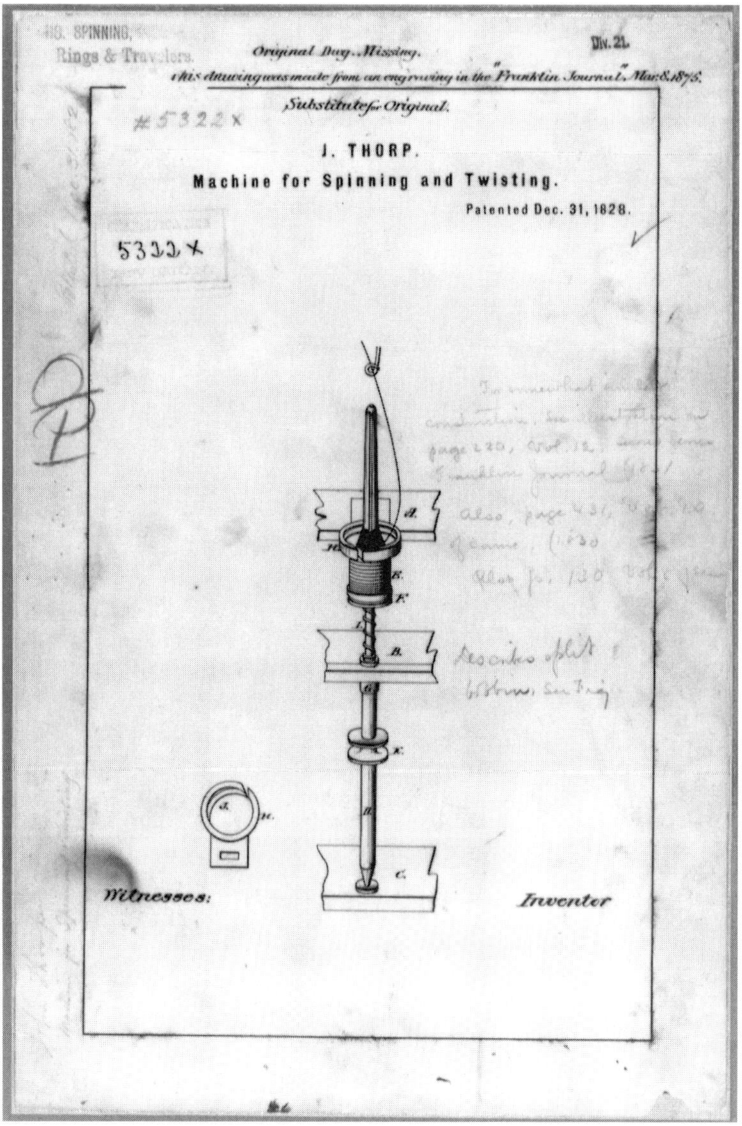

Drawing no. 6 Machine for Spinning and Twisting
(Courtesy National Archive)

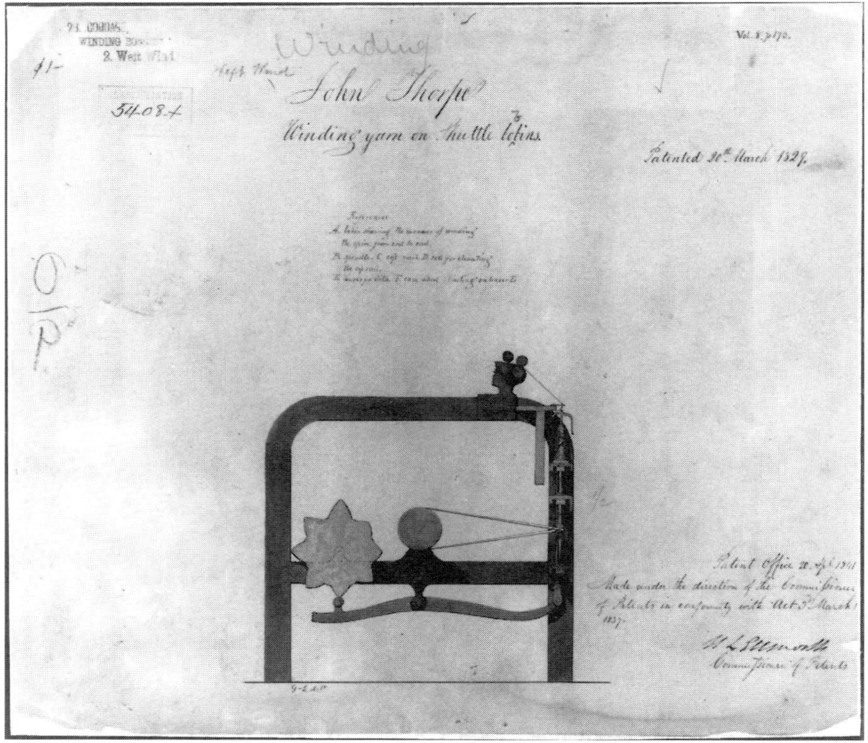

Drawing no. 7 Winding Yarn on Shuttle Bobbins (Courtesy National Archives)

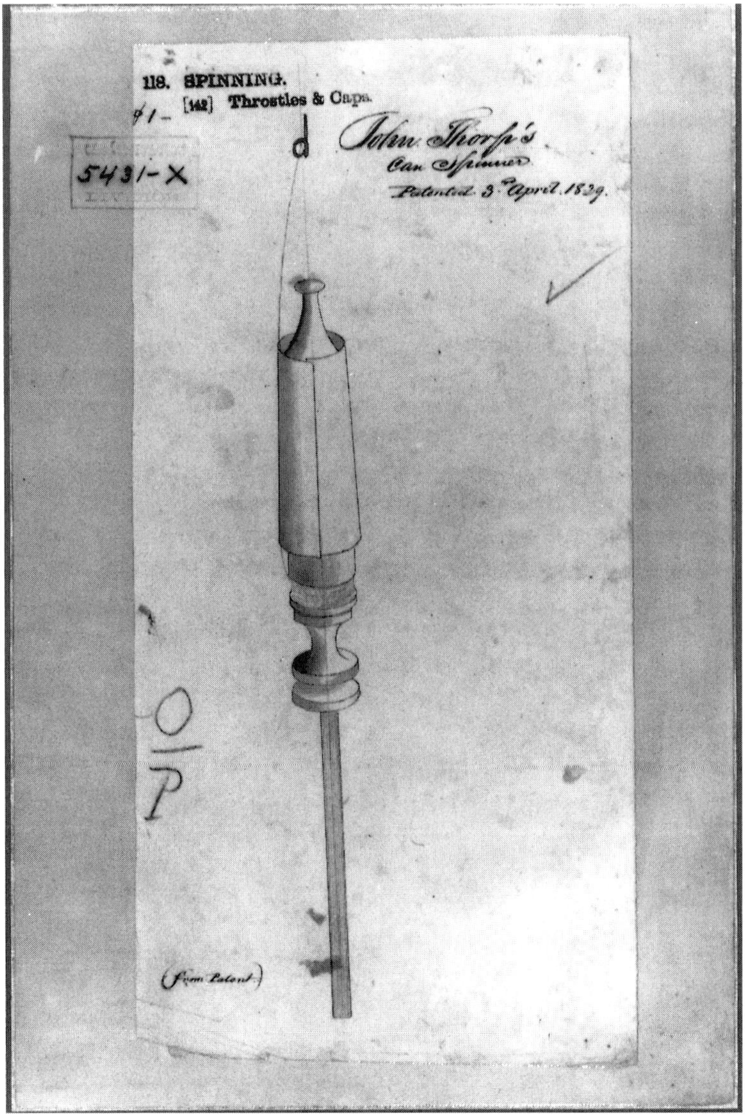

Drawing no. 8 Can Spinner (Courtesy National Archives)

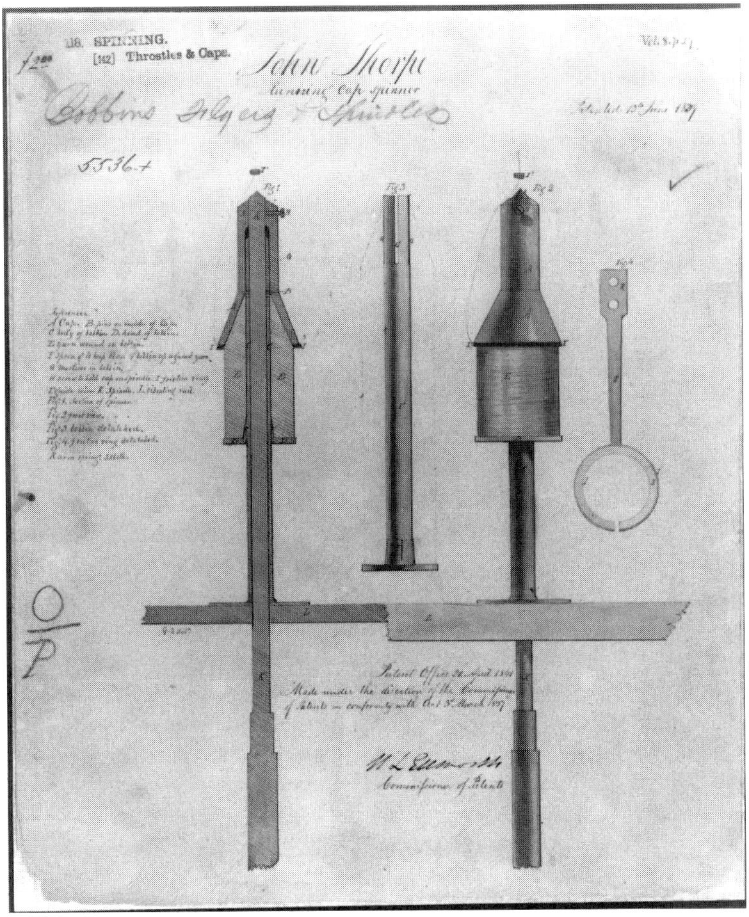

Drawing no. 9   Running Cap Spinner (Courtesy National Archives)

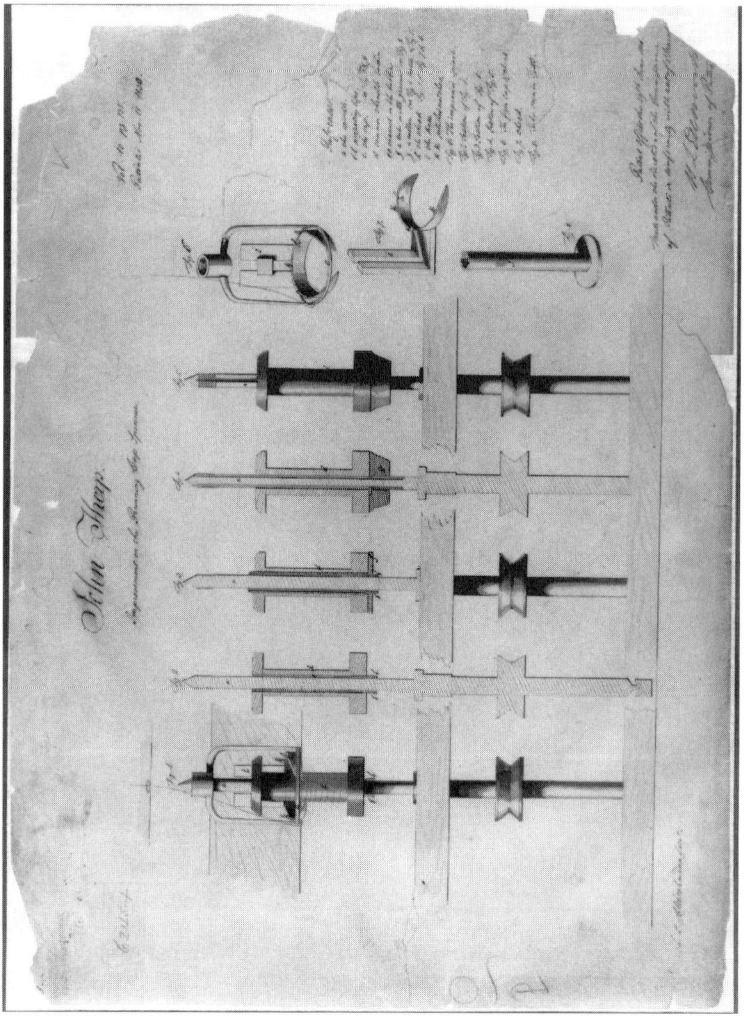

Drawing no. 10 Improvement in the Running Cap Spinner (Courtesy National Archives)

Appendix II

Appendix II, which follows, shows the results of a careful comparison between the subject and chronological lists (see Table of Thorp's inventions for the bibliographic details of these lists). It demonstrates the following points:

1 Only 12 (out of 432) patents are possibly wrongly dated in the subject list (Appendix IIa).
2 Another 24 of the 432 patents in the subject list do not appear in the chronological list (see Appendix IIb) and this could be due to apparent gaps in the chronological list, e.g., 12–13 April 1819.
3 A further 10 patents not in the subject list (or at least its "Class III, Manufacturers of Fibrous and Textile Substances, including Machines for preparing Fibres of Wool, Cotton, Silk, Fur, Paper, &c") were found in the chronological list (Appendix IIc).

Assuming that the 24 extra patents in the subject list were in fact awarded, then we are faced with a total degree of error of 22 out of 432 + 10, which is a probability of error of 4.98 per cent. In short, the tables compiled for the above article are much more reliable than I at first feared when I discovered the discrepancies between the two U.S. patent lists.

Appendix IIa

Discrepancies in Dates (of more than a week) between the U.S. chronological and subject patent lists in mechanical textile inventions, 1790–1830 (Eastern seaboard states, N.H. to MD. only)

State	Name of patentee	Invention	Subject list date	Chronological list date
MA (Excl. ME)	Draper, Ira	In loom	7 January 1816	7 June 1816
	Phelps, Russel	Spindle	21 December 1830	31 December 1830
RI	Stillman, William	Shearing cloth	11 June 1812	11 June 1811
	Brown, John	Spinning wheel	12 June 1813	12 June 1818
	Brown, John	Spinning machine	24 February 1814	24 February 1815
	Thorp, Jphn	Yarn, rolls, shuttle bobbin	20 November 1819	20 March 1829
	Hazard, JP & RG	Cloth manufacturing	6 December 1823	6 December 1828
CT	Levalley, William	Figure loom	26 March 1819	20 March 1829
NY	Wheeler, JR & JB	Spinning machine: “fancy spinner”	3 November 1826	3 March 1826
PA	Bacon, DM	Carding machine	10 April 1828	10 September 1828
MD	Morgan, James	Improvement in warping	5 April 1831	5 April 1821
Discrepancy in States				
	Slater, Samuel	Vertical spinner	4 April 1825 but VT	4 April 1825 but CT

Appendix IIb

U.S. patents in the U.S. subject list but not in the U.S. chronological list, for inventions in the mechanics of textile processing, 1790–1830 (Eastern seaboard states, N.H. to MD. only)

State	Name of patentee	Invention	Subject list date
MA (Excl. ME)	Bradley, Zalmon	Spinning wheel	21 April 1818
	Decker, Israel	Twine spooling	3 April 1819
	Lamb, Joshua	Carding machine	21 August 1819
	Lamb, Joshua	Cards, cylinder	21 August 1819
	Wright, Duncan	Cloth drying, by steam rollers	31 August 1820
RI	Waterman, John	Loom temple	28 May 1818
	Brown, John	Spinning, roping	22 July 1819
	Whitehead, John	Powerloom lathe	30 August 1819
	Thorp, John	Yarn, rolls	20 November 1819
	Thorp, John	Spinning machine	25 November 1828
	Williams, Thomas Robinson	Felt manufacture	22 May 1830
CT	Olney, James	Napping cloth	1 March 1813
	Forrester, Lot	Loom	15 May 1819
	Blackmar, John	Loom	7 August 1824
	Arnold, Peleg	Spinning, flier	6 October 1824
	Griggs, John	Spinning wool	30 December 1824
NY	Armstrong, Stephen	Loom	7 April 1818
	Hatch, M	Spinning wheel	10 April 1819
	Mudge, I & Hatch, M	Spinning wheel heads	10 April 1819
	Fowler, William & Germond, Richard	Spinning roping	24 November 1824
	Otis, Elisha	Loom, wire harness	31 December 1824
	Gill, John	Family spinning machine	19 February 1825
	Wheeler, John W	Spinning cotton and wool	24 July 1826
PA	Prescott, Levi & Jedediah	Napping cloth	5 November 1818

Appendix IIc

U.S. patents in the U.S. chronological list but not in the U.S. subject list, for inventions in the mechanics of textile processing, 1790–1830 (Eastern seaboard states, N.H. to MD. only)

State	Name of patentee	Invention	Chronological list date
NH	Steward, Stephen	A mode of weaving millinet, muslin, lace	31 December 1813
MA (Excl.	Blydenburgh, Samuel	Weaving machine	12 September 1810
ME)	Goulding, John	In mode of scouring & washing cloth	24 August 1827
	Goulding, John	In mode of manufacturing wool	24 August 1827
RI	Thorp, John	In can spinner patented 25 November 1828	3 April 1829
	Thorp, John	Weaving narrow goods without shuttles	22 December 1829
NY	Drummond, John	In weaver's lathe	24 July 1813
PA	Hess, John	Revolving weaving loom	2 October 1813
MD	Clarkson, Joseph	Machine for laying and twisting cotton seine twine	16 August 1826
No state given			
	Harris, William	Lathe for loom	15 March 1800

IX

Innovation in American Textile Technology during the Early 19th Century

Introduction

As technological historians from John L. Hayes down to George S. Gibb and W. Paul Strassmann have shown,[1] the most seminal period in the history of American textile technology occurred during the three decades succeeding the War of 1812. Following suggestions in H. J. Habakkuk's stimulating comparative study of 19th-century British and American technology,[2] this article attempts to relate the influence of product market pressures and organizational forms to innovation in textile technology during those seminal decades.

Between the Revolution and the end of the War of 1812, American proponents of capitalist manufacturing concentrated on acquiring the technology of Britain's system of textile manufacturing, which by 1812 had been basically mechanized at all stages in the cotton industry (see table 1); on overcoming Jeffersonian resistance to industrialization; on discovering and establishing feasible and

Reprinted from TECHNOLOGY AND CULTURE
Vol. 14, No. 1, January 1973

[1]John L. Hayes, *American Textile Machinery* (Cambridge, Mass., 1879); George S. Gibb, *The Saco-Lowell Shops: Textile Machinery Building in New England, 1813-1949* (Cambridge, Mass., 1950); W. Paul Strassmann, *Risk and Technological Innovation: American Manufacturing Methods during the Nineteenth Century* (Ithaca, N.Y., 1959). The most recent study is confined to an analysis of cotton spinning patents, 1790-1836, and hinges upon a comparison of all those issued before the Patent Office fire of 1836 with those reconstructed after the fire (see Robert R. MacMurray, "Technological Change in the American Cotton Spinning Industry, 1790-1836" [Ph.D. diss., University of Pennsylvania, 1970]).

[2]H. J. Habakkuk, *American and British Technology in the Nineteenth Century: The Search for Labour-saving Inventions* (Cambridge, 1962).

profitable product markets; and on diverting capital investment into textile-manufacturing enterprises. During this period of technology acquisition, the direction of technology adaptation seems to have been most firmly set by two kinds of limiting pressure: crowded product markets and the organizational structures created at the time of the formation of a factory-based industry, especially those parts of these structures developed in response to severe shortage of unskilled labor.

Product Market Influences on Innovation

In locating secure markets, American manufacturers sought those areas in which competition from British imports on one hand and from household manufactures on the other were most vulnerable. This meant, in the cotton industry, concentrating on coarse and medium yarn spinning before the War of 1812 and, thereafter, developing an efficient power loom (which within two decades eclipsed the household hand loom), and on making coarse and medium cotton goods like shirtings, sheetings, and bed ticking.[3] In the woolen industry safe product markets appeared first in supplementing household manufacturing by building for sale implements or machines such as improved spinning wheels or spinning jennies and by operating carding, spinning, and fulling mills. Military demands for coarse cloths during the War of 1812 turned American woolen manufacturers toward a market which they increasingly served as cities grew, transportation networks ramified, and power looms multiplied. Whereas many early woolen manufacturers had ambitiously (and frequently suicidally) aimed at displacing their English rivals in the fine cloth (broadcloth) market, in the 1820s they and their successors diversified production to include inferior fabrics like cassimeres, kerseymeres, satinets, cassinets, and flannels. This progressive change reflected the prudence of reducing production costs by applying simpler technology in manufacturing to meet the demands of expanding lower-class and urban mass markets. Cassimeres and kerseymeres needed less finishing than broadcloths; satinets and cassinets had a cotton warp which immediately made them far cheaper than broadcloth; and flannels required little more than washing and pressing in their finishing.[4] In short, British competition in product markets disposed American cotton and woolen

[3]Strassmann, p. 83.

[4]Arthur H. Cole, *The American Wool Manufacture,* 2 vols. (Cambridge, Mass., 1926), 1:195–207.

TABLE 1

BRITISH TEXTILE TECHNOLOGY (INDUSTRIAL): MACHINES AND DEVICES USED OR DEVELOPED, 1700–1812

	Wool			Cotton		
Stage	Process	Fiber State	Machine	Process	Fiber State	Machine
Preparatory	Sorting	Wool (raw)	. . .	Sorting	Raw Cotton	. . .
	Scouring	Wool (raw)	. . .	Picking	Loose Cotton	Picker
	Dyeing	Wool (raw)	. . .	Mixing	Loose Cotton	. . .
	Picking	Wool (raw)	Picker	Dyeing	Loose Cotton	. . .
	Oiling	Wool (raw)	. . .	Carding-breaker card	Sheet or lap	Card (carding machine)
	Carding-breaker card	Sheet or lap	Card (carding machine)	Carding-finisher card	Sliver (endless tubular length)	Card
	Carding-finisher card	Short rolags	Card	Drawing	Doubled and twisted end or sliver	Drawing frame
	Slubbing	Continuous sliver	Billy	Roving	Drawn and twisted end or sliver	Roving can and double speeder†
Spinning	Intermittent spinning DT W*	Yarn	Jenny, Mule	Intermittent spinning DT W	Yarn	Jenny, Mule
	Reeling	Yarn	Reel	Continuous spinning DT W	Yarn	Water frame, Throstle
		. . .	. . .	Reeling	Yarn or thread	Reel
				Twisting DT W	Plyed yarn (thread)	Twisting frame
	Twisting	Plyed yarn (thread)	Twisting frame	Bleaching	Yarn or thread	. . .
Weaving	Spooling/winding	Yarn or thread	. . .	Spooling/winding	Yarn or thread	. . .

Prep. Warp	Warping	Yarn or thread	Warping mill	Warping	Yarn or thread	Warping mill
	Dressing	Yarn or thread	. . .	Dressing (slashing)	Yarn or thread	. . .
	Drawing-in	Yarn or thread	. . .	Drawing-in	Yarn or thread	. . .
	Reeding	Yarn or thread	. . .	Reeding	Yarn or thread	. . .
Prep. filling	Quill winding	Yarn	. . .	Quill winding	Yarn	. . .
Weaving	Weaving	Cloth	* Fly shuttle	Weaving	Cotton goods	Power loom:† Shedding, Picking, Beating-up, Letting off, Taking up, Stop motions: Warp, Weft, Warp-protection
Finishing	Fulling	Cloth	Fulling mill	Printing	Cotton goods	Roller printing machine†
	Tenter drying	Cloth	. . .	. . .	. . .	. . .
	Raising	Cloth	Gig Mill	. . .	. . .	. . .
	Shearing	Cloth	Shearing frame (non-rotary)†	. . .	. . .	. . .
	Burling	Cloth	. . .	. . .	. . .	. . .
	Pressing	Cloth	Screw press	. . .	. . .	. . .

*D = drafting; T = twisting; W = winding. In intermittent spinning, drafting, and twisting are simultaneous and winding is separate; in continuous spinning, drafting, twisting, and winding are simultaneous. The latter is faster and cheaper but was not appropriate for fine cotton yarns or any woolen yarns at this time. See n. 16.

* *All except these four were introduced into the United States before 1794 (see David J. Jeremy, ed., *Henry Wansey and His American Journal, 1794* [Philadelphia, 1970]).

manufacturers toward coarser and cheaper fabrics while the rise of mass markets focused their concern on increasing production rather than developing product quality or variety.

The one important exception to this general trend occurred in the 1820s in Rhode Island. Here, competition from the northern New England giants persuaded many cotton manufacturers to switch to "high quality specialty goods, lines that could be manufactured profitably on a relatively small scale."[5] So Rhode Island cotton-mill operators would have become interested in machinery capable of manufacturing finer yarns and cloths and in reliable equipment of greater versatility able to spin a wide range of yarn finenesses or to weave a variety of fancy patterns as well as plain weaves.

Organizational Impact on Innovation

The limits set to technological innovation by crowded product markets were further modified by the emergence of two distinct organizational structures as the factory system was applied to the American cotton industry. The more traditional industrial organization appeared in Rhode Island, southern Massachusetts, Connecticut, and states to the south. Here, between 1790 and 1812, a structural pattern arose of small partnership firms horizontally specialized in factory spinning and utilizing the family as the fundamental labor unit.[6] The partnership derived from mercantile models, but horizontal specialization seems to have owed much to Samuel Slater, who dominated the organizationally formative period of the industry. The British system of cotton manufacturing which he succeeded in making practicable in Rhode Island in 1790 was in fact mechanized only at the carding and spinning stages of production. Although a functional power loom was pioneered in Britain during the two decades after Slater's emigration, and was brought to the United States by 1815, Slater resisted it until as late as 1823 apparently. For reasons which need to be explored, he fixed upon a hybrid manufacturing system: mechanized factory spinning in conjunction with put-out or contracted household handloom weaving.[7]

Slater's example was widely followed, and the typical Rhode Island cotton business in 1820 (when there were less than 100) was a

[5]Peter J. Coleman, *The Transformation of Rhode Island, 1790-1860* (Providence, R.I., 1963), p. 92.

[6]Caroline F. Ware, *The Early New England Cotton Manufacture* (Boston, 1931), pp. 19-38; Coleman, pp. 77-86; Mack Thompson, *Moses Brown, Reluctant Reformer* (Chapel Hill, N.C., 1962), pp. 202-33.

[7]Ware, p. 74.

small one capitalized for well under the 1832 average figure of $40,000, and running a spinning mill of less than 1,000 spindles. Its factory labor force, of families moved into company-built single or row houses located close to the mill, also reflected current British practice.[8] Given the general shortage of unskilled labor, the focus on machine spinning inherent in Rhode Island's industrial organization of cotton manufacturing must have concentrated the local searches for labor-saving devices into efforts to increase the productivity of carding, roving, and spinning equipment, that is, until the 1820s when the change in product markets noted above would have widened the search again toward machinery used at all processing levels and for fine as well as coarse products.

In contrast to the Rhode Island pattern of industrial organization stood that established at Waltham in 1813 and extended into much of northern New England before 1830. The Waltham model, the Boston Manufacturing Company, was conceived by the merchant entrepreneur Francis Cabot Lowell (1775–1817) as the investment alternative to the Boston mercantile community's declining financial interests in trade.[9] By means of incorporation, Lowell's enterprise was given a generous injection of Boston capital, its original authorized capital being $400,000 (that of pre-1840 Waltham-type companies would rise to as much as $2,000,000). Meantime, Lowell acquired, through some personal industrial espionage in Britain, a power-loom design which in 1814 materialized as an efficient machine through his own exertions and the skill of an outstanding New England mechanic, Paul Moody (1779–1831), who had earlier worked for the Scholfield brothers and had been recommended to Lowell by another employer, the highly respected engineer Jacob Perkins (1766–1849). Once weaving was successfully mechanized, Lowell vertically integrated all the stages of manufacturing within a single mill complex and then confined production to one or two standardized cheap cloths, sheetings, and shirtings, from no. 14 yarn in the case of the Boston Manufacturing Company. In 1820 the Boston Manufacturing Company ran 5,376 spindles and 175 power

[8]Coleman, pp. 89, 93; Ware, pp. 198–200; George R. Taylor, *The Transportation Revolution, 1815–1860* (New York, 1951), pp. 270–74.

[9]Ware, pp. 60–78; Gibb, pp. 4–14, 33–34; Kenneth P. Mailloux, "The Boston Manufacturing Company of Waltham, Massachusetts, 1813–1848: The First Modern Factory in America" (Ph.D. diss., Boston University, 1957), pp. 73–84; Robert V. Spalding, "The Boston Mercantile Community and the Promotion of the Textile Industry in New England, 1813–1860" (Ph.D. diss., Yale University, 1963), pp. 6–15; *Dictionary of American Biography* (*D.A.B.*).

looms which then annually processed 450,000 pounds of cotton.[10] The other manufacturing giants which spawned after Lowell's death, in the town named for him, each produced different grades or types of cheap cotton goods so that product markets were assured for, if not monopolized by, the respective cotton manufacturing corporations.

Another bold organizational experiment made at Waltham was Lowell's remedy for the shortage of unskilled labor. He recruited teen-age girls from New England farms and housed them in dormitories or boarding houses close to the mill and supervised by matrons of the highest respectability. As an alternative to domestic service and as a change from teaching school, factory work, with its comparatively high wages and opportunity for urban adventure, attracted girls from all over Massachusetts, Vermont, New Hampshire, and Maine, increasingly so as new promotional corporations set up more mills. Of the 264 workers employed at Waltham in 1820, 225 were women and girls, thirteen boys, and twenty-six were men (this excludes forty mechanics and thirteen bleaching workers).[11]

The Waltham organizational structure had two important effects on innovation in textile technology. First, by assembling a labor force of almost maximum physical and economic vulnerability to operate British cotton manufacturing equipment, it precisely defined the areas needing labor-saving improvements, and it made those innovations imperative. Second, by concentrating optimum organizational and material resources, it gave technological innovative effort a very good chance of succeeding.

The latter effect had several facets. Large capital resources permitted the best engineers to be hired and the most well-equipped workshop facilities to be built. Corporation structure would relieve the engineers, and to some extent the managers, of the anxieties of the broadest financial policy and decision making. Commitment to mass production also had liberating consequences for the matrix of innovation. On the one hand it reinforced the need for well-equipped workshops. On the other, it reduced the number of engineering problems that would arise in modifying British-type equipment because large-scale production combined with product specialization considerably limited the number of operations any given machine would be required to perform.

[10]National Archives, Washington, D.C., Record Group 29, "Records of the 1820 Census of Manufactures: Massachusetts," p. 33.

[11]Ware, pp. 200–202; Taylor, pp. 274–77; 1820 Census (n. 10), p. 33.

Most determinative for technological innovation was the unique composition of the Waltham labor force. The general shortage of unskilled labor, felt most strongly at the beginning of the enterprise before factory employment was known to be socially respectable and financially remunerative, would have encouraged the search for higher operating speeds to increase productivity. The limited physical strength of women and girls had to be compensated by a wider use of inanimate power and a more efficient movement of materials. A rapid turnover rate in female employees minimized operative experience and put a premium on increasing mechanical control and performance. The pursuit of an urban experience, rather than a subsistence wage, in seeking factory employment, easily distracted educated rural middle-class daughters from the monotony of machine minding; some operatives at Lowell were even allowed to read recreational literature and to relax on the job! Since factory discipline could not be ruthlessly imposed without seriously jeopardizing an already unstable labor force, automatic fault-detection devices became necessary at some stages of processing. The critical influence of the mill girls in early American textile manufacturing is attested by the experience of John Williams, agent of the Dover Manufacturing Co. in New Hampshire, who complained (1825):

> We have had to encounter many serious difficulties in consequence of being obliged to make frequent changes in our Girls who have very frequently been induced to go to the neighbouring establishments. This has not only had a tendency to cause a great deficiency in quantity but quality also, for new hands must be learnt & we cannot expect good Goods from them until they have had some experience.[12]

[12]Despite the ban on books in all the machine rooms in the Lowell mills, books were taken into cloth rooms and pages of books were pinned to machine frames or pasted in windows anywhere. In this way the boredom of machine minding was relieved by memorizing scripture verses, hymns, and poetry. Furthermore, female weavers and drawing-in girls, who worked long hours, were allowed to take rest periods as long as thirty minutes whenever they pleased—piece rates discouraged excessive rests. For the ten- or twelve-year-old doffer girls life was even more pleasant. Only half their time was actually spent doffing (removing loaded bobbins from roving or spinning frames). In between doffings they either played or, in summer, were permitted to run home. Of course treatment varied with the overseer (see Lucy Larcom, *A New England Girlhood Outlined from Memory* [Cambridge, Mass., 1891], pp. 153–54, 175–76, 233; Harriet H. Robinson, *Loom and Spindle, or Life among the Early Mill Girls* [New York, 1898], pp. 32, 46, 71; Hannah Josephson, *The Golden Threads: New England's Mill Girls and Magnates* [New York, 1949], pp. 77–84; John Williams to George Bond, October 1, 1825, Letterbook I, Dover Mfg. Co. Papers, New Hampshire Historical Society [Concord, N.H.]).

Resultant Labor-saving Principles and Innovations

Together then, market forces and organizational structures in the Rhode Island and Waltham textile manufacturing systems during the early 19th century promoted a number of labor-saving principles which were expressed in a variety of modifications to British manufacturing equipment. The modifications constitute the core of American improvements during the 1814–40 period of technology adaptation in the textile industry. As such they can be most conveniently surveyed by subsuming them under their respective labor-saving principles which may be enumerated as follows: (1) large-scale production, (2) production of standardized low or medium quality goods, (3) vertically integrated manufacturing, (4) flow production, (5) maximum use of inanimate power, (6) high operating speeds, (7) mechanical vertical integration, (8) maximum mechanical control and work, (9) automatic fault detection. It will be observed that these labor-saving principles *almost* add up to a system of modern mass production (the further division and specialization of labor being implicit in the factory system). Almost, because the product did not lend itself to interchangeability of parts (although there was standardization, and hence interchangeability in the sense of a standardized yarn being used in a variety of ways to make a number of different cloths) and because vertical integration in an advanced economy is not essential to mass production. Equally obvious is the fact that these principles were most completely and vigorously expressed during the early 19th century within the Waltham system of cotton manufacturing. Some of the innovations each incorporated more than one of these labor-saving principles, so that one device can sometimes be used to illustrate several principles.

1. Large-Scale Production

In 1834 the eight mammoth textile manufacturing corporations at Lowell (the Merrimack, Hamilton, Appleton, Lowell, Suffolk, Tremont, Lawrence, and Middlesex) ran 116,804 spindles, 3,933 looms; employed 5,051 females and 1,512 males; and processed weekly 235,700 pounds of cotton and about 9,000 pounds of wool into 704,000 yards of cotton sheetings, shirtings, drillings, and prints; 43,270 yards of carpeting, rugs, and Negro cloths; and 6,000 yards of broadcloths and cassimeres.[13] Apart from the motivation of cornering a large share of the product market, large-scale production,

* [13]"Statistics of Lowell Manufacturers as Published Jan. 1, 1835," *Annual Statistics of Manufactures in Lowell and Neighboring Towns No. 45* (Lowell, Mass., 1882), pp. 21–23.

as practiced under the Waltham system, effected well-known economies in the use of the factors of production.

The major labor-saving economy of scale was gained technologically in one way simply by increasing the size of a single machine but not the number of its operatives. This economy was certainly practiced at Lowell in the 1820s. For example, the throstle spinning frames built for the Merrimack Manufacturing Company's first two mills (built and equipped 1822–25) held sixty-four spindles each. However, when Mill No. 2 was rebuilt after a conflagration in 1828 it was reequipped with forty-eight 128-spindle throstle frames, of the same size as the fifty installed in Mill No. 5 (built and equipped 1828–30).[14] It is safe to assume that both sizes of throstle frame needed only one operative, for in New England mills in 1840 one operative per 166-throstle spindles was the norm,[15] and no applicable stop motions or new patterns of machine management are known to have been introduced in the meantime. It would be pertinent to know whether there were other increases in machine size such as taller bobbins (in roving and spinning frames) with greater traverses which, because a larger bobbin would hold more roving or yarn, would diminish the number of doffing (bobbin changing) operations in a fixed period of time.

2. Production of Standardized Low- or Medium-Quality Goods

The prime consideration in manufacturing standardized coarse or medium goods, as undertaken first at Waltham, was to capture mass markets. But in cotton manufacturing another auxiliary but significant reason for specializing in low-quality yarns was to avoid the use of mules which, spinning on the intermittent principle, were less productive than throstles and, before 1830, required a man's physical strength to operate.[16]

[14]Directors' Minutes, vol. 1 (1822–43), pp. 82–85, Merrimack Manufacturing Company MSS, Baker Library, Harvard University, Cambridge, Mass.; Gibb, pp. 71–72.

[15]James Montgomery, *A Practical Detail of the Cotton Manufacture of the United States of America* (Glasgow, 1840), pp. 116–19.

[16]Spinning consists of three distinct actions: drafting, or stretching the roving to give it the small diameter that characterizes yarn; twisting, which binds the fibers together within the yarn, and winding, or packaging the yarn after it has been formed. In intermittent spinning (as on the spindle and whorl, the spindle wheel, and the jenny, jack, and mule) drafting and twisting are simultaneous, and winding is a separate operation. In continuous spinning (as on the bobbin and flyer wheel, the water frame and the throstle, cap, and ring frames) drafting, twisting, and winding are carried on simultaneously in three zones of the spinning machine. During the 19th century continuous spinning was much faster and cheaper than intermittent

Given mass markets to support standardized products, certain labor savings could be made simply by narrowing the range of goods produced. The narrower the range of product, the less versatile and therefore less complex each piece of equipment in a production line needed to be. This made an operative's job simpler, for example, in reducing the number of times change gears had to be removed to change the speed of roving or spinning frames. Simpler operative procedures shortened the new operative's learning period and reduced the chances of operative error. A less-complex machine would also be less demanding in its maintenance and easier to modify than a more complex one.

Furthermore, production of a standardized yarn encouraged the construction of standardized machines at the weaving and finishing levels. One yarn fineness could be used in different looms to make different cloths varying according to width (sheeting differed from shirting in that it was wider) or weave (whereas sheeting and shirting were plain weave, drilling was twill weave). Again, cloth woven in one loom could be transformed in several different fabrics by applying different finishes (bleaching, dyeing, calendering, printing). Standardized equipment designs would have reinforced the kind of labor savings accomplished already in using less versatile machinery.[17]

---

spinning but not appropriate for fine cotton or any woolen yarns because of the limitations imposed by fiber length and structure. For spinning these kinds of yarns, twisting had to be concurrent with drafting (or else the roving would break), as in the mule. However before the 1830s the mule required a man's strength both to back off (reverse spindles prior to winding and lower yarn from spindle top to cop top) and to put up the mule (wind on the yarn while pushing back the carriage). For a useful discussion of spinning principles, see W. E. Morton and G. R. Wray, *An Introduction to the Study of Spinning* (London, 1962), pp. 133-75; for the mule, see Harold Catling, *The Spinning Mule* (Newton Abbot, Devon, 1970), pp. 1-51. That continuous spinning was more productive (spindle for spindle for the coarser cotton counts) than intermittent spinning in the early American factory system is clear from Montgomery, pp. 69-82.

[17]Discussing desirable production lines for the projected Hamilton Manufacturing Company at Lowell, its first superintendent, Samuel Batchelder (who had just succeeded at New Ipswich, N.H., in adapting the Waltham power loom to make twilled goods, i.e., weaves requiring three harnesses), nicely expressed the possibilities of standardized yarn: "I should think the machinery for the first factory should be built one half to spin about No. 20, the other half No. 40 to 44. The No. 20—warps and filling—would be suitable for the twilled stripes you have seen, and for stout bleaching and printing. The No. 40 would make very good dimity or fine jean—and would be suitable for doubling, for warps for Derrys, to be filled with No. 20. These would be wove in the common loom for plain goods, and a loom, constructed to weave a proper twill, would weave either of the other articles. So that a factory on this plan would be

3. Vertically Integrated Manufacturing

When all the stages of processing were concentrated in one mill or mill complex (vertical integration), transportation costs were lowered, quality control became more uniform, flow production appeared logical, and the labor force could be more rationally deployed. Vertical integration in textiles only became possible in the United States after cotton weaving and printing were mechanized in the period 1810-15; beginning at Waltham in 1813-15, the vertically integrated mill evidently remained a basic model in the cotton industry for the rest of the century. Vertically integrated woolen factories appeared by the 1830s after the adoption of power-loom weaving and the maturation of napping and shearing machines.[18]

4. Flow Production

In 1814 flow production was a 300-year-old principle. It had been brought to America in the 17th century and in 1787 was first mechanized by Delawarean Oliver Evans.[19] Though the flow-of-materials concept was not original with Lowell, he was the first to apply it to textile manufacturing, probably because it was an obvious corollary to vertical integration. In the Waltham system mills, flow production

---

capable of making a considerable variety of goods." However the finer goods (made of no. 40 yarn) fetched little more than the coarser ones so in 1827, two years after the company was organized, the first mill was retooled (for about $300) for the standardized spinning of no. 20 yarn and variety was gained at the weaving and finishing levels only (see Samuel Batchelder to Nathan Appleton, September 25, 1824 [copy], Minutes of the Proprietors [1824-64] and Directors [1825-57], inside front cover and pp. 24-27, vol. a, Hamilton Manufacturing Company MSS, Baker Library). Montgomery observed in 1840, "The machine makers [in the USA] seem to proceed upon the supposition that their machinery is already perfect. The machines are calculated for one kind of goods, and only one system of working. Hence, in making alterations in the adjustment of any machine, so as to . . . suit a different quality of goods, it is frequently necessary to make some new patterns" (see Montgomery, pp. 109-10). Since hardly any pre-1840 machine drawings, wooden patterns for parts, or series of complete machines of one maker and type have survived, it is difficult to prove the point about standardized designs. However one example does suggest this to be so. Three extant twisting frames built for the Merrimack Manufacturing Company, 1830 to about 1840 are so identical as to be distinguished only by the number plates originally affixed by the makers (see below, "Automatic Fault Detection," for further details).

[18]Paul F. McGouldrick, *New England Textiles in the Nineteenth Century: Profits and Investment* (Cambridge, Mass., 1968), pp. 12-13; Cole, 1:253-58.

[19]John B. Rae, "The Rationalization of Production," in *Technology in Western Civilization*, ed. Melvin Kranzberg and Carroll W. Pursell, Jr., 2 vols. (New York, 1967), 2:38-40.

was physically vertical: the water wheel and the machine shop were housed in the basement; carding equipment was located on the first floor, spinning on the second, and weaving on the third and fourth floors. Because of fire hazard, picking was done in an outhouse.[20]

5. Maximum Use of Inanimate Power

At Waltham there were several examples of the maximum employment of inanimate power. Spinning on the throstle frame was one.[21] Admittedly the prime reason for an exclusive reliance on continuous spinning was that it was more appropriate and more productive than intermittent spinning for low-count yarns. But another virtue of the throstle was that, unlike the hand-powered mule, it required little physical strength for its operation. Another instance of the application of power was the Waltham warping frame. This will be examined in more detail later. The relevant point here is that the European warping mill, turned by hand, was replaced with a powered frame.[22] The power loom, too, was an example of this principle.

6. High Operating Speeds

This principle may be defined as the pursuit of higher processing speeds in existing equipment. Together, labor scarcity and the introduction of the power loom stimulated the search for means of raising spinning speeds. Speed in continuous spinning on the throstle, the most economical cotton spinning machine, could not be increased much above 3,000-4,000 rpm without encountering the

[20]Montgomery, pp. 13-17. The point deserves further investigation (in insurance surveys for example), for it is clear that this machinery arrangement was not rigidly preserved throughout the 19th century in Waltham system mills. The Merrimack Manufacturing Company, for example, in 1877 had looms in the lower (but above-basement) stories, spinning frames and mules in the middle stories, and drawing and roving frames and carding machines in the top stories of its four mills. It would be interesting to learn why this reversion to the English system occurred, sometime between the 1840s and 1870s (see "Merrimack Manufacturing Company, Buildings & Machinery, 1877," Merrimack Manufacturing Company MS 58-A, Baker Library).

[21]Essentially the throstle was identical with Arkwright's water frame; the only difference was that in the water frame the spindles were powered from a vertical driving shaft while in the throstle they were powered from a horizontal tin roller called the throstle. The latter machine allowed more spindles than the water frame but not so flexible in that emptying bobbins or making repairs necessitated stopping the entire machine (see Abraham Rees, ed., *The Cyclopaedia; or Universal Dictionary of Arts, Sciences and Literature,* 45 vols. [London, 1802-19], s.v. "Manufacture of Cotton").

[22]Ibid., s.v. "Weaving."

problem of flyer vibration.[23] Hitherto the flyer had been fixed to the top of the spindle by which it was powered. Since the spindle had its bearings in three rails below the flyer and since the flyer legs could not be perfectly balanced by the machine technology then available, centrifugal forces caused the flyer legs to spread and the flyer to wobble at high speeds. In the dead spindle used at Waltham, the flyer was pivoted at its eye, or top, and the flyer legs were brought down and riveted to a plate welded to the upper side of the wharve or power pulley. By this means the flyer could reach speeds of 4,700 rpm and was used at the Boott Mills at Lowell to spin cotton yarns as fine as no. 50 in 1840.[24]

Whereas Moody seems to have been satisfied with the dead spindle, the pursuit of higher speeds preserving yarn quality continued in the 1820s under the Rhode Island system, as might be expected, given its history of horizontal specialization and factory spinning. We know, for example, that Gilbert Brewster, mechanic of Norwich, Connecticut, and later Poughkeepsie, New York, was then working on an improved spindle (as well as a self-acting mule) which he patented in 1824.[25] This was suited to work with the trumpet flyer patented by Oliver G. and Nathan Rogers of Whitestown, Oneida County, New York, also in 1824. Though the specifications and drawings of the Rogers patent are ambiguous, they do indicate a groping toward a substitute for the flyer.[26] This, the real breakthrough in spinning productivity, came in 1828 when Charles Danforth of Ramapo, Rockland County, New York, patented the cap throstle[27] and John Thorp, of Providence, Rhode Island, the ring spinning principle.[28] Cap spinning was an immediate success for it

[23]Andrew Ure, *The Cotton Manufacture of Great Britain,* 2 vols. (London, 1836), 2:132.

[24]Montgomery, pp. 73, 168-69.

[25]Gilbert Brewster (at New Hartford, N.Y.) to John R. Watkinson (at Middletown, Conn.), November 3 and November 8, 1824, Sanseer Manufacturing Company Papers, Connecticut State Library, Hartford; U.S. Patent Office, *A List of Patents Granted by the United States from April 10, 1790, to December 31, 1836* (Washington, D.C., 1872), p. 267 (March 13, 1824). This is cited when the patent specification and/or drawing has not survived.

[26]For the Rogers's patent of August 28, 1824, see National Archives, Record Group 241, "Restored Patents, Specifications," 5:181, and Patent Drawing no. 3931.

[27]*D.A.B.;* for Danforth's cap spinning patent of September 2, 1828 see "Restored Patents, Specifications," 7:231-32 and Patent Drawing no. 5214.

[28]*D.A.B.;* for Thorp's ring spinning patent of November 20, 1828, see "Restored Patents, Specifications," 7:87-89 and Patent Drawing no. 5279.

made possible continuous spinning speeds of up to 7,000 rpm.[29] Unknown problems in ring spinning delayed its more general adoption until the late 1850s, though Montgomery appreciated its high speeds and relatively low power need in 1840 when both cap and ring spinning were spreading in Rhode Island.[30] Since then ring spinning has come to dominate the whole of the coarse and medium cotton and woolen yarn field.

In the dressing frame (considered more fully in the next two sections), drying the sized warp, hitherto achieved by fans, was greatly accelerated by passing the yarn over steam-filled cylinders. For this considerable improvement, Samuel Batchelder (for whom, see below) was awarded a patent on June 22, 1832; by 1840 the modification, which enabled warp preparation to keep pace with large-scale power weaving, was extensively used in New England cotton mills.[31]

Speeders, for roving, were essentially a matter of mechanical control, so are dealt with under that heading. Higher speeds, of course, made fault detection devices more imperative wherever economical.

7. Mechanical Vertical Integration

By mechanical vertical integration is meant either the elimination of an operation or the mechanical continuity of two consecutive operations previously separated in the mill. At Waltham there were two good examples of this: the filling frame and the dressing frame used in conjunction with the warping frame. Up to about 1817, yarn destined for the filling or weft had been unwound from spinning frame bobbins and wound onto quills for the shuttle. The process was laboriously done on the old quiller or hand wheel, one quill at a time. In October 1816, Silas Shepperd of Taunton, Massachusetts, and John Thorp patented a mechanical winder which Lowell and Moody went to inspect. In haggling over the terms of a lease, Moody bluffed with the threat to spin the filling directly onto quills. The idea materialized in his filling frame, patented in 1819.[32]

29Ure, 2:145–47; Evan Leigh, *The Science of Modern Cotton Spinning*, 2d ed., 2 vols. (Manchester, 1873), 2:208–10.

30Montgomery, p. 70.

31"Restored Patents, Specifications," 14:67–68 and Patent Drawing no. 7120; Montgomery, pp. 96–97; Samuel Batchelder, *Introduction and Early Progress of the Cotton Manufacture in the United States* (Boston, 1863), p. 86. Montgomery and, surprisingly, Batchelder give incorrect dates—which emphasizes the need to double check even these first-hand contemporary sources.

32*List of Patents, 1790–1836* (n. 25 above), p. 205 (May 6, 1819); Gibb, p. 37.

Before this, in 1816, Moody patented a warping frame, the characteristics of which are established below in the consideration of automatic fault-detection devices. Since it appeared nearly two years before Moody patented his dresser (or slasher, which sized the cotton warp) and since its main feature was a stop motion, it is most likely that the warping frame wound the warp directly onto a loom warp beam, which, when filled, would be removed from the warping frame and positioned in the loom. Once inside the loom it would have been slashed (sized) in the old laborious way, a few feet at a time. After the dressing frame was developed in 1817, the warp would have been built up on a back beam which, when full, would have been taken to the dressing frame. Here the yarn was sized as it passed from the back beam to the loom warp beam. For a warp with more than a few hundred ends, several back beams loaded with warp would be set in the dresser; then all the ends would be sized and combined on the loom warp beam. Whether one or several back beams were used in the dresser, the three operations of warping, sizing, and beaming, hitherto performed in three distinct stages, were achieved in two steps by the warping and dressing frames.[33]

These machines and their labor-saving principle were taken up outside the Waltham system. The Rogers's trumpet flyer of 1824, for example, was claimed to be applicable for "spinning shuttle bobbins."[34]

Railway carding, patented by William B. Leonard of the Matteawan Machine Shop, Fishkill, New York, incorporated the so-called railway drawing head by which the ends produced by a line of cards were brought to the center of the line on traveling bands or railways and there fed into a drawing head. By this means one transference of cans of card ends (between carding machine and drawing frame) became redundant. The cans of doubled ends filled by the railway drawing head still had to be removed to a drawing frame, of course.[35]

The most famous example of American mechanical vertical in-

[33]*List of Patents, 1790–1836*, pp. 164, 189, for Moody's two patents of March 9, 1816 and January 17, 1818. The warping frame is discussed further under "Automatic Fault Detection" and the dressing frame under "Maximum Mechanical Work and Control." For the latter see also Montgomery, pp. 91–101 and Plate VII.

[34]Restored Patents Drawing no. 3931 of Oliver G. and Nathan Rogers's flyer.

[35]For Leonard's patent of September 16, 1833, see "Restored Patents, Specifications," 16:127–48, Patent Drawings nos. 7760, 7761; and Montgomery, pp. 44–47 and Plate IV. Leonard's device is remarkably similar to the delivery system patented in England by John George Bodmer on October 14, 1824 (Great Britain, Patent no. 5016).

tegration occurred in the woolen industry: the ring condenser. The problem it solved was the production of a continuous sliver from the delivery end of the carding machine. Unlike the convoluted structure of cotton fibers, the scaley, tubular structure of wool fibers prevented the formation of a continuous roll by passing the carded sheet or lap through a trumpet. Wool-carding engines turned out short rolls or rolags, the length of the machine's doffer and roller bowl, from which the rolags fell laterally. These rolags then had to be taken from the carding machine and pieced end to end on the apron of the slubbing billy. The billy, which had an action similar to that of the jenny (except that the spindles were mounted on the carriage and not fixed in the frame) stretched, reduced and slightly twisted the slubbing ready for spinning on the jenny or mule. But the billy was labor-intensive: it required a boy to stand at the apron end to piece the rolags onto the slubbed ends and a man who would power and control the machine.[36]

* As arthur H. Cole has shown, since 1810 American manufacturers had been seeking a device which would eliminate the billy.[37] John Goulding of Dedham, Massachusetts, after five years of experimenting came up with the answer in his 1826 condenser patent.[38] By combining two devices designed by others—the ring doffer (or final clearing cylinder in a carding machine with the card clothing fixed in rings around it), patented by Ezekiel Hale of Haverhill, Massachusetts, in 1825, and revolving tubes (as many as the rings of card clothing on the doffer and positioned horizontally close to each ring) patented in the cotton tube roving frame by George Danforth of Norton, Massachusetts, in 1824[39]—and adding them to the carding machine, Goulding produced a continuous wool roving. Two ring doffers were needed to separate the carded wool into strips or ends, and two sets of revolving tubes, through which the ends passed, gave them enough twist to make the fibers adhere to an endless sliver or tubular form.

8. Maximum Mechanical Work and Control

Since mechanical control invariably includes mechanical work in

[36]Rees, s.v. "Woollen Manufacture."

[37]Cole, 1:101-2.

[38]For Goulding's December 15, 1826 patent, see National Archives, Record Group 241, "Restored Patents, Reissues," 2'141-49, and Restored Patents, Drawing no. 4605; Cole, 1:102-7.

[39]Cole, p. 102; *Lists of Patents, 1790-1836,* p. 284 (February 18, 1825), and p. 272 (September 2, 1824). Danforth's tube roving frame is described in Montgomery, pp. 60-61, and tube twisting is clear from the English patent on the device taken out in the name of Joseph C. Dyer (Great Britain, Patent no. 5217, July 16, 1825).

these early innovations, the devices may be grouped according to the means by which this work and control were extended: either by increasing the versatility and/or productivity of an existing machine or else by mechanizing an operation previously performed by hand.

One of the most challenging obstacles to maximum mechanical control in an existing machine was presented by the roving frame. At this stage the delicate cotton ends produced on the drawing frame had to be further drafted (attenuated), slightly twisted and then wound onto bobbins ready for spinning. Two alternative machines had been developed in England: Arkwright's roving can frame and, later but before 1812, the double speeder. The former had two serious disadvantages: the degree of twist could not be controlled (it depended on the point at which the roving came to rest inside the rotating can), and rewinding the rovings onto bobbins was both damaging and laborious.[40]

Both disadvantages of the roving can frame were obviated in the double speeder, but another appeared. Roving was so weak that both flyer and bobbin had to be powered (the reason it was called the double speeder), but unless the roving was wound onto the bobbin at the same rate that it emerged from the delivery rollers of the drafting element, it would stretch or break, disastrous in either case. Since the acting surface of the bobbin increased in size after each traverse, some method had to be found for decreasing the speed of the bobbin at this point. In the English double speeder this was accomplished by two cones, one on the shaft driving the flyers (rotating at a constant speed) and the other, to which the first cone transmitted power by means of a belt, on the shaft driving the bobbins. Since the cones were set in opposite directions the shifting of the belt altered the speed of the bobbins. This belt movement, governed by a rack-controlled shipper, was the heart of the double speeder problem. Whenever drafting needed changing, to produce a finer or coarser roving, the rack and perhaps cones had to be changed also—each tooth of the rack corresponded to one layer of roving, and the difference in cone sizes equaled the difference between the empty and full bobbin. There was yet another disadvantage. The finer the roving, the smaller the increments of movement of the belt along the cones needed to be; and with high counts the belt and cones device was not sufficiently sensitive.[41]

[40]Rees, s.v. "Manufacture of Cotton."

[41]Ibid.; James Montgomery, *The Carding and Spinning Master's Assistant; or the Theory and Practice of Cotton Spinning* (Glasgow, 1832), pp. 109–12.

Shortly before his death, Lowell mathematically devised a gearing mechanism, which Moody patented in 1820 and 1821, that was based on the English double speeder and which Gibb believes brought the double speeder into American use for the first time.[42] Undoubtedly the Lowell-Moody gear train simply compensated for the increasing size of the bobbin. As the Waltham company made only a few yarn counts (no. 14 for yard-wide shirting being one)[43] to avoid English competition, there was little need to invent further adjustments. The double speeder problem provoked American innovation in other directions, however. Simpler machines were built, like George Danforth's tube roving frame (1824) and Gilbert Brewster's eclipse roving frame (1829) for example.[44] These were more productive than the double speeder and a lot cheaper to buy or build (the Waltham double speeder cost $950 to build and sold for $2,000),[45] but their qualitative results were less satisfactory.

On the other hand, attempts were made, particularly in Rhode Island (no doubt in response to the cotton industry's organizational form in that state and to market competition in coarse goods from northern New England), to build an English-type double speeder more versatile in its range of counts with a minimum of adjustment. These efforts succeeded in 1822 when Asa Arnold (1788–1865) of North Providence, Rhode Island, invented the differential or equational gearbox (patented January 1823)—now most familiar in the rear wheel axle of the automobile. By inserting a differential gear between the bobbin cone and the shaft driving the bobbins, various and finer rates of change in bobbin speed could easily be accomplished with the replacement of the change gear engaging the differential—an enormous advance over the old method of machining and installing a new rack and sometimes different cones, or even building a new machine altogether (to accommodate longer cones). Arnold claimed that his differential made speeders 20–30 percent more productive than those without it.[46]

[42]Gibb, p. 36.

[43]Mailloux, p. 89.

[44]*List of Patents, 1790–1836,* p. 272 (September 2, 1824) and p. 380 (April 18, 1829); Montgomery, *Carding and Spinning Master's Assistant,* pp. 127–32; idem., *Practical Detail,* pp. 60–61; Ure, 2:101–15. See also n. 39 above.

[45]Gibb, p. 47.

[46]*D.A.B.* for Arnold; *List of Patents, 1790–1836,* p. 251 (January 21, 1823), a patent which MacMurray (n. 1 above, pp. 106, 447) notes was renewed on September 10, 1836; Montgomery, *Carding and Spinning Master's Assistant,* pp. 109–26; Ure, 2:63–100, Plate IV; idem, *Dictionary of Arts, Manufacturers, and Mines* (London, 1839), pp. 362–64; Charles Tomlinson, ed., *Cylopaedia of Useful Arts, Mechanical and Chemical, Manufactures, Mining and Engineering,* 2 vols. (London, 1854), 1:458–65. I am grateful

Leonard's railway drawing head card also extended the mechanical work of an existing machine by increasing its productivity. In common carding machines only one third of the main cylinder was actually carding at any one time during processing. In the cards Leonard built at Matteawan, this proportion was as high as two-thirds.[47]

Three major examples of extended mechanical control and work achieved by mechanizing a manual task were developed by Americans at the weaving stage of manufacturing. The self-acting temple, the first of these, was a mechanism for automatically keeping the cloth laterally stretched as the filling was beaten up by the lay or batten (a hinged frame which carried the reed and shuttle race and boxes in the loom). In the handloom the freshly woven cloth was stretched by two pieces of hardwood tied together (so as to be adjustable to a variety of cloth widths). Each of these was tipped with a set of metal points that gripped one edge or selvage of the cloth. Every time the cloth was taken up (wound onto the cloth beam), the temples had to be reset.[48] Such an interruption would seriously diminish production rates in a power loom. For example, an 1830 power loom running at 120 picks per minute and weaving cloth with eighty picks (representing eightly filling yarns) per inch, would make 90 inches of cloth per hour.[49] Without self-acting temples the loom would have to be stopped and the temples reset every time 6 inches of cloth were woven—6 inches being an average distance between the fell of the cloth (the edge where the last pick or filling yarn was beaten up) and the breast beam. Should each stoppage last only thirty seconds, fifteen stoppages per hour would represent a drop in loom productivity of 15 percent. In developing the British power loom, it was therefore imperative for the Boston Manfacturing Com-

to Mr. William B. Strzelewiecz, textile engineer of Andover, Mass., for a telephone conversation on the principles of roving frame bobbin mechanisms.

Arnold's patent specification is in National Archives, "Restored Patents, Reissues," 2:157-59, and a copy of the patent drawing is in the Miscellaneous Papers Box of the Zachariah Allen Papers in the Rhode Island Historical Society (Providence). The latter also contains Arnold's account of his invention and its early adoption in the Waltham system; the account is also in the *Transactions of the Rhode Island Society for the Encouragement of Domestic Industry* for 1861, pp. 80-85. I am indebted to Paul E. Rivard, director of the Old Slater Mill Museum, Pawtucket, for bringing these Rhode Island sources to my attention.

[47]Montgomery, *Practical Detail*, p. 44.

[48]Clinton G. Gilroy, *The Art of Weaving, by Hand and by Power* (New York, 1844), p. 83.

[49]This was the speed of an English Sharp & Roberts power loom in 1836 (see Ure, 2:323).

pany to acquire a self-acting temple. Some kind of improved temple it did purchase from one J. Stimpson in 1814, and this, or some modification of it, sufficed for ten years.[50]

However, the first thoroughly successful self-acting temple was invented and then patented in 1816 by Ira Draper (1764-1848), a textile machinery manufacturer of Weston, Massachusetts.[51] After experimenting with nipper temples, Draper, as his patent shows, turned to the rotary principle. His temple consisted of a freely rotating ring of spiked teeth set in a casing which was screwed to one side of the breast beam. The teeth gripped the cloth from underneath and, to prevent the cloth from moving in opposite directions at the same time, the ring was partially enclosed so that only a semicircle of teeth caught hold of the cloth. To keep the selvage from slipping off the teeth, these were angled outward and a hinged clasp that came over the selvage pressed the cloth down onto them. No manual adjustment was necessary because as the cloth was taken up the teeth of the temples (one on each side of the breast beam) revolved and in so doing seized a new section of selvage.[52] Draper slightly modified this design in 1829 by setting the temple on a spring beam (secured to the breast beam) in order to reduce some of the strain exerted on the filling as it was beaten up by the lay.[53] The success of Draper's temples was evidenced by their popularity in power weaving, and in 1824 the Boston Manufacturing Company purchased patent rights to use them.[54]

A second example of mechanizing a manual operation was the perfection of the dressing frame. In handloom cotton weaving, dressing (toughening the warp yarn by impregnating it with a size, like starch) was performed after the warp was set in the loom. The weaver would push back the lease rods (placed in the warp to separate alternate yarns and to maintain an even tension), comb any burrs or knots off the warp exposed between the warp beam and the harnesses (24-36 inches, perhaps more, depending on the size of the loom), size that portion of the warp, let it dry, move back the lease rods, weave 2 or 3 feet of cloth, and then repeat the dressing cycle.[55] If this method was used in a power loom with a production

[50]Gibb, p. 14.

[51]*D.A.B.*

[52]Ibid.; Draper's patent of June 7, 1816 is in "Restored Patents, Specifications," 3:525-33, and Patent Drawing no. 2608. It is discussed in Gilroy, pp. 410-12.

[53]Draper's patent of April 1, 1829 is in "Restored Patents, Specifications," 8:365-66, and Patent Drawing no. 5419 (see also Gilroy, pp. 412-15).

[54]Gibb, p. 739, n. 26.

[55]Rees, s.v. "Weaving."

rate of 90 inches an hour, four dressings would be needed every two hours. Assuming that dressing a yard of warp 30 inches wide took ten minutes (a low guess), that loom's productivity would drop 30 percent. This estimate seems conservative because one early 19th-century handloom weaver claimed that dressing the warp occupied a third of his time.[56] In Britain an English cotton manufacturer, William Radcliffe, in 1803 had invented a mechanical dressing frame which sized the warp before it went into the loom.[57] Radcliffe's dresser, though sound in design, was defective in materials. Water in the sizing caused wooden rollers to crack or warp and iron rollers to rust. The American innovation was to solve this materials problem. Following a suggestion from his brother, Paul Moody at Waltham made the rollers from soapstone. This simple answer he patented in 1818 and the material continued to be utilized for this purpose until after 1850.[58]

The last major American mechanization of a manual task in the weaving process was the perfection of a versatile shedding device. In early power looms, cams actuated the shedding (lowering and raising the harnesses to form a shed through which the shuttle was propelled), one cam controlling one harness. Though accomplishing an ideally smooth movement in raising and depressing the harnesses, and so minimizing stress on the warp, cams had two major drawbacks in shedding. First, if they were positioned inside the loom, directly below the harnesses, the height of the harnesses above the ground (generally 2-3 feet) limited the size of the cams to within a diameter of about 6 or 7 inches; this in turn limited the potential number of cams (and therefore harnesses) generally to four or five, because within this range of diameter any more cams would create an unacceptably jerky harness movement. This difficulty could be reduced but not eliminated by using larger cams positioned outside the loom, either to one side or above. But here the second defect of multiple cam shedding became apparent. A change in weave pattern frequently entailed a change in the number of harnesses, which meant installing a new set of cams; and this was a job which could take hours if there were half a dozen or more cams whose timings had to be set carefully in relation to one another.[59] So, until a

[56]Duncan Bythell, *The Handloom Weavers: A Study in the English Cotton Industry during the Industrial Revolution* (Cambridge, 1969), p. 67.

[57]Walter English, *The Textile Industry: An Account of the Early Inventions of Spinning, Weaving and Knitting Machines* (London, 1969), pp. 102-3.

[58]Gibb, p. 35; Moody's dressing frame was patented January 17, 1818.

[59]International Correspondence Schools (International Textbook Co.), *International Library of Technology*, vol. 80 (Scranton, Pa., 1906), sec. 56, pp. 1-15, and sec. 58, pp.

substitute for cam shedding was found, power looms would be confined to plain, twill, and satin weaves; though satin weaves, needing five or more cams, were not a practical proposition unless a very long production run was assured to offset loom fixing costs. Consequently until the 1820s, in Europe as well as the United States, complex weaves demanding up to thirty or forty harnesses continued to be done on the handloom with the aid of the drawboy (a weaver's assistant who, working from a squared and colored chart, could select and move harnesses by means of levers and rigging).[60]

In the 1820s, the dobby or witch appeared in England.[61] Its characteristic principle was harness selection by means of a rotating chain of bars, in which each bar represented one pick of the pattern and each hole in the bar represented one harness. The chain was positioned close to a set of levers and cords which raised the harnesses. For a given pick, a pin was bolted into each of the holes (in the bar) corresponding to each of the harnesses which had to be raised at that pick. Thus when at that pick the bar reached the actuating position, the pins would depress the levers controlling the appropriate harnesses and so lift them. In this manner almost fifty harnesses might be fixed to form very complicated weaves; equally important, to change a weave (as dictated by fluctuations in fashion), the dobby chain could be taken apart and rebuilt in a matter of minutes rather than hours. This dobby principle, a variation of the jacquard principle, was patented in England in 1824 and by 1830 was applied to handloom silk weaving at Spitalfields, London.[62] What English-born William Crompton (1806-1891)[63] did in the

---

1-16. I am grateful to Professor John L. Merrill, lately of the Textile Department, Lowell Technological Institute, for answering my questions on cam looms.

[60]Robert Patterson, "Spinning and Weaving," in *A History of Technology*, ed. C. Singer and others, 5 vols. (Oxford, 1954-58), 3:164-65.

[61]English, pp. 105-7.

[62]G.B., Patent no. 4951 (May 13, 1824), awarded to John Potter; Ure, 2:313 (which gives the wrong date); Anon, *A Treatise on the Origin, Progressive Improvement, and Present State of the Silk Manufacture* (in *The Cabinet Cyclopaedia*, ed. Dionysius Lardner [London, 1831]), p. 260.

[63]*D.A.B.*; George Crompton, *The Crompton Loom: William and George Crompton* (privately printed, Worcester, Mass., 1949), pp. 30-31, where Crompton's U.S. patent of November 25, 1837 is reproduced. The first application of the power loom to the weaving of damask linen tablecloths, which would necessitate the use of a jacquard mechanism, was supposedly achieved in 1828 or 1829 by William Mason (1808-83), the textile machinery manufacturer and inventor. Primary corroborative evidence is really needed to confirm this claim, and to discover Mason's reasons for complete abandonment of his device (see J. D. Van Slyck, *New England Manufacturers and Manufactories*, 2 vols. [Boston, 1879], 2:431).

United States was to build a metal dobby and to incorporate it into the power loom. This he patented in 1837 and three years later had his first machine working in a woolen loom for the Middlesex Manufacturing Company at Lowell.[64] Within twenty-five years the production of fancy cassimeres (at first four harness weaves; always characterized by variety of design) in Massachusetts rose 15 percent, accounting for 30 percent of all woolen cloth manufactured in the state in 1865.[65] Unable to compete against British broadcloths with their superior finishes, American woolen manufacturers found in the Crompton dobby a new means of capturing mass domestic markets. It permitted complex weaves which could be easily changed to keep pace with fluctuations in fashion. In turn, the dobby stimulated the addition of multiple shuttle boxes for manufacturing fancy-colored as well as fancy-structured cloths. Cotton manufacturers also switched to the dobby to diversify production but after 1860.[66]

9. Automatic Fault Detection

The final labor-saving principle which motivated major American contributions to textile technology was automatic fault detection in equipment other than power looms. Mechanical stop motions, to operate when warp or filling yarns broke or the shuttle jammed in * the shed, were a prerequisite of power loom weaving and had been incorporated in some of the first power looms in England. There, the labor-intensive situation apparently retarded the application of stop motions to other textile machines in which they would have been economically feasible, apart from doubling machines used in the silk industry.[67]

In early 19th-century American cotton mills, the attempts to make one operative responsible for more than one machine and/or to

[64]Cole, 1:306-8.

[65]Ibid., p. 309.

[66]According to Ware (p. 84), few fancy cottons were made in the Waltham system until after the Civil War. Statistics are difficult to obtain for individual cloth types, and 19th-century American cotton cloth production really needs closer investigation.

[67]The silk doubling frame stop motion (drop wire type) is described and illustrated in Rees, s.v. "Silk Manufacture of," and Anon, *Treatise on Silk Manufacture,* pp. 203-5. This stop motion first appeared in Britain in Lombe's silk doubling frame (patented 1718) according to W. English, "A History of Winding," *Journal of the Textile Institute* 49 (1958); Proceedings, 169-70. If so, the device probably came from Italy or the East. The classic early 19th-century technological accounts of the British cotton manufacture in Rees, *Cyclopaedia* and Ure's *Cotton Manufacture* mention no stop motions outside the power loom. Evidently those patented for twisting machines after 1770, e.g., by Samuel Unwin (G. B. Patent no. 1009, March 23, 1772) or by Thomas Margrave (no. 2802, December 19, 1804), were not in widespread use by the 1830s.

increase the speed of machines revealed two kinds of operation in which productivity was impeded without automatic stop motions. The first kind involved the fusion of two or more strands, as in the drawing frame or the twisting frame. Failure to detect a broken sliver on the former altered the fineness of the resultant doubled sliver, and consequently of the roving and of the yarn. Failure on the twister, which plyed yarns, meant that in a frame making 3-ply yarn, breakages would yield 2-ply, single yarn, or nothing at all. The second kind of operation had a unique example, warping. In this, strands were not joined, but a large number (perhaps 200-400 or more) were closely juxtaposed and wound onto the loom's warp beam or onto a back beam (as explained above under "Mechanical Vertical Integration"). If one yarn broke, the warp beam or section beam (back beam) had to be stopped immediately, turned back, the severed yarn had to be found, and then the two ends rejoined. Unusual vigilance was therefore demanded of the operator attending the powered warper. But above a certain speed even the most alert operator would fail to detect a breakage until many feet of warp had been wound onto the beam. With several hundred ends in the warping frame, it is not surprising that the twin impulses of American manufacturers, to increase productivity and to diminish labor-intensiveness, should have led to the invention of a warping frame stop motion before those for drawing and twisting frames.

Credit for the American development of both kinds of stop motion (there were only two, apparently, before 1840 – not including devices to halt a machine when those elements being fed, like bobbins in the fly frame, were full) belongs to engineer-manufacturers in the Waltham system. Though no evidence I have seen explicitly states the point, it is reasonable to suppose that the composition of the Waltham-system labor force of unskilled, easily distracted young girls encouraged, if not compelled, the early invention (or adoption, if the drop wire device derived from British silk doubling) and use of stop motions outside the loom.

The warping frame stop motion (of the drop wire and rocker type) was developed by Paul Moody at the Boston Manufacturing Company's mill at Waltham and was perfected and patented by him in early 1816. The evidence for this is scattered but can be quickly summarized as follows. On February 7, 1815, one Caleb Pratt received $2.67 from the company for two days' work on the warping mill.[68] Nearly sixteen months later, on May 28, 1816, the company

[68]Ledger, vol. 23 (1813-17), p. 85, Boston Manufacturing Company MSS, Baker Library.

paid out $30 for "patent for warping machine."[69] This must refer to the March 9, 1816 patent awarded to Paul Moody for "winding yarn from bobbins or spools."[70]

That the Waltham warper was an unusual one is evident from the prices the company charged for it. In the first sales of machinery in 1817, before market prices were settled, power looms were sold for $90 and warpers for $250. From 1817 until 1823 warpers, which cost $200 to build, were sold for $600 while power looms, which cost $90 to build, were sold for $125. Only the dresser and the double speeder cost more ($1,000 and $2,000, respectively) than the warper.[71] Yet the warping frame was a simple machine. It consisted of a creel (or rack to hold loaded bobbins), a heck (or grid by which yarns from the bobbins were kept from tangling) and a cylindrical frame rotating on a vertical axis onto which the warp or chain was wound.[72] A warping mill like this (probably second hand) fetched $12 in 1816—this was the Philadelphia price a Brandywine cotton mill then paid for such a warper.[73] Now, in the period 1817-23 the Boston Manufacturing Company sold sixteen warping frames at $600 each and three warping frame patent privileges at $300 per machine built.[74] The frame thus brought the company a clear profit of $7,300 in six years. Its main attraction was, it seems, the stop motion.

The nature of the stop motion design is disclosed in the list of forgings made by Samuel C. Oliver for warping frames built by the Boston Manufacturing Company for the Merrimack Manufacturing Company's Mill No. 2 (built 1823-25 and its machines 1822-25). This list includes a rocker shaft, top arm for rocker shaft, plate for rocker, long and short spring rods, pieces to hold same, steel drop wires, catch for spring rod, and piece to hold catch.[75] Earlier entries in the company's Accounts Current mention drop wires—1,560 for seven warpers or about 220 per machine on the average.[76] Besides

[69]Ibid., Accounts Current, vol. 34 (1816-22), p. 12.

[70]*List of Patents, 1790-1836,* p. 164.

[71]Gibb, p. 47.

[72]Rees, s.v. "Cotton," "Mill," and "Weaving."

[73]Account with Archibald McCall, June 16, 1817, Box 4, Duplanty, McCall and Company Papers, Accession 500, Eleutherian Mills Historical Library, Greenville, Delaware.

[74]Gibb, p. 47.

[75]Ibid., pp. 57-58; Samuel C. Oliver's Forging Patterns List, item A-1, Case S-2, Proprietors of the Locks and Canals on the Merrimack River MSS, Baker Library.

[76]Accounts Current (n. 69 above), p. 192 (June 1819); Samuel Batchelder ordered 250-thread warpers for the Hamilton Company in 1826. Blotter (1825-29), vol. 20, December 4, 1826, Hamilton Manufacturing Company MSS, Baker Library.

suggesting that sectional warping was practiced at this early date (see "Mechanical Vertical Integration" above), these parts indicate that the stop motion was of the drop wire and rocker type described and illustrated by James Montgomery (see fig. 1).[77] As Montgomery, a Scottish cotton mill manager, had emigrated to America in late 1835 or early 1836 to become superintendent of the York Manufacturing Company at Saco, Maine, and spent a number of years there,[78] it is safe to assume that the warping frame stop motion he describes was very little different from those built by Moody after 1816. Montgomery, it should be noted, attributed the invention of the device not to Moody but to Moody's mentor and one-time employer, the eminent engineer Jacob Perkins.[79] What is also interesting is that an 1820 inventory of the Dover Cotton Factory (New Hampshire) lists "2 warping machines with stop motions with 2 dressing of Adams & Markhams plan," all valued at $1,000.[80] As the Dover company paid $6,000 the next year to the Boston Manufacturing Company for the right to use and build all machinery used or patented by the Waltham company,[81] it is likely that these warping frame stop motions were not the Perkins-Moody type. This would indicate that the American search for fault detectors was more widespread than a first glance might suggest.

The other kind of stop motion, the latch type shown in figure 2, was invented by Samuel Batchelder (1784–1879) specifically for the drawing frame in 1832.[82] As manager of the New Ipswich Cotton Manufactory (New Hampshire), about 1812–25, Batchelder had purchased a Waltham warping frame in 1821. From 1825 through 1831 he was agent or manager of the Hamilton Manufacturing Company at Lowell (yet another of the Boston-promoted land-and-power development *cum* cotton manufacturing corporations).[83] He was therefore quite familiar with the automatic fault-detection principle. In 1831 he was hired as the principal agent of the York Manufacturing Company at Saco, Maine, and within

[77]Montgomery, *Practical Detail*, pp. 85–90 and Plate VI.

[78]Gibb, p. 749, n. 12.

[79]Montgomery, *Practical Detail*, p. 85. Greville and Dorothy Bathe, *Jacob Perkins: His Inventions, His Times and His Contemporaries* (Philadelphia, 1943).

[80]Agreement between Williams & Wendell and A. & J. Wendell, June 29, 1820, Dover Cotton Factory file, Williams and Wendell, 1815–31, Case 11, Wendell Collection, Baker Library.

[81]Gibb, p. 43.

[82]Batchelder, pp. 84–85. Batchelder's stop motion was patented in Britain by John Houldsworth in 1835 (see G. B. Patent 6951 [December 9, 1835]).

[83]*D.A.B.*; Gibb, p. 110; Accounts Current (n. 69 above), p. 308.

twelve months had designed but did not patent his stop motion. This, too, is described in Montgomery's *Practical Detail* and is best understood from that source.[84]

The drop wire and rocker stop motion proved appropriate also for the twisting frame. Here the evidence seems confined to three surviving sixteen-spindle throstle twisters, donated in 1939 to the Smithsonian (twister no. 3, see fig. 3) the Old Slater Mill Museum (twister no. 1), and the Chicago Museum of Science and Industry

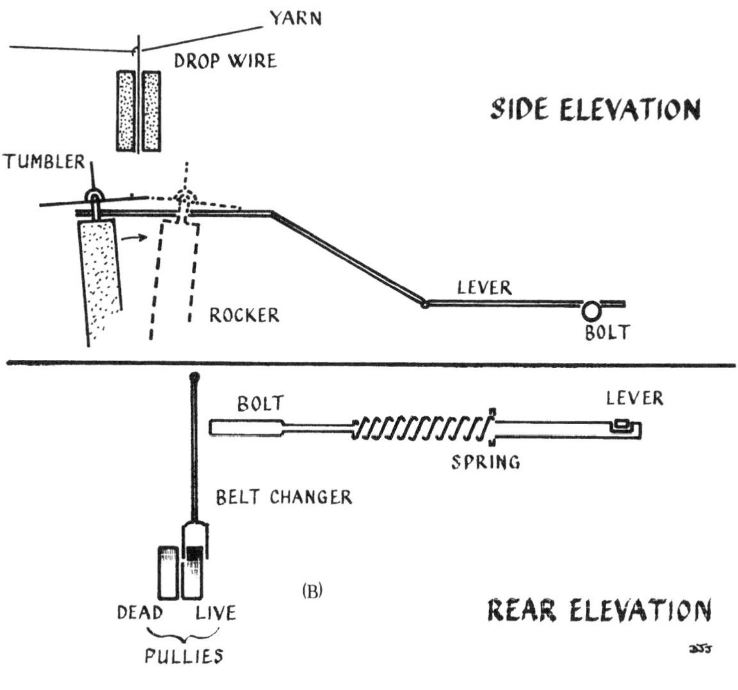

FIG. 1—Warping frame stop motion of the drop wire and rocker type found in American cotton mills before 1840. The breakage of a yarn end releases a drop wire which strikes against a pivoted tumbler on an oscillating rocker. The tumbler turns and depresses a pivoted lever, the other end of which rises from a retaining groove in a spring bolt. When thus released, the spring bolt strikes the belt changer (shipper) moving it from the live (fixed) to the dead (loose) pulley, halting the machine for the breakage to be repaired. A good description and some productivity figures for the warping frame stop motion are found in Montgomery, *Practical Detail,* pp. 85–90 and Plate VI. This stop motion closely compares with that on contemporary American twisting frames (see figs. 3–5).

[84]Montgomery, *Practical Detail,* pp. 56–59.

(twister no. 2) by the Merrimack Manufacturing Company (incorporated 1822; liquidated 1958-59).[85] Two aspects of these twisters are curious. The donors dated them to 1822 and attributed them to Moody, but a search of the records of the Boston Manufacturing Company (which built the Merrimack machines before 1825) and those of the Proprietors of the Locks and Canals on the Merrimack River (which built most of them from 1825 until 1845) indicates that the Merrimack Company's first twister (called a twine machine) was constructed in 1835—and only one at this time.[86] Furthermore, an 1830 valuation of the Merrimack Company which

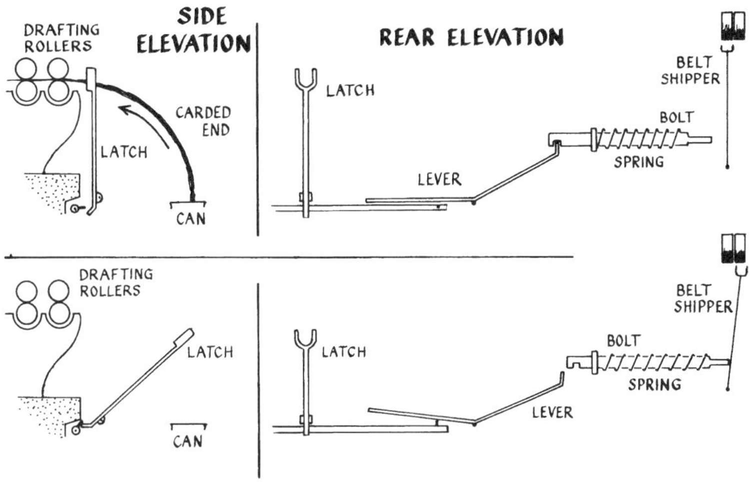

FIG. 2.—Drawing frame stop motion of the latch type found in American cotton mills before 1840. In this, the movement of the carded slivers passing from the can to the drafting rollers supports a pivoted latch, so balanced as to fall back when the sliver breaks. In falling back, the lower end of the latch turns a rod which has a radial catch at its far end and this, as the rod rotates, raises a pivoted lever. The other end of this lever is thereby depressed and removed from a retaining groove in a spring bolt. The drawing frame is then stopped in the same manner as the warping frame. For a good description and some productivity figures, see Montgomery, *Practical Detail,* pp. 56-59.

[85]Correspondence received from the various involved parties by Dr. Frederick L. Lewton (curator), February 9-June 20, 1939, Artifact accession file no. 152,776, Smithsonian Institution.

[86]Day Book, vol. LN-2 (1832-36), Merrimack Manufacturing Company account, opp. p. 441, Proprietors of the Locks and Canals on the Merrimack River MSS, Baker Library.

lists all the machinery in the company's five mills does not mention a twisting frame or twiner.[87] An 1877 inventory of the company shows that there were five twisters on the fifth story of No. 1 Mill (each had eighty spindles) which were built in 1846, 1849, 1852, 1867, and 1869 respectively.[88] However, this evidence cannot be regarded as conclusive yet. The Hamilton Manufacturing Company (in-

FIG. 3.—Merrimack Manufacturing Company twisting frame no. 3, probably built in the period 1830—ca. 1840; now in the custody of the Division of Textiles, National Museum of History and Technology, Smithsonian Institution, Washington, D.C. (Smithsonian accession no. 152,776 and catalogue no. T. 10,890). The frame measures approximately 5 feet 9 inches long, 4 feet wide, and 3 feet 6 inches high and has sixteen spindles. The creel at the rear of the frame can hold four rows of sixteen bobbins each, so the twister could have plyed sixteen threads of up to four yarns each. In the photograph, the yarns from three loaded bobbins have been drawn through the machine to indicate the yarn path. From its bobbin, each yarn passes under the hook of a drop wire (thereby holding up the wire), through guide pins, between a pair of rollers acting as a clasp (shown in an open position), and then under and over two pulleys at which the several yarns are brought together before passing through the eye of the flyer to be twisted into thread. The spindles are dead throstle spindles, that is, the flyer is powered but the bobbin around which it turns is not; as the flyer drags the bobbin around, the twisted thread is wound onto the latter. (Courtesy of the Smithsonian Institution.)

[87]Directors' Minutes (n. 14 above), pp. 82-86.

[88]"Merrimack Manufacturing Company Buildings & Machinery, 1877," vol. 58-A, Merrimack Manufacturing Company MSS.

corporated in 1825 as the second of the Lowell manufacturing mammoths) received amounts of 2-12 or more pounds of "harness twine" from the Merrimack Manufacturing Company at intervals ranging from a few days to a few weeks from 1826 through the early 1830s. Harness twine was 3-ply cotton thread twisted from coarse yarn (about no. 8) and used to knit heddles. In view of the 1830 valuation it is possible that the Merrimack Company obtained its twine from an outside supplier and this might be a safe guess except for an early March 1832 entry in a Hamilton Manufacturing Company waste book which reads, "Rec'd of J. B. Howard . . . 32 Bobbins for twister." Would the Hamilton Company have been the first Lowell corporation to utilize a twister? If so, it failed to meet the demand for harness twine, because even after this date this company bought twine from the Merrimack Company, other mills at Lowell

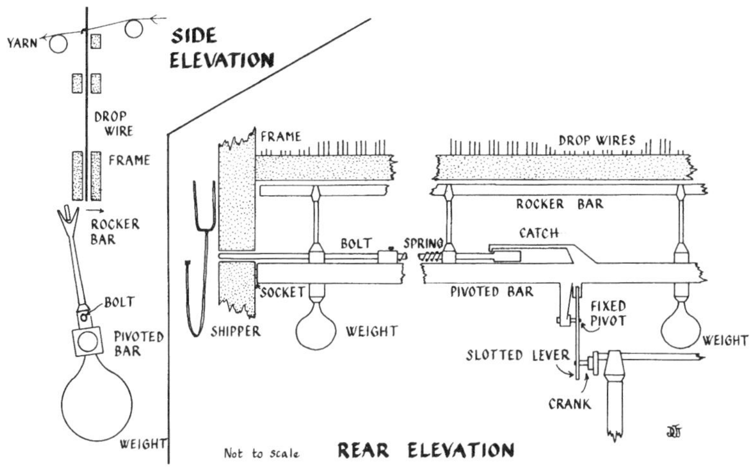

FIG. 4.—Diagram of a stop motion on the back of the Merrimack Manufacturing Company's twisting frame. Under the drop wires (arranged in groups of three, corresponding to the plying of three yarns), is an oscillating steel bar (the rocker bar). When a yarn breaks, its drop wire falls to impede the rocker bar with a sudden jolt. This releases the delicately held spring bolt positioned on a socketed wooden bar which is the axis for the rocker bar. When released, the bolt strikes the shipper, the fork of which moves the driving belt from the live to the dead pulley and so stops the machine for the yarn to be mended. A turning movement for the wooden pivoted bar (and hence oscillation for the rocker bar) is gained by transforming rotary motion into laterally reciprocating movement through the rocker arm, a variation of the Scotch yoke familiar to engineers. Essentially the rocker arm movement consists of a crank and a lever with a center pivot. The crank pin runs in a slot at one end of the lever, and, since the center pivot is fixed, the rotary movement of the crank pin becomes oscillating movement at the other end of the lever. I am grateful to Mr. James Stevenson of Andover, Mass., for identifying this mechanical movement.

(like the Appleton), and at least one outside supplier (Taylor & Souther). So much for the manuscript evidence. It seems to suggest that the three surviving twisters were built in the 1830s; further interpretation or unequivocal additional evidence may place them in the 1820s or, equally possible, in the 1840s.[89]

The other curious aspect of these twisters is the method by which the jammed rocker released the bolt that knocked the belt shipper from the live to the dead pulley. When a Smithsonian curator (Dr. Philip W. Bishop) and I went to the Suitland storage area on December 5, 1969 to collect a delivery roller and a drop wire from the Smithsonian twister for metal analysis, we tried to coax the stop motion into action, We learned, when I accidently triggered it and the curator nearly lost a finger between the bolt and the shipper, that it was the sudden jarring of the impeded rocker which released

FIG. 5.—Close-up view of the drop wire and rocker stop motion on the back of the Merrimack Manufacturing Company's twisting frame (see fig. 4 for an explanation of the stop motion mechanism). (Courtesy of the Smithsonian Institution.)

[89]Blotter (1825-29), vol. 20, and Blotter (1829-38), vol. 21, passim. Hamilton Manufacturing Company MSS, Baker Library; Montgomery, *Practical Detail,* p. 104.

the bolt (see figs. 4 and 5). This mechanically clumsy but physically compact arrangement also suggests an early date for the machines.[90]

Efforts to apply fault detection devices to other machines in this period appear to have been either uneconomical (as with the mule), unnecessary (as with the belt-driven throstle), or superseded by a more efficient operation of the machine in question (like the roving frame, for which T. & J. O. Lewis of Wilkinsville, Mass., invented a stop motion in 1836).[91]

The Momentum and Results of Innovation

If the experience of the Hamilton Manufacturing Company was at all typical, in the vertically integrated Lowell mills of the 1830s at least two production influences gave momentum to these innovations springing from the interaction of American market forces and factor endowment. The first arose from the early and unavoidable reliance on empirical rather than scientific knowledge in textile manufacturing. Because productivity at each processing stage could not be precisely predicted, imbalances appearing between production capacities of machines at the various stages impelled improvements at the levels falling behind. So, in Hamilton Mill No. 1 in spring 1831, when unspecified alterations in the card room promised to supply roving for four extra spinning frames, ten new double spinning frames were substituted over a two-year period for twenty old single ones, thereby increasing cloth production by 1,500 pounds per month. Three years later, however, the preparatory machinery was found to be lagging behind the new throstle frames in terms of productivity. The agent therefore recommended that more speeders and stretchers be installed with the possibility of acquiring improved cards if the looms could be run faster. Theoretically the imbalance problem might have been solved simply by adding to the number of old-style machines wherever productivity dropped. But this, if it involved, for example, extending plant floor space, could prove more uneconomical than introducing the new and more productive equipment.

At this point, the second production influence became relevant. Although there was negligible competition between the Lowell cor-

[90]The mechanical principles behind these two kinds of stop motion continued to be utilized into the 20th century, though mechanical stop motions were displaced by electrical ones.

[91]*Franklin Journal*, n.s. 17 (1836): 317; evidently it was not patented. Since the reference is so vague, there is the possibility that this stop motion halted the machine when the bobbins were full, rather than when an end broke.

porations in product markets, the Hamilton records shows that by 1831, when nine companies existed at Lowell, agents had become aware of competition in terms of production, and consequently cost—how many thousand pounds of cotton the other mills were manufacturing per month and cost per pound. Such productivity and cost comparisons provided measures of performance which, in the mind of the Hamilton agent, appeared to be related to improved equipment.[92] Making this connection, it was natural to turn to technological innovation as the internal answer both to production imbalances and to the interlocking directorates' pressure for rising dividends when a major capital expansion could not be justified. How long this factor persisted in Lowell is difficult to estimate. Elsewhere product competition would have operated, imperfectly at first except, perhaps, in urban markets, as a spur to continued innovation.

The application, conscious or unconscious, of these foregoing labor-saving principles produced some remarkable results. Whereas one fairly typical American cotton factory of the period 1810-20 (Samuel Batchelder's New Ipswich mill) employed forty-eight persons and processed 25,000 pounds of cotton annually, the Boston Manufacturing Company employed 264 persons and processed 450,000 pounds of cotton annually; put another way, though the Waltham company employed five and one half times more persons than Batchelder's factory, it annually processed eighteen times more cotton. By this rough yardstick the Waltham, and hence later American, system of cotton manufacturing, was three and one half times more labor saving than that used before 1812.[93] One Rhode Island cotton factory, Union Mills in Johnston (established 1805), which in 1831 had 1,600 spindles, forty power looms, plus cards, warpers, spoolers, and double speeders, estimated that over its twenty-five-year history, the cost of manufacturing had decreased by 33-50 percent.[94] How much of these savings were due solely to the power loom and how much to American innovations is unknown.

Comparing American (Waltham type) and British vertically integrated cotton mills in 1840, Montgomery calculated that an average British mill with sufficient carding and spinning equipment to

[92]Minutes of the Proprietors and Directors, vol. a, pp. 197-99, 202, 212-13, Hamilton Manufacturing Company MSS.

[93]1820 Census (n. 10 above), Massachusetts, p. 33; New Hampshire, p. 35.

[94]Louis McLane, *Report of the Secretary of the Treasury, 1832: Documents Relative to the Manufactures in the United States* (22d Cong., 1st sess.; *House Executive Document* no. 308, 2 vols. [Washington, D.C. 1833], 1:949).

support 128 looms would in two weeks produce 35,200 yards of cloth (plain weave shirting, 35 inches broad, with 2,000 warp ends of no. 16 yarn, and with sixty-three picks of no, 18 yarn per inch). An equivalent 128-loom American mill would, over the same period, make 51,300 yards of cloth (three harness twill, 30 inches broad, with 2,400 warp ends of no, 18 yarn, and with sixty-two picks of no, 18 yarn per inch).[95] In other words, discounting capital and labor costs, a line of American equipment on this scale had an output per unit time of about 30 percent higher than that of an equivalent British production line. Montgomery suggests that the respective looms were comparable (by 1836 self-acting nipper temples were built into British looms and would have been as effective though less smooth in operation than Draper's temples)[96] and his figures (which unfortunately are not wholly consistent)[97] indicate that American production was increased at the carding, spinning, warping, and dressing stages of processing. Here the Americans were using larger cards, faster throstle spindles (and no mule spindles at all), and more efficient warping and dressing frames.

This paper has looked at the technological details of the major American labor-saving innovations which were built into these machines and the other basic equipment of the British system of textile manufacturing before 1840. If further production and labor data relating specifically to the equipment of both the British and American systems could be found, then a more certain estimate of the increased production and productivity attributable to American innovations could be made.

Conclusion

The mechanized system of textile manufacturing developed in Britain before 1812 and transmitted to the United States by that date was modified there before 1840 in specific directions in response to the pressures of crowded product markets and the structures of industrial organizations interacting with the shortage of unskilled labor.

Generally, market pressures compelled the large-scale production of cheap, standardized, low-quality goods, at first. Attempts to compete against British imports, as in the woolen industry, brought

[95]Montgomery, *Practical Detail,* pp. 112-25.

[96]Ure, *Cotton Manufacture,* 2:317-21.

[97]The total number of hanks processed in winding in the American mill does not agree with the total number spun as weft, there being a difference of 42,048 hanks (Montgomery, *Practical Detail,* pp. 120, 123).

failure. Only when technological innovation supplied versatile equipment (like the cotton roving frame differential or the fancy loom dobby) for manufacturing cheap and medium-to-fine goods, could British products be challenged. In the cotton industry, standardized products tended to stimulate standardized manufacturing machinery which, because it was simpler and therefore presented fewer mechanical or manufacturing problems than more versatile or specialized equipment, reduced operative and maintenance problems and was more amenable to innovation.

In determining the direction of innovation, the shortage of unskilled labor was the important underlying factor. Alone, however, it is not sufficient to explain the remarkable difference between the narrow range of improvements made under the Rhode Island system and the wide range of innovations achieved under the Waltham system. The difference, as this article has implied, derived from the two kinds of organizational structure adopted by the two systems.

In the Rhode Island system, horizontal specialization provided the matrix for the pursuit of labor-saving innovations at the preparatory and spinning stages of processing only. Since the basic labor unit, the family, was the same as that operating identical equipment in Britain, increased productivity was sought by means of the conventional mechanical principles of higher operating speeds, maximum mechanical work, and greater mechanical control. The first was exemplified in the cap and ring spindles; the second in the railway drawing head; and the third in the tube and eclipse speeders and the roving frame differential (the pursuit of mechanical control in developing a self-acting mule was relatively unsuccessful until the late 1830s, over a decade after such a machine had been perfected in Britain).

Under the Waltham system, vertically integrated and massive textile manufacturing corporations allowed the concentration of optimum material resources and engineering skills for adapting a complete line of British equipment to meet the general shortage of unskilled labor. Most significant for innovation was the choice of unskilled teen-age females as the nucleus of the labor force, for it demanded several radical mechanical and organizational compensatory devices to balance the operatives' physical weakness, inexperience, high turnover rate, and easily distracted mentalities. So, besides higher operating speeds and greater mechanical work and control, Waltham engineers sought to implement a more efficient movement of materials, maximum use of inanimate power, mechanical vertical integration, and automatic fault detection. Maximum mechanical work and control were exemplified in the development of the dressing frame, the self-acting temple, the dobby head, and the installation of larger throstle frames. Higher speeds were achieved with the dead spindle and, in the dressing frame, with steam-filled cylinders. A more efficient movement of materials was gained by means of flow production. The throstle, the warping frame, and the power loom evidenced a greater dependence on inanimate power, while the filling frame and the warping frame used in conjunction with the dressing frame were two examples of mechanical vertical integration. Automatic fault detection outside the power loom was applied through the drop wire and rocker stop motion in the warping and twisting frames and the latch stop motion in the drawing frame. The Goulding condenser for the wool carding engine, an example of mechanical vertical integration, may be viewed as a spinoff from both the Waltham and Rhode Island systems.

These and other innovations, sustained not least by production imbalances, cost competition (at Lowell at least), and imperfect product competition, spread with varying degrees of rapidity from one system to the other. So, large-scale production, production of stan-
* dardized or medium quality goods, vertically integrated manufacturing, flow production, maximum use of inanimate power, high operating speeds, mechanical vertical integration, maximum mechanical control and work, and automatic fault detection eventually became characteristic features of textile manufacturing in New England. As components of the American system of manufacturing, these principles would later be absorbed into the concept and practice of early 20th-century American mass production.

Errata:

p. 41, line 13, British imports *should read* British and Asian imports
p. 43, column 4, Fly shuttle *should read* Fly shuttle loom
p. 43, notes,*All... *should read* †All
p. 48, line 2 from the bottom, Manufacturers *should read* Manufactures
p. 52, line 4 from the bottom, the waterframe but not so flexible *should read* the water frame but was not so flexible
p. 56, line 17, As arthur H. Cole *should read* As Arthur H. Cole
p. 63, l ine 22, warp or filling *should read* ever filling
p. 76, line 20, standardized or *should read* standardized coarse or

X

Technological diffusion – the case of the differential gear

1. Chronological outline

The differential, known to the ancient Greeks and claimed for the originators of the ancient Chinese south-pointing chariot, seems to have been reinvented in modern times first by clockmakers. Joseph Williamson, a London clockmaker, between 1719 and 1725, constructed the earliest surviving example of a differential (apart from the ancient Greek one rescued from the Antikythera wreck), consisting of a mobile wheel engaging two flat-toothed wheels in the manner of the motorcar differential. Clearly developed from earlier epicyclic gears, the differential served as an equation mechanism, driving two clock faces, one showing solar time and the other mean or apparent time. Ingenious versions of the differential appeared in the technical literature as early as 1741, especially in French works, and an analysis of the mathematical theory represented by the gear was tackled by two Austrian priest-horologists in Vienna, Aureliano and David a San Cajetano, between the 1770s and 1790s.[1]

The earliest industrial application of the differential appears to have been in the windmill. In the late 1780s a millwright, James White, used it for automatic tentering (or adjusting the gap between the millstones) in variable

---

[1] H von Bertele 'The Origin of the Differential Gear and Its Connection with Equation Clocks', *Newcomen Society Transactions*, 30 (1955–1957), 145–155. I am grateful to Professor Derek de Solla Price for bringing to my notice his definitive study of the Greek differential: 'Gears from the Greeks. The Antikythera Mechanism – a Calendar Computer from ca. 80 BC', *American Philosophical Society Transactions*, new ser. 64, 1974).

It is clear from this present study that the invention, re-invention or diffusion of the differential in the eighteenth and nineteenth centuries was more labyrinthine than Professor Price suspected, ibid., 61.

For the Chinese antecendents see Joseph Needham, *Science and Civilization in China* 4 (Cambridge. 1965), 286–303, a reference kindly brought to my attention by Joe Marsh of UMIST.

winds. White claimed that it worked with some success in at least one windmill in Kent.[2]

Over 30 years later, White, now a civil engineer in Manchester, published a new application for the differential in the design of a dynamometer. His *New Century of Inventions* (1822) illustrated both his mill governor and his dynamometer, though he emphasised that the mill governor came first. In both, the differential consisted of four bevel gears, the two pinions being held by a pulley carrier, much like the crown wheel in the motorcar differential. In the dynamometer, power was measured by weights attached to the circumference of the pulley carrier, (see Plates I and 2) though White did suggest various arrangements of levers which might be employed for this purpose.[3]

That same year, 1822, a Rhode Island mechanic, Aza Arnold, built a differential into a speeder or cotton roving frame and in the following year took out an American patent for it (Plate 3). In old age, he recalled that he assembled the differential itself as early as 1818. In the roving frame the gearing allowed the bobbins to slow down as the cotton roving, being wound on to their circumference, increased their effective diameter.[4] Before the end of the 1820s this application of the differential had been taken up by the Lowell corporations and other cotton manufacturing firms in northern New England, and also in Europe by engineers and manufacturers in Manchester. Glasgow and Mulhouse.[5]

The use of the differential in the road vehicle was first recorded in a French patent awarded in 1828 to a Parisian, Onésiphore Pecqueur. Pecqueur saw the differential as the best solution to the driving wheel problem posed by all road vehicles with two or more driving wheels on a common axle linc. Pecqueur placed his differential (consisting, like Arnold's, of three bevel gears) on the rear axle of his steam carriage but to one side rather than in the centre.[6] Four years later Richard Roberts, the eminent Manchester engineer, gained a British patent in which he placed the differential (again comprising

---

[2] James White, *A New Century of Inventions, Being Designs and Descriptions of One Hundred Machines Relating to Arts, Manufactures and Domestic Life* (Manchester, 1822), 223–225 and plate 27. For James White (1762–1825) see H. W. Dickinson, 'James White and His "New Century of Inventions"', *Newcomen Society Transactions*, 27 (1949–51), 175–179. The *Manchester Mercury* for 24 December 1825 recorded White's death 'on the 17th Inst. aged 63 years'. I am grateful to Dr R. S. Fitton of Manchester Polytechnic for these references to White,

[3] White; *New Century* 15–25, 223–225, plates 1, 3–6, 27.

[4] David J. Jeremy, T*ransatlantic Industrial Revolution: the Diffusion of Textile Technologies between Britain and America, 1790–1830s* (Cambridge, Mass., and Oxford, England, 1981), 58, 206–208.

[5] ibid., 206–208, 244, 249; James Montgomery, *The Carding and Spinning Master's Assistant, or the Theory and Practice of Cotton Spinning* (Glasgow, 1832), 112–116.

[6] French patent, new no. 5915 (old no. 3524) (25 April 1828).

Plate 1 James White's design for a differential dynamometer, published in 1822 in his *New Century of Inventions*.

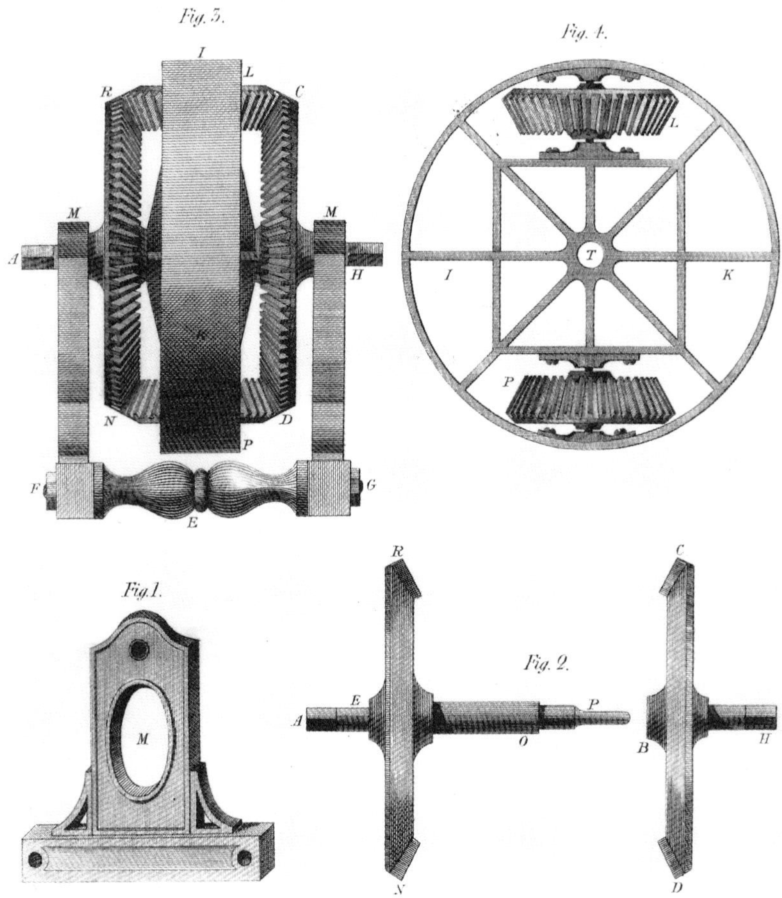

Plate 2 James White's design for a differential gear train, utilising cast iron bevel gears, published in 1822 in his *New Century of Inventions*.

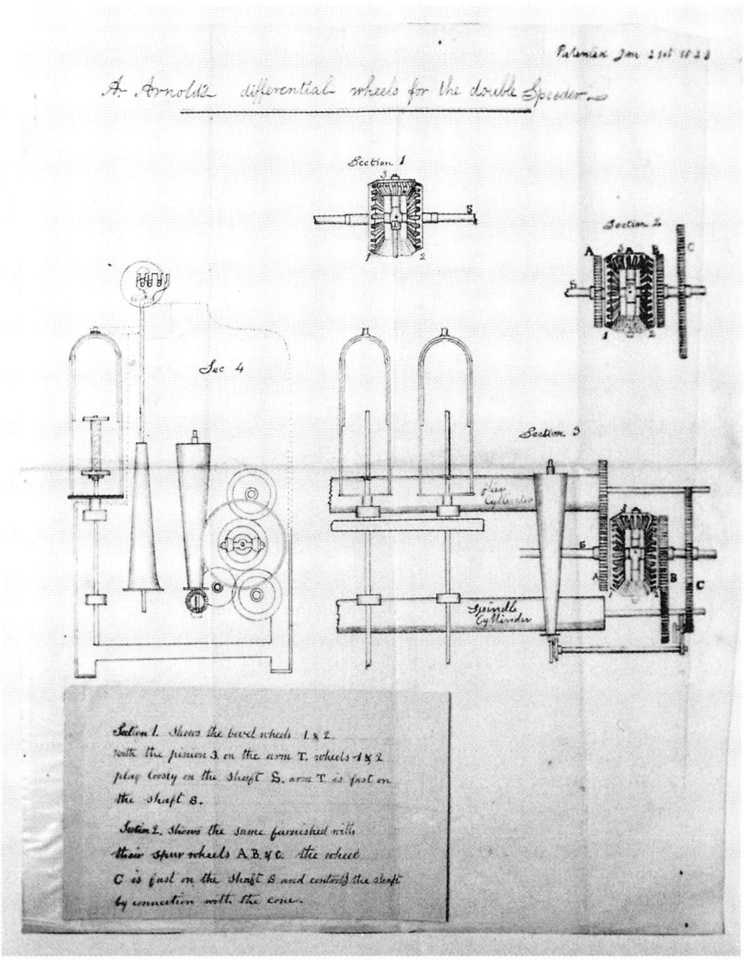

Plate 3 The drawing of Asa Arnold's American patent for a differential gear, dated 21 January 1823; drawing in the Zachariah Allen papers, Rhode Island Historical Society, Providence, Rhode Island.

As can be seen, the differential gear train divided the driving force entering the cotton roving frame between the cylinder powering the flyers and the cylinder prowering the spindles. The outer circumference of the bobbins on the spindles increased as the delicate roving was wound onto the bobbins, requiring the bobbin and therefore spindle speed to be slowed progressively as each layer of roving was completed with the up-and-down movement of the bobbins inside the flyers. This was achieved as the weight of the bobbins held back the speed of the spindles leaving the excess of power to be absorbed by the epicycle pinions in the differential gear train.

three rather than four bevel gears) on a drive shaft, the two halves of which were then linked to the driving wheels by chains: a more cumbersome arrangement than Pecqueur's.[7] The first English patent to place the differential on the centre of the driving axle of a road vehicle and to employ four bevel gears (thus giving a balanced movement within the differential train) – as in the cotton roving frame or in the modern motorcar – was taken out by Frank Hills, a Deptford industrial chemist, in 1843.[8] While the Pecqueur patent remained at a design stage, both Roberts and Hills developed their patents into operating steam carriages.

Meanwhile in the late 1830s an American cotton manufacturer, Samuel Batchelder, developed the differential dynamometer; it gained acceptance in both the USA and Europe in the following decade.[9]

When new powered road vehicles appeared after 1860 the differential of the steam carriage was slowly transferred to them. James Starley of Coventry built a differential in the rear axle of his tricycle, patented in 1877,[10] Benz installed one in his gasoline-driven car patented in Britain in 1886.[11] But other motor vehicle engineers and builders, faced with the differential driving wheel motion problem, sought mechanical solutions other than the elegant design known to eighteenth-century horologists. Jaw clutches, ratchet clutches and friction devices (belts, rollers, or the leather discs initially favoured by Daimler) were all tried and recorded in numerous British and American patents between the 1860s and the 1880s.[12]

2. General questions about the approach to diffusion of technology studies

Several of these arise:

2.1 Should we even try to study the diffusion of a single implement, machine or technique? A number of earlier historical diffusion studies have examined single devices, for example Draper's automatic power-

---

[7] British patent no. 6258 (13 April 1832).

[8] British patent no. 9684 (30 May 1843).

[9] David J. Jeremy *Technology and Power in the Early American Cotton Industry: James Montgomery, the Second Edition of His 'Cotton Manufacture' (1840), and the 'Justitia' Controversy about Relative Power Costs* (Philadelphia: American Philosophical Society, 1990) 201, 241, 247, 293.

[10] British patents for 1877, no. 3388 (5 September 1877).

[11] British patents for 1886, no. 5789 (28 April 1886).

[12] Leonard H. Dyer, 'The Evolution of the Motor Vehicle as Shown by Patents: Part 1, the Compensating Gear,' *The Horseless Age* 4 (28 June 1899). I am grateful to Professor Lynwood Bryant for directing me to this.

loom, or even a single occasion, as in the history of tubular bridge building.[13] The choice of a narrow topic like these, or of the differential, runs certain risks (and the authors in question have negotiated them with varying degrees of success):

2.1.1 We are likely to miss what Rosenberg calls 'complementarities' – the technical bottlenecks whose removal permits the diffusion of an innovation.[14] Thus the diffusion of the differential to industrial machinery obviously hinged upon the invention of bevel gears and the skill to design and cast them in iron.[15]

2.1.2 We may overlook the subtle interactions between a whole technological system and its economic and cultural frameworks. Without knowing about Rhode Island manufacturers' pursuit of versatile equipment, making a range of yarn finenesses over short production runs, we should be puzzled by the independent invention and diffusion of the differential to the roving frame in New England.[16]

2.1.3 While case studies move us towards a broader understanding of circumstances favouring the diffusion of technology, we are unlikely to hasten towards a general theory of the diffusion of technical innovations unless we reach beyond a strictly intrinsic understanding of the innovation. Indeed, as the history of the differential illustrates, those who invented and later adopted the innovation had little or no scientific understanding (i.e. mathematical analysis) of the mechanisms. In this instance a purely intrinsic appreciation fails to uncover the circumstances of diffusion.

2.2 How do we handle the two rival theoretical approaches to the diffusion of innovations? Diffusion theory has tended to favour either economic or sociological explanations of the diffusion of innovations. Rosenberg succinctly expressed the views of economists; 'the diffusion of inventions is an essentially economic phenomenon, the timing of which can be largely explained by expected profits.'[17] On the other hand, histori-

---

[13] Irwin Feller, 'The Draper Loom in New England Textiles, 1894–1914: a Study of the Diffusion of an Innovation,' *Journal of Economic History* 26 (1966), 320–347. Nathan Rosenberg and Walter Vincenti, *The Britannia Bridge: the Generation and Diffusion of Technological Knowledge* (Cambridge, Mass., 1978).

[14] Nathan Rosenberg, 'Factors Affecting the Diffusion of Technology,' *Explorations in Economic History* 10 (1972–1973), 21.

[15] I am grateful to Mr M. T. Wright of the Department of Mechanical and Civil Engineering, Science Museum, London for comment on this point.

[16] Jeremy, *Transatlantic Industrial Revolution*, 204–206.

[17] Rosenberg, 'Factors,' 6.

ans from John E. Sawyer in the 1950s (when the transfer of American technology under Marshall Aid programmes met unexpected European resistance) to Merritt Roe Smith in the 1970s (when non-economic explanations attracted fresh interest, especially after Smith's demonstration of the force of cultural factors delaying the adoption of new technology at the Harpers Ferry Armoury) have been aware of the inadequacy of wholly economic explanations of diffusion.[18] Kenneth Arrow suggested that economists have been looking at demand, and sociologists at problems in the supply of communication channels. This is perhaps the best reconciliation of the two perspectives.[19] Certainly historians of technology must take account of both, as the differential case study illustrates. While the rapid spread of the differential roving frame in New England can be attributed to its ability to increase mill productivity by around 30 per cent, with an impact on profits of significant if not identical magnitude, the profit motive surely cannot explain the contribution of the Cajetanos to the spread of the clock differential: they were under monastic vows and profited (in an economic sense) little or nothing from their work.[20]

2.3 Then there are problems of evidence, common to most historical investigations. In the case of the differential, we simply do not have enough evidence to construct the economist's demand and supply schedules; nor to trace the elements and stages exemplified by the sociologists' communication model (Source, Message, Channel, Receiver, Effects); nor to chart thc slopc of the S-shaped curve which the diffusion of an innovation invariably follows.[21] In this situation we are obliged to grasp any partial or surrogate measures as do survive and to look closely at the wider economic or social conditions and circumstances restraining or encouraging diffusion and the kind of agents or vehicles by which an innovation was diffused. In the light of these broader questions and the perspective they bring to the study of the diffusion of technology we can now turn to specific aspects of the diffusion of the differential.

---

[18] John E. Sawyer, 'The Social Basis of the American System of Manufacturing,' *Journal of Economic History* 14 (1954), 361–379. Merritt Roe Smith, *Harpers Ferry Armory and the New Technology: the Challenge of Change* (Ithaca, NY., 1977).

[19] Kenneth J. Arrow, 'Classificatory Notes on the Production and Transmission of Technological Knowledge,' *American Economic Review* 59 (1969), 33.

[20] Jeremy, *Transatlantic Industrial Revolution*, 206; von Bertele, 'Origin', 151.

[21] Everett M. Rogers and F. Floyd Shoemaker, *Communication of Innovations: a Cross-Cultural Approach* (2nd. edn., New York, 1971), chapters 1 and 2.

3. Specific issues raised by the differential's diffusion

The history of the differential, from clock to motorcar, exhibits several interesting features with respect to invention and diffusion.

3.1 It appears to have been invented at least thrice over, first by a London clockmaker in 1720 or thereabouts, then by a Kentish millwright in the 1780s, and then by a Rhode Island mechanic around 1820. While the distances of time and space made it possible for three identical but independent acts of invention to occur, we should pause before leaving the matter at that. The American invention implies that either American mechanics were cut off from European industrial developments or, if they were not, that European mechanics were removed from European clockmakers. In view of the movement of skilled artisans and of technical publications across the Atlantic, and of American visitors to Europe, it seems unlikely to me (especially in the light of a close study of the subject) that Americans would not have known about European industrial applications of the differential had they been extensively developed before the 1820s. Arnold's roving frame differential was in fact plagiarized in America by the Lowell corporations who, in defending their piracy in American courts, sent agents to Britain to try to discover prior invention. Their enquiries, presumably confined to industrial practice in cotton manufacturing, discovered no precedent.[22]

Arnold evidently assembled his differential by the empirical method, much as Williamson had done a century earlier. Presumably Arnold had been tinkering with gear arrangements, for not until two years later, in 1820 (if we may trust his memory), did he perceive its applicability to the cotton roving frame, an insight that came to him as he was repairing one. We may wonder that he did not see the possibility much sooner for he had been apprenticed to Samuel Slater and was surely well aware of the nature of the roving frame problem and the commercial bottleneck that it posed to higher count spinning. At any rate, when he described to some of the most able New England engineers and mathematicians what he had done, he met incredulity and the initial reaction that such a gear train was 'impracticable.'[23]

The repeated invention of the differential therefore turns our attention back to the interface between European clockmakers and European mechanics. Did James White derive his windmill governor differential from the clockmakers or did he develop it without reference to the horologists? His first patent certainly shows that by the 1780s he was

---

[22] Jeremy, *Transatlantic Industrial Revolution*, 216.

[23] ibid., 206.

familiar with clockwork. Until we know a lot more about the connections of friendship and business and a lot more about the intellectual horizons of clockmakers and mechanics, we shall not be able to surmise whether we are looking at a series of independent inventions (1720s, 1780s, then Arnold in the 1820s) or a series of stages in the process of diffusion. This in turn will affect the way in which we try to understand the development of this 'building block' in mechanical technology. Was it the product of inventors working from scratch or was it the achievement of lateral thinking? In White's case there was obviously some lateral thinking, as he reapplied the differential from the windmill to the dynamometer – slow thinking judging by the intervention of three decades. Had White been a really sharp inventor he might have seen the differential as the key to the roving frame problem, as Arnold did; instead when he turned to cotton spinning machinery, White invented a cumbersome combination of roving frame and spinning frame, to resolve the problem of weak roving, for which he made high claims unmatched by extent of adoption.[24]

3.2 The known aspects of the diffusion of the differential are interesting in several respects:

3.2.1 It seems to have taken a long time for the mechanism to gain universal acceptance as the best mechanical solution to a particular kind of mechanical problem, whether that problem arose in clockmaking, windmill wrighting, cotton spinning, power measurement or road transport. If we think only of the mechanical problem, of dividing one uniform motion into two variable motions, rather than specific applications, then the differential's diffusion took about 170 years, from Williamson in the 1720s to Franz Reuleaux's published analysis of the kinematics of epicyclic gears in the 1870s and the acceptance of his ideas over the following two decades by mechanical engineers.[25]

---

[24] British patent no. 4485 (11 July 1820). James White (see note 2) was one of the more literate millwright-engineers of the early industrial period, spending some time in Revolutionary France where, he later claimed, he supplied ideas which were taken up by French engineers. Presumably he knew about, and may have inspired or been inspired by, the differentials illustrated in Hachette, *Traité Elementaire des Machines*, (Paris 1811), and in Lanz et Bétancourt, *Essai sur la Composition des Machines*, (2nd.edn., Paris, 1819).1am grateful to Joe Marsh of UMIST for bringing these drawings to my attention and to Michael Wright of the Science Museum for also discussing them with me.

Although governors were employed in windmills to tenter the millstones, White's ingenious combination of the two operations made little headway if we may rely on the silence on the matter in Rex Wailes, *The English Windmill* (London, 1954), 93, 98–99, 136–138.

[25] Franz Reuleaux, *The Kinematics of Machinery: Outlines of a Theory of Machines* (first English edn., 1876 new edn., intro. by Eugene S. Ferguson, New York, 1963), 425–435.

Two hiatuses appear to exist during this period, unless we are looking at an illusion due to lack of evidence: first between the eighteenth-century horological use of the differential and its post-1780 industrial applications; and second between the steam carriage differential and the motorcar differential. These hiatuses raise questions about constraints on the movement of knowledge or limitations on the agencies by which the differential was or might have been carried from the miniature world of the clockmaker to the industrial-scale world of the cotton factory, from the world of the steam engineer to the world of the tricycle and light car manufacturer.

On the other hand a series of rapid inter-industry and international transfers occurred in the two decades after 1823. The differential was carried eastwards across the Atlantic to the Lancashire cotton industry from whence it was transferred by Richard Roberts into the steam carriage and then, back in the USA, by Batchelder into the dynamometer. The timing was clearly related to the spread of early industrial forms, and expectation of higher profits resulting from higher productivity (to say nothing of higher yarn qualities and greater technical versatility, which also contributed to higher profits) in cotton spinning.

Richard Roberts's transfer of the differential over to the steam carriage reflected western Europeans' hopes for road transportation in contrast to Americans who relied much more on the railway and, to a lesser extent, river and canal boat, to open up their far greater expanses of territory. Those hopes receded by the late 1840s and steam road vehicles survived mostly in European agriculture as traction engines.[26]

3.2.2 A variety of carriers or vehicles of diffusion, the supply side of the diffusion equation, achieved the transfer of the differential from the USA to Britain and from the cotton industry to other industries. Between America and Britain a connection of manufacturers facilitated the transfer of Arnold's differential (pirated without his knowledge) which was secured by a visit to England and the protection of an English patent. Charles Richmond, manufacturer of Taunton in southern Massachusetts, took the differential to Manchester in 1825 and exchanged it there for a printing machine – a deal that allowed the Taunton manufacturers to gain some technical leadership and renewed profit margins, through factory-method calico printing and quality cotton-goods-finishing, in New England. The English patent was taken out by the Manchester cotton spinner, Henry Houldsworth. Jr. who had visited the USA himself in 1822 and 1823, when presumably he estab-

---

[26] See William Jesse Hughes, *A Century of Traction Engines* (London, (1959), 69–71, *passim*.

lished the trust placed in him by Richmond and the other directors of the Taunton Manufacturing Company.[27]

Although Richard Roberts did not patent his steam carriage differential until 1832, he would have learned about the differential through White's book, to which he subscribed, and his knowledge of the developments occurring in the cotton industry, rather than through any acquaintance with Pecqueur's steam coach French patent of 1828. Within a year of Houldsworth's 1826 patent for the cotton roving frame differential, Richard Robert's firm of Sharp, Roberts & Co apparently exported one of these improved roving frames to Mulhouse in, France.[28] We may safely assume that he derived some inspiration from it. To be sure, Robert's steam carriage was reported to have run down the Oxford Road in Manchester in spring 1834 but then its boiler tubes collapsed and Roberts soon gave up his attempts to compete with horses and railways.[29]

In short, manufacturers and capital goods manufacturers acted as carriers for the diffusion of the differential in the early nineteenth century. Patents guaranteed the various applications of the differential. What is not entirely clear is whether the patents also secured claims to the invention of the differential itself. Both Houldsworth in 1826 and Roberts in 1832 laid patent claims to the differential or variations of it, if used for the purposes stated in their patents.[30] But Hills's patent of 1843 clearly infringed Robert's claims in this respect.[31] And Starley in 1877 claimed the invention of the differential as well as its use in 'traction engines, velocipedes, or other self-impelled vehicles',[32] which was nonsense because traction engines were using the differential in their power transmission systems. In part this confusion can be ascribed to the inadequacy of the British patent system.[33]

---

[27] Theodore Z. Penn, 'The Introduction of Calico Cylinder Printing in America: a Case Study in the Transmission of Technology', *Technological Innovation and the Decorative Arts*, ed. by Ian M. G. Quimby and Polly Anne Earle (Charlottesville, Va., 1974), 235–255; Jeremy, *Transatlantic Industrial Revolution*, 109, 244.

[28] ibid., 249 and W. H. Chaloner, 'New Light on Richard Roberts, Textile Engineer,' *Newcomen Society Transactions* 41 (1968–1969), 35.

[29] H. W. Dickinson, 'Richard Roberts, His Life and Inventions,' ibid., 25 (1945–1947), 128–129.

[30] British patents, nos. 5316 (16 January 1826) and 6258 (13 April 1832).

[31] British patent no. 9684 (30 May 1843).

[32] British patents for 1877, no. 3388 (5 September 1877).

[33] Not until 1905 was novelty investigated by the British patent-awarding authorities, under an act of 1902. See Barbara M. D. Smith, 'Patents for Invention: the National and Local Picture,' *Business History* 4 (1961–1962), 110.

Throughout the history of the differential it is not clear either, what part was played by publications as vchicles of technology diffusion. French horologists, from Thiout l'aine in 1741 to F. Berthoud, the contributor to Diderot's *Encyclopaedia* in 1765, recorded the differential in French works.[34] The English encyclopaedias appear to have been slow in recording the various industrial applications of the mechanism, judging by a search of articles on the dynamometer.[35] At the same time, White's work was certainly widely disseminated, as the list of subscribers shows.[36]

Subscribers to James White, *A New Century of Inventions* (1822), by location of address

Area	*Number of subscribers*	*Percentage*
Lancashire and Cheshire	346	54.32
Yorkshire	110	17.27
London	76	11.93
Midlands	47	7.38
Scotland	37	5.81
Foreign	9	1.41
South	7	1.10
Ireland	3	0.47
North East	1	0.16
East Anglia	1	0.16
Total	637	100.01

---

[34] Thiout l'âine, *Traité de l'Horlogerie Méchanique et Pratique* (Paris, 1741), plate 25, which shows a version of the differential in which the mobile pinion moves in the same plane as the other gears comprising the differential train. I am grateful to Mr R. J. Law, Department of Mechanical and Civil Engineering, Science Museum, London, for this reference. See also von Bertele, 'Origin,' 150.

[35] I searched the 'Dynamometer' articles in Abraham Rees (ed.), *The Cyclopaedia: or Universal Dictionary of Arts, Sciences and Literature* (45 vols., London, 1802–1820); the *Encyclopaedia Britannica* 6th. edn. (1823), 7th. edn. (1842), 8th. edn. (1855–1860); David Brewster (ed.), *The Edinburgh Encyclopaedia* (20 vols., Edinburgh, 1830); Charles Tomlinson, *Cyclopaedia of Useful Arts and Manufactures* (2 vols., London, 1852–1854). The information in these entries almost wholly related to absorption dynamometers such as those of George Grahame, Leroy, Regnier and Poncelet. The Rees entry also recorded Robert Salmon's weight and pulley arrangement for measuring the force exerted by oxen in turning a horse-wheel.

[36] White, *New Century of Inventions*, end of volume.

3.2.3 The question of the appropriateness of a given technology for differing sets of economic and social circumstances, obviously present in each of the transfers (or inventions) of the differential, may be illustrated with reference to its application in the dynamometer. Although the differential dynamometer first appeared in Britain, being announced by James White in 1822, it does not seem to have come into use for one or two decades. The *Edinburgh Encyclopaedia* article on dynamometers, dated 1830, described several devices none of which could have been applied with ease to individual machines in a production line, for none could have been attached to a rotating shaft.[37] Prony's brake, publicised in France in 1821–1822, could not accurately measure the power applied through one machine in a production line because it was an absorption type of dynamometer, which converted the energy involved into heat by means of a friction mechanism.[38]

In contrast to existing alternatives, the differential dynamometer offered several distinctive advantages. Firstly it was of the transmission type and could, with reasonable accuracy, measure the power passed through a machine driven by rotating shafts, without altering that machine's rate of power consumption. Secondly it expressed power exerted in terms of weight over distance and therefore facilitated conversion of power measures to horsepower units. Thirdly it was a rugged device, well suited for mill practice.

The differential dynamometer illustrated by White, in which weights for measuring power were hitched to the differential pulley crown wheel, would have been less satisfactory than his alternative of a lever pivoted across the carrier pulley because the pulley weights would have to be much greater than lever or steelyard weights. Use of the steel yard (already employed in at least one eighteenth-century absorption dynamometer) came when Samuel Batchelder, a leading northern New England cotton manufacturer, adapted the differential for use as a dynamometer, sometime before September 1838. The device was awarded a silver medal in the second exhibition of the Massachusetts Charitable Mechanic Association in Boston in September 1839 and received high praise and wide spread use in both America and Europe between the last 1830s and the 1870s.[39]

---

[37] See note 35.

[38] S. B. Hamilton, 'The French Civil Engineers of the Eighteenth Century,' *Newcomen Society Transactions*, 22 (1941–1942), 156.

[39] Sources in note 9.

4. Leaders and Followers

The diffusion of the differential to the transmission type of dynamometer raises a very interesting question: why in a 'leader' industrial economy reputed for dearer power (Britain) were power measuring instruments adopted more slowly than in a 'follower' industrial economy reputed for cheaper power (the USA)?[40] We can rule out imperfections in the dissemination of knowledge because the machine was well advertised in Britain in the heart of the country's leading manufacturing and machine-making region.[41] We are therefore left with the following alternative explanations.

4.1 The evidence considered is incomplete and therefore illusory: British manufacturers did adopt the differential dynamometer advertised by White in 1822 and therefore did know how much power their equipment was consuming. This possibility awaits much more investigation.

4.2 Our presuppositions about comparative British and American power costs in the early nineteenth century are incorrect. If so, then so is all the testimony of contemporary commentators before 1840, an unlikely possibility in view of the unanimity of opinion among American manufacturers and foreign visitors alike from the 1820s onwards.[42]

4.3 British manufacturers, increasingly reliant on steam power, did not need to measure power consumption of individual machines in a production line all that accurately. If modification necessitated more power, then steam engine pressure was simply increased.[43] In contrast, American manufacturers, almost wholly reliant on water power, could not easily increase power supply: instead they would be more inclined to check power demand.

---

[40] If we regard cotton machine spindles as a surrogate measure of industrial growth, Britain was (as is well known) well ahead of the USA up to 1850: in 1820 Britain had nearly 8 million cotton spindles, compared to 350,000 in the USA; in 1850 Britain had 20 million compared to the USA's 3.63 million. See Stanley D. Chapman. 'Fixed Capital Formation in the British Cotton Manufacturing Industry,' in J. P. P. Higgins and Sidney Pollard (eds.), *Aspects of Capital Investment in Great Britain, 1750–1850* (London, 1971), 75; Brian R. Mitchell and Phyllis Deane, *Abstract of British Historical Statistics* (Cambridge, 1962), 185; Jeremy, *Transatlantic Industrial Revolution* 277; US, *Eighth Census,* (1860), 3, *Manufactures* (Washington, D.C., 1865), 1.

[41] See Table in text of article.

[42] See, for example, the testimony of Kirkman Finlay, Joshua Bates and James Kempton in GB, *Parliamentary Papers (Commons)* 1833 (690) VI, 'Report from the Select Committee on Manufactures, Commerce and Shipping.'

[43] I am grateful to Mr R. J. Law for providing this insight.

4.4 Americans were anxious to measure the power consumption of their machines for reasons other than the cheapness of power in the USA as compared to Britain. In textile manufacturing at least two cases existed in the late 1830s and early 1840s in the USA where it would be important for Americans to learn how much power individual machines were consuming:

4.4.1 Where alternative power sources presented themselves, in particular steam or water power. Thus in the so-called Justitia debate of 1841, steam and water protagonists argued with reference to the number of different types of spindles driven by one horsepower of water or steam.[44] Until measuring equipment of accuracy was available such measures lacked universality.

4.4.2 Where differing types of manufacturing equipment were available, lower power consumption could well be the factor to tip the balance in favour of one kind. At this time, in the late 1830s, ring throstles and self actor mules were beginning to come onto the market in New England.[45] And in this context an accurate dynamometer could help to resolve disputes about relative costs and efficiencies of spinning machines. The multiplication of mechanical substitutes for earlier machines intensified the need for accurate measuring devices across a growing spectrum of mechanical technology after 1830.

Conclusion

Diffusion theory, the duplicated invention of the differential and its protracted spread to new machine technologies, all direct our attention back to the communities and connections of clockmakers, millwrights and mechanics. To establish and understand multiple invention, or its alternative of prolonged diffusion, we must learn much more about the early shops of clockmakers, millwrights and machine makers, their pupils, their professional or institutional groups and means of spreading new information. Only then can we offer tentative theories about the linkages between say the French horologists of the Enlightenment, James White's Kentish windmill governor, the seven clockmakers at the Gregs' Styal mill near Manchester in 1790, and the Manchester mechanics of the 1820s. When we have collective biogra-

---

[44] See Jeremy, *Technology and Power* Part II.

[45] See John W. Lozier, 'Taunton and Mason: Cotton Machinery and Locomotive Manufacture in Taunton, Massachusetts, 1811–1861' (Ph.D. dissertation, Ohio State University, 1978), chapter 4.

phies of these key machine making groups (as opposed to Smilesian biographies of their leaders) spanning the transition from pre-industrial to industrial society, we will also be able to say a lot more about the carriers of early industrial technology and perhaps more about the varying receptions given to the differential dynamometer in Britain and the USA.[46] This last aspect will probably not be clarified until we have a thoroughgoing comparative study of British and American power costs. Given the work of Louis Hunter, G. N. von Tunzelmann, Jeremy Atack and others, this cannot be long delayed.[47] Meantime the differential invites continued investigation as a 'building block' in mechanical technology which, I argue, can best be understood by studying the operating systems, and the economic and social contexts, in which it was adopted.

Acknowledgements

For comments which have helped me in the preparation of this paper, I am very grateful to Professor Lynwood Bryant of the MIT, Messrs R. J. Law, M. T. Wright and Peter Mann of the Science Museum, London, Jonathan Liebenau, my colleague in the Business History Unit at the LSE, and Joe Marsh of UMIST. I am also indebted to the organizers of the conference on 'New Perspectives on the History of Technology,' held at Manchester University in late March 1981, for the opportunity to present a version of this article which has been improved by the consequent discussion – though what remains is my responsibility, of course

---

[46] Examples of what can be learned from collective biographies are Ian Inkster, 'Studies in the Social History of Science in England during the Industrial Revolution, circa 1790–1850' (Ph.D. thesis, University of Sheffield, 1977) and Arnold Thackray, 'Natural Knowledge in Cultural Context: the Manchest Model, *American Historical Review*, 79 (1974), 672–709.

[47] Louis C. Hunter, *A History of Industrial Power in the United States, 1780–1930*, volume one, *Waterpower in the Century of the Steam Engine* (Greenville Delaware, 1979). G. N. von Tunzelmann, *Steam Power and British Industrialization to 1860* (Oxford, 1978). Jeremy Atack, 'Fact in Fiction? The Relative Costs of Steam and Water Power: a Simulation Approach,' *Exploration in Economic History*, 16 (1979), 409–437.

XI

Cotton Mills in Developing Regions 1820-1840

Massachusetts, and Some Comparisons with Lancashire

While cotton manufacturing became an urban phenomenon based on steam power in Lancashire, in Massachusetts, where very similar machine technologies were originally used, it thrived in a rural environment utilising water power. The purpose of this essay, which uses familiar but hitherto uncollated data, is to re-examine the Massachusetts experience, making some comparisons with Lancashire, and to re-open questions about alternative strategies for early industrial growth.

I. *Massachusetts and Lancashire populations and cotton industries in the late 1830s Description*

1. Populations

By 1840-41 the population of the USA (at 17.1 m) had, for the first time, overtaken that of England and Wales (at 15.9 m). However, the rates of growth of the two populations in the 1830s differed with that of England and Wales moving at 1.38 % per annum (simple interest formula), while the USA's was more than twice that, at 3.27 % per annum, of which 2.8 % was due to natural increase and 0.46 to immigration. The means of subsistence, as Malthus noted, were so much vaster in the USA and could support this high rate of population growth. The American population was also very young : 55 % of the white population (83 % of the whole) were aged under twenty.(1)

In the cotton manufacturing regions, Lancashire and Massachusetts, population growth rates were closer : at 2.48 % per annum in Lancashire in the 1830s (which represented a falling trend, after 2.71 per annum 1811-21 and 2.69 %, 1821-31), compared to 2.09 % per annum in Massachusetts (which marked a rising trend, following 1.08 % per annum 1810-20 and 1.66 per annum 1820-30). More interesting, from the viewpoint of labour supply, are the two regions' shares of the populations of their respective economies. While both regions increased their populations between 1800 and 1840 (Lancashire by 148 %, Massachusetts by 74 %), Lancashire's share of the population of England and Wales grew from 7.42 to 10.46 % while Massachusetts' share of the USA's population halved from 9.65 to 4.31 %. In other words the cotton industry in Lancashire was strongly associated with population growth ; in Massachusetts any such relationship was much weaker.[(2)]

In Lancashire population growth meant increasing urbanisation, as can be seen from Table 1 (see Appendix). This lists the population densities of Lancashire's cotton parishes (identified from the Factory Inspectors' Returns on power in cotton mills in 1835). These twenty-five parishes accounted for 74.7 % of Lancashire's population in 1811, 75.3 in 1821, 74.8 in 1831 and 74.4 in 1841. (The major non-cotton parishes in 1841 were Liverpool, Manchester's port with 33 000 and its satellites of Ormskirk, Toxteth Park and Walton with 94 000 ; and Kirkham parish near Preston with 12 000.) Noticeably the number of parishes with densities greater than the mean for Lancashire dropped from fifteen in 1811 and 1821 to fourteen in 1831 and thirteen in 1841, implying a growing concentration of persons in those parishes above the mean. The densities cited in this table were greater than the parish areas suggest because people clustered in townships within the parishes ; however the absence of area statistics for townships precludes a calculation of population density for anything less than the parish unit. The density of population in Manchester, the region's largest city with 136 000 in 1811 and 353 000 in 1841, draws attention to the fact that whereas Lancashire's population grew twice as fast as that of England and Wales between 1811 and 1841, Manchester's population grew faster still. While Lancashire's population grew at 23-27 % per decade 1811-41, Manchester's rose sharply from 37 % per decade during 1811-21, to 45 %, 1821-31, and settled back at 30 % per decade during 1831-41. Manchester in truth was the expanding centripetal hub of all Lancashire's urban populations.

In Massachusetts population growth started from a much lower base — nearly 60 % that of Lancashire — but could not maintain Lancashire's growth rate, rising by 56 % 1810-40, compared to Lancashire's 101 % increase 1811-41. Just four of the fourteen counties increased their shares of the state's population between 1810 and 1840 : Bristol, Middlesex, Norfolk and Suffolk. The last-named and fastest growing was almost totally occupied by the city of Boston, the largest entrepôt in the North East (Table 2). No strong correlation between cotton manufacturing and population growth, as evidenced in Lancashire (where the correlation coefficient R between the size and distribution of population in 1841 and the size and distribution of steam power in cotton mills in 1835 is 0.94, implying a very high positive relationship between population and cotton manufacturing), developed in Massachusetts before 1840. In the late 1830s four counties claimed 85 % of the state's capital investment in cotton manufacturing – Bristol, Hampden, Middlesex and Worcester. Yet together they accounted for only 40 % of the state's population. Equally, the four cotton manufacturing counties were not all among the most densely populated (Table 3). Six counties in 1840, on average, had over 100 persons per square mile but only two cotton manufacturing counties, Bristol and Middlesex, were among them. In short industrialised forms of cotton manufacturing in the USA, which had been spreading for fifty years, and thus were only a decade or two younger than in Lancashire, had not had the English vortex effect, of sucking population into a highly localised urban region.

2. Cotton industries

The relative situation of Lancashire, which had "no unique range of physical advantages for the growth of the manufacture,"[(3)] is briefly summarised in Table 4. Lancashire by all available measures, contained 70-75 % of the whole cotton industry of England and Wales by the late 1830s. The distribution of productive resources within the Lancashire cotton industry is suggested by Table 5. Within the county the single most important physical factor determining location was the Lancashire coalfield on which every cotton town except Preston was sited. As may be seen in Table 5, 87.7 % of the Lancashire cotton industry's horsepower needs were generated by steam engines (presumably all coal-fired) in 1835-36. Over a quarter of all cotton manufacturing horsepower was located in Manchester, no other district accounting for as much as 12 %. Behind Manchester ranked, by horsepower, the parishes of Oldham, Bury, Ashton, Rochdale and Bolton, each with 5-12 %. In view of Manchester's extreme shortage of waterpower sites, the city's parish held nearly 31 % of all the steam power in the county. In powerloom weaving Bury outranked all but Manchester whose share was relatively smaller than its share of horsepower, implying that it held a larger proportion than 31 % of the county's spindles. Powerloom mills, by the rule of proportions applied to Table 5, were larger in Manchester, on average, than in Bury. By 1840 the long-persisting pattern of regional specialisation had been laid down, with weaving in the northern Lancashire towns of Preston, Blackburn and Whalley (Burnley), and also Wigan ; spinning in the southern towns of Bolton, Oldham and Ashton ; and a balance between processes in Manchester, the merchant base of the industry.[(4)]

Some idea of the relative size of mills in the cotton parishes in the mid 1830s can be seen in Tables 6 and 7, which present the numbers of steam engines and water wheels in each range of sizes in the cotton parishes. What is most noticeable, and well known, is that Manchester's cotton mills contained the greatest proportion of the largest steam engines — half those over 50 hp and a third of those over 30 hp. Again as might be expected, the vast majority of water wheels were of small power, half under 10 hp ; the vast number of these were in the parishes of Bury, Rochdale and Whalley, each partially on the Pennines, high ground drained by fast streams.

The regional distribution of power in the Lancashire cotton industry in 1841 is summarized in Table 8 which includes doubling, smallwares and waste firms as well as the more common types of spinning and weaving firms. Unless it conceals discrepancies in data collection from the 1835 survey, it suggests that, relatively and overall, the gravitational strength of Manchester was weakening, having fallen from 27 to 20 % of the county's steam and water power capacity in the cotton industry. Absolutely, Manchester's cotton power rose significantly from 6 291 hp in 1835 to 7 850 in 1841. Within the cotton industry in 1841 Manchester's relative strengths were fine spinning (a branch where it had 45 % of Lancashire's steam power), powerloom weaving (where it had 31 %), yarn doubling (61 %) and smallwares like ribbons (82 %). Not much of the county's power was employed in these smaller branches, however, and here Manchester's share of steam power totalled only 1 396 hp. In the cotton manufacturing branches which consumed most power, coarse spinning and integrated spinning (in both of which Manchester's share was nearly 21 %), the shifts were towards Ashton, Blackburn, Prestwich, Oldham and Whalley, as may be seen from Table 9. By 1841, of the 37 860 hp employed in the Lancashire cotton industry (excluding the minor branches of doubling, smallwares and waste), 6.6 % was used in fine spinning firms, 34.3 % in coarse spinning firms, 55.5 % in integrated spinning and weaving firms and 3.6 % in power weaving firms.

While we have no census of mills comparable to those for Massachusetts, the work of V.A.C. Gatrell and others has shown that by the early 1830s, although the largest steam powered mills might cost £80 000 to build and equip (compared to £10 000 in the 1790s) and costs of entry to the industry were rising, it was still possible for small producers to enter the industry.(5) Profitability was achieved by means of numerous and familiar capital-saving techniques — power sharing, greater use of labour-intensive methods, reliance on water power — and by shifting working to fixed capital (by holding small stocks or by increasing reliance on specialist merchants for marketing functions). Gatrell also found no evidence in the 1830s of continued growth in the numbers and size of the giant firms, nor of a pursuit of vertical integration, both of which might be expected if economies of scale were to be achieved.

The best view we have of the structure of the Lancashire cotton industry in the late 1830s is found in Gatrell's work, based on the Factory Inspectors' Returns of 1841. The dimensions of the industry as plotted by Gatrell are summarised in Table 10. Just over 20 % of the county's 37 800 horsepower was located in Manchester, as were most of the largest mills, employing over 200 hp (or upwards of 100 000 spindles) and over 1 000 workers each. Only a third of the industry's firms were vertically integrated (i.e. combining spinning with weaving) though they used over half (56 %) of the industry's power and employed 70 % of the workforce. These firms were, individually, the largest in the industry. The largest group of firms were the coarse spinners, who comprised nearly half of all firms in the Lancashire cotton industry, but used only a third of the industry's power and employed a third of the industry's workforce. The Lancashire cotton industry in the late 1830s was therefore characterised thus :

- Half its firms engaged in coarse spinning and, on average, employed 28 hp (ca 15 000 spindles) and 110 workers.
- A third of its firms combined spinning with weaving and on average employed 65 hp (over 30 000 spindles) and 350 workers ; among these integrated concerns were most (eighteen) of the giants of the industry with, on average, 235 hp (100 000 plus spindles) and over a 1 000 hands.
- Some 10 % of firms specialised in power weaving and another 10 % in fine spinning.

Whereas the data on Lancashire for 1841 relates to firms, that for Massachusetts pertains to mills, which therefore makes comparison difficult. However while one firm might own several mills in the largest centers like Lowell or Waltham, or outside these two places mills may sometimes have been divided on a power-sharing basis, we may safely assume that the great majority of mills in Massachusetts were owned by single businesses.

To set the Massachusett cotton industry in context, Table 11 shows how in the 1820s it moved from third to first place doubling its share from 13 to 27 % of the nation's cotton spindleage. By 1836, as Table 12 demonstrates, cotton manufacturing claimed the largest share of industrial capital investment in the state, ahead even of fisheries, the traditional leading industry. The historical development behind the rise to prominence of cotton in Massachusetts has long been familiar through the pioneering work of Caroline Ware and many others.(6) Few attempts have been made to assemble aggregate measures of cotton manufacturing activity within the state during the years of most rapid growth, 1820-40, and none to compare the business strategies played out in Massachusetts with those in the contemporary English cotton industry. The focus here is on the American experience.

The first significant feature of the Massachusetts cotton industry was its rural location throughout the rapid growth period, as may be seen from Table 13. The projectors of the early cotton manufacturing concerns clearly planned the vast majority of their mills outside the most densely populated counties of Suffolk (and the city of Boston) and Essex (see Table 3). By 1820, following the buoyant War of 1812 years and then the post-war depression with its renewed British competition, the less efficient firms in the industry had been weeded out and the industry's location settled in Bristol, Middlesex, Worcester and Hampden counties (in that rank order). By 1832 Middlesex and Worcester had overtaken Bristol, Middlesex having a third of all the state's capital investment in cotton manufacturing machinery. By 1837 Middlesex had nearly a half of the state's capital investment in cotton manufacturing while Hampden drew level with Bristol, with Worcester county a few percentage points ahead of them both.

A comparison of population with investment and spindleage in the cotton townships of the four leading counties in 1837 summarises and underlines the rural character of the industry in Massachusetts (Table 14). Of the fifty-five cotton townships only fourteen had populations over 2 500, and of these fourteen only five were above 5 000 : Lowell (nearly 21 000), Springfield (nearly 11 000), Taunton (nearly 7 650), Worcester (nearly 7 500) and Fall River (nearly 6 750). By comparison, of Lancashire's twenty-five textile parishes only eight had populations under 20 000 and (excluding Halton and Cartmel) none less than 5 000 in 1841 (Table 1). With average spindleages between 1 500 and 5 000 per mill (frequently per firm) these American firms, on average, were no larger than the smallest range of Lancashire firms, operating under ten horse power and representing the bottom 3 % of the Lancashire industry by power and the bottom 4 % by labour. Yet there were a few exceptionally large firms in Massachusetts : those at Lowell. It must be emphasised however that they were the exception.

At the end of 1836 Lowell's ten textile companies comprised one firm making textile machinery and locomotives (the Locks and Canals Co.), one woollen company (the Middlesex Co.) and eight cotton manufacturing firms. The latter each spun their own yarn and wove it into prints, sheetings, shirtings, drillings and, in one case (the Lowell Co.), carpets, rugs and negro cloths. Altogether $8.05 million was then invested in the ten companies, of which $6 950 000 was the capital of the eight cotton firms. Together they ran 22 mills (excluding printworks), 141 508 cotton spindles and 4 467 cotton powerlooms

— on average 17 688 spindles and 558 powerlooms per firm and 6 432 spindles and 203 powerlooms per mill. The largest firm, the Merrimack Manufacturing Co, had 5 mills, 35 704 spindles, 1 253 powerlooms and a capital investment of $1.5 million. It also had the largest workforce — 1 400 women and 437 men.(7) Thus the largest firm at Lowell ranked with the combined spinning and weaving firms which comprised a third of Lancashire's cotton industry but, by spindleage, it was only a third of the size of giant Lancashire firms. Clearly the Lowell companies, promoted by Boston merchants, were pursuing the same economies of scale which had been attained in England. Yet not all the Massachusetts firms pursued high growth and the region produced the phenomenon of the persisting rural cotton mill. It flourished throughout the pre-Civil War period alongside the urban giants which emerged first at Lowell and then at other mill cities on the Merrimack (Lawrence, Ma ; and Manchester, NH, for example) and Connecticut Rivers (Holyocke for example). How can we explain this phenomenon of the rural cotton mills ? Why in the first place were they located at a considerable distance from existing urban population centers ? How did they manage to survive against the growing competition of the new urban-based giant firms ?

II. *Massachusetts rural mills : explanations*

1. Why go into the country ?

The New England cotton industry grew rapidly in the 1820s and 1830s and for a complex of economic reasons, as Zevin has shown. On the demand side (he argues), the tariff, the growth of population (at 3 % per annum), its redistribution into the West or the cities, a rise in real per capita income (under 1 % per annum) and transportation improvements increased demand between 1815 and 1833 to a level of 8 or 9 % per annum but, after 1824, at a decelerating rate of 7 or 8 % per annum. "After 1833 the influence of these forces became even less pronounced."(8) Rather, the cotton industry grew as a result of supply changes : the powerloom replacing the handloom, a secular decline in the price of raw cotton, labour-saving technology which reduced total wage costs, and an expansion of the industry's capacity. While all this explains the industry's total growth rate (compound) of 15-16 % per annum 1815-33, it does not explain why the New England cotton industry eschewed the Lancashire example and moved to water power sites rather than concentrating in existing urban centers using available steam engine technology — which was currently employed by Americans in river steamboats, railroads, numbers of urban manufacturing firms and a very few New England cotton textile mills in the 1820s and 1830s. By how much would a steam engine technology have raised the costs of New England cotton manufacturers in the 1820s and 1830s ? Like Habakkuk, G.N. von Tunzelmann argues that water power, being cheaper in America than Britain, allowed Americans to run their equipment at much higher speeds.(9) Until 1840, when James Montgomery and Charles T. James (as Justitia) debated New England's relative costs of steam and water power in the Boston newspapers, we have no hard data on whether a steam mill was more expensive.(10)

The data provided by both sides of the Justitia debate allow the computation of a measure of steam and waterpower costs in two specific locations, Newburyport and Lowell, Massachusetts, in 1841. While a reading of the documents shows that prejudice surfaced more easily with Justitia than with his opponents, Justitia's data must be accepted for the steam power costs in the mills he managed. On the other side, we can now use the contributions of "B" (probably Samuel Batchelder, an ex Lowell manager and Montgomery's superior in the York Manufacturing Co of Saco, Maine) and the correspondent from the Boot mill at Lowell, as well as those of Montgomery, to calculate waterpower costs for the Boot mill at Lowell.

Justitia made a mistake by failing to calculate power costs separately from total manufacturing costs. By expressing the cost of power per yard of goods he fell into several accounting errors. He totally ignored labour costs, which immediately falsified his result for the cost of manufacturing cotton goods. His ratios of manufacturing capacity to power for the various spindle types were inaccurate. And any savings he actually made and then attributed to the use of steam might in fact have derived from other savings in labour or capital costs. If we take power costs alone, add transportation charges to Justitia's steam mill figures, regard Lowell waterpower charges as an operating rather than a capital cost, and make the other adjustments claimed by Justitia's opponents, then a rather different picture emerges as Tables 15 and 16 show. The figures confirm Temin's finding that capital costs formed a higher proportion of overall costs in water than in steam driven mills.(11) But while the figure for steam costs (the same as Temin's) falls just below the wide range specified in Atack's simulation model for New England power costs, that for waterpower costs is just over half of the lower level of Atack's narrower simulated range.(12) And at $34 per horsepower per annum it is 30 % lower than Temin's figure. The omission of depreciation and insurance, much higher for steam engines than for waterwheels, would have raised steam costs but these in turn would have been offset by the use of larger engines. Clearly waterpower costs at Lowell have been seriously overestimated, largely because a rate of $4 per spindle, which included rent on four acres of land as well as waterpower, was quoted by Justitia and has been repeated uncritically ever since. When he came to prepare the second (unpublished) edition of his *Cotton Manufacture*, Montgomery assembled costs that were nearly double these present ones for steam power at Newburyport and just under half these present ones for waterpower at Lowell. But without knowing the method by which he reached these figures, it is impossible to tell what biases were built into his cost accounting.

These calculations pertain only to large American cotton mills of the Lowell type. On a much smaller scale, the cost advantages of using a water power site may have been much greater — but empirical work remains to be done on this, particularly on differentials in rents and water wheel efficiencies between Boston and up country sites for small mills. What is certain is that, to accommodate the state's 340 000 spindles in 1830 the city of Boston would have had to offer 4 250 horsepower (80 dead throstle spindles per hp) — a requirement which doubled by 1840. But the Charles River drained only 215 square miles and had a minimum flow of only 44 cubic feet per second (at Newton Upper Falls) compared to the Merrimack River's drainage basin of 4 085 square miles and a minimum flow of 1 275 cubic feet per second at Lowell (which was regularly measured from 1840 onward by James B. Francis, the English immigrant resident engineer at the Locks and Canals Co.). By 1880 the whole length of the Charles River had been utilised to the extent of producing only 3 362 horsepower, compared to the Merrimack which then was used to produce 88 818 horsepower. On the other side of the state the Connecticut River at Hartford, twenty miles south of the Massachusetts line, had a drainage basin of 10 154 square miles and a minimum flow of 5 208 cubic feet per second ; by 1880 the whole Connecticut River was utilised to produce 118 026 horsepower.(13) This cheap source of power the Americans proceeded to exploit vigorously : much more vigorously than their English counterparts who preferred the capital intensive steam engine on city sites and neglected available water power in England (according to the recent work of Gordon).(14) While this does not answer the counterfactual question (by how much would a switch to steam have raised American costs ?), it does demonstrate the rationale behind individual manufacturers' preferences for water power.

2. How did smaller rural mills survive against competition from the new giants at Waltham and Lowell ?

It might be argued that the small and medium rural cotton mill represented a passing phase in the cotton industry's development, following Alfred Marshall's analogy of firm growth to "trees in the forest" with smaller ones overshadowed by larger ones (until aging displaced these too).[15] Sooner or later widening transportation systems would remove local market imperfections and expose the small mills to the price competition of the giants whose large turnovers permitted them to pare their unit margins more finely than those of smaller mills. This might be the situation in the Massachusetts cotton industry of the 1830s when the giant mills of Middlesex county (at Waltham and at Lowell) were increasing their shares of the industry's capital investment, from 30 to nearly 50 % (see Table 13). In 1832 their share of the industry's throughput of 21.7 million pounds of raw cotton was just over 29 % and presumably this share also increased in the 1830s. The logical conclusion was the extinction of the state's rural cotton mills well within the space of the rest of the nineteenth century. Yet *Davison's Textile Blue Book* (the industry's trade directory) in the late 1940s still listed a number of cotton manufacturing firms in Massachusetts sprinkled through the rural counties, as well as numbers of firms clustering in the major centers of Lowell, New Bedford and Fall River — and this when the bulk of the industry had migrated to the South.[16] Rural cotton mills were not rapidly eliminated by the large firms. To be sure, they found new functions (specialty work for example) but they persisted as an economic and business phenomenon long after they are supposed to have disappeared.

Clearly part of the answer must lie in setting the smaller mills in the context of the changing economy of Massachusetts. Declining returns (profit rates of 3-4 % per annum) in agricultural activity in the state during the first decades of the nineteenth century made small cotton mills look profitable investments. However here I have started by focussing on the situation *within* the state's cotton industry.

In an attempt to discover how rural manufacturers in Massachusetts survived against their giant competitors at Waltham (the forerunner of the Lowell promotions, established in 1813 on the Charles River near Boston) and Lowell, I have taken two of the four leading cotton counties, Worcester and Bristol, and a much smaller but very rural cotton manufacturing county, Berkshire, and abstracted as much uniform cost information as possible about their cotton manufacturing firms from the compendious *McLane Report*.[17] It has not been possible to use the Hampden County Returns because they are too incomplete or conflated. In a very few instances I have had to estimate some information from a known variable about a firm (e.g., it has been assumed that cotton cost eleven cents a pound and on this basis, knowing the annual cotton bill, a few measures of throughput have been calculated). The only costs I have not included are merchants'commission and the cost of transportation. Thus as far as I can tell, the figures in Table 17 give a picture of factory gate costs. I have also assumed, contrary to McGouldrick, that most of these country mills used powerlooms, which might have been supplemented by handlooms when margins warranted. (See Appendix.)

The first point to emerge from Table 17 is that entry to the industry was relatively easy. For a small rural mill fixed costs in 1832 amounted to under $3 000 a year (with an exchange rate of $4.8, this, incidentally, looked like under £600-700 from Lancashire perspectives). Overheads could be reduced in various ways. The cash flow problem posed by labour costs might be offset by barter payments or by a company store. And money for the cost of the cotton was usually borrowed at the prevailing rate of 6 %. Certainly with $5 000 one could minimally get into business at the beginning of the season (which lasted July-November) in a rural cotton mill in Massachusetts in the early 1830s — $3 000 for a year's rent on land and machinery, $400 for cotton loan interest, and $1 600 for three months' wages and manufacturing costs. For an average mill in Worcester and Bristol counties, twice that sum was needed. Easy entry meant that throughout the 1820s and 1830s, whenever depressed conditions eliminated the less efficient manufacturer, a fresh entrepreneur could step forward and take his place — which is exactly what was happening according to the McLane evidence.

Having got into business, how did the small rural mills survive in competition with the giants of eastern Massachusetts or Lancashire ?

One point to emphasise is that the rural mills were fairly widely spread across the state. In 1836-37 in Worcester County for example seventy-four mills were spread through twenty-six towns, all but two with populations under 3 000 (Table 14).(18) To some extent this suggests an attempt to serve very local markets and thereby reduce competitive pressures. This seems to have been true for Athol in the north west of Worcester County : in 1832 it recorded its markets as Massachusetts. However, higher up Millers River (a tributary of the Connecticut River) at Winchendon, another firm was supplying Boston in 1832, and in the north east of the county, where the two branches of the Nashua River attracted two clusters of cotton mills, at Fitchburg (4 mills, 3 880 spindles in 1836) on the northern branch and at West Boylston (7 mills, mean size 1 150 spindles in 1836) on the southern branch, the explicit markets in 1832 were Boston, "all the cities in the Union" and South America — national and world markets.(19) In the south of the county the markets were also national : Boston, New York, Philadelphia or the South were recorded by sixteen of twenty-one mills in the Blackstone valley in 1832 ; the other five mills were sending print cloths to printeries at Dover, New Hampshire, Taunton, Massachusetts, or Providence, Rhode Island.

The wide markets of the country cotton mills are not surprising in view of the transportation networks which opened up the remote districts of Massachusetts in the 1820s and 1830s. There were three canals :

the Middlesex (opened 1803), linking the Merrimack River near the site of Lowell to Boston ; the Blackstone (opened in 1828), linking Worcester to Providence in Rhode Island ; and, least successful, the New Haven and Northampton (opened 1835), linking Hampden County to Long Island Sound. By 1835 three railroads radiated from Boston, to Providence, to Worcester and to Lowell. An extension from Worcester to Springfield opened in 1839 and would shortly reach Albany, New York as Bay State merchants tried to divert up-country trade toward Boston and away from New York, an effort which was not successful in the long run.[(20)]

In the 1830s, then, the rural mills of central Massachusetts were securing national rather than merely local markets and were not protected from Lowell competition by regional product market imperfections. (Their products were all or some of the following few kinds : shirtings, sheetings, bedtick, cotton warp, i.e. were in the same product markets.)

The analysis of relative costs in Table 17 contains some clues to understanding how the smaller mills survived. The eighteen Berkshire firms, in the remotest part of the state, represent the smallest mills in the industry, with costs averaging \$16 312 (excluding stock). The average firm in Worcester County in the centre of the state was similar in size to that in Bristol County, adjacent to Rhode Island, with average costs of around \$35 000 (excluding stock). The Boston Manufacturing Company at Waltham was nearly seven times larger and the average company at Lowell fifteen larger than these medium firms in Worcester and Bristol Counties. Proportionately, costs of land and power do not seem to have varied greatly throughout the state on the water power sites chosen. Nor do the costs of machinery. (The inference being that the ratio between land and capital was remarkedly stable despite capital intensive techniques.) The greatest disparities are plainly in the cost categories of labour, stock and raw cotton. Labour formed the highest proportion of costs in Worcester, Berkshire and Bristol Counties. At Waltham and Lowell they were comparatively much lower. This is not in the least surprising since the Waltham and Lowell promoters invested heavily in a labour-saving technology, which I have described and analysed elsewhere.[(21)] The relative closeness of the proportionate fixed costs of land and machinery at Waltham may be due to the depreciation of Waltham's machinery (which was, in 1832, getting on for twenty years old). The smaller mills used numbers of these improvements pioneered at Waltham but not on Waltham's scale and, apparently, the smaller mills of Massachusetts after 1824 produced relatively few innovations, at least patented ones.

What is interesting is the inverse relationship of stock and cotton between the Lowell and non-Lowell firms. Smaller or medium firms could not afford to have large amounts of finished or semi-finished goods in progress or unsold at any one time : they bought their cotton, processed it and moved most of it quickly. The impetus to do this was that many smaller firms had to buy the cotton on six months' credit at a rate of 6 %. By moving their raw or finished materials they lowered their variable costs in the short term. In the Waltham-Lowell mills the vastly increased scale of operations required production runs at a constant rate of output as well as large scale buying of cotton. Since the cotton season came in summer and fall the very large mills had to buy huge quantities of cotton for processing over the remainder of the year. Evidently they held onto stocks of yarn or uncoloured or unprinted cloth until the next season's fashion preferences (dictated by European taste and echoed in New York and Boston) were known. Thus the giants had to lock about 60 % of their finances into raw material, compared to 45-50 % for the medium and small mills. In the case of the Boston Manufacturing Company at Waltham it looks as though costs have shifted from the fixed to the variable sort. While this may reflect the ageing of the company's plant (much nearly twenty years old in 1832), it may have been deliberate policy on the part of the giant company to keep stocks ready for immediate use in the Boston market. In this situation the smaller firms could take advantage of falls in raw cotton prices the trend of which was downwards, so this could have been a considerable positional advantage. On the other hand, competing in the same market as the giants, they were obviously more exposed if finished goods prices fell, since their margins between costs and receipts (Table 17 far right columns) were proportionately a third or a quarter those of the giants. (Some portion of stocks must be added to raw cotton costs and some portion to the value of finished goods in order to calculate a profit level ; profits were said to have fluctuated between 6 and 15 % in Worcester County's Blackstone Valley 1829-31 ; at Lowell the Merrimack Company's profits on net worth were at least 14 % per annum 1828-35 according to McGouldrick.)(22)

While labour costs looked proportionately high for the smaller mills in Berkshire, Worcester and Bristol Counties, these mills could be more flexible in responding to demand changes because of the presence of (to use Clapham's phrase) "a reserve army of labour."(23) In New England a portion of that reserve army was female. Table 18 analyses the data recorded in the McLane and Bigelow censuses. In 1832 the remotest mills, in Berkshire County, relied least on women, having a ration of 1.9 females to every male while at the Waltham system centers, Waltham and Lowell, it was almost 3.5 females to every male.(24) By the late 1830s some interesting changes had taken place. Not only had the Waltham system centers vastly increased their female workforces without increasing their male cohorts — possible because males were more or less confined to overseer, manager, mechanic, machine shop, labourer positions, a small minority compared to machine operatives — but Worcester County displayed a distinct erosion of its proportion of females to males. The Worcester figures may record only full time mill hands (Bigelow did not define his categories). In this case there may indeed have been more reliance on part time female and child labour when business in the mill demanded. Alternatively, the rural cotton mills were indeed being operated by a workforce in which males and females were more closely balanced, as in the Rhode Island system, and which therefore was likely to have been more family based and possibly utilising a less automated technology than at Waltham. In this case perhaps the technology was being accommodated to suit the local population structure (on which work has yet to be done).

Interaction between industry and agriculture in the rural industries of course allowed mill operators to use numerous familiar wage saving devices : payment in kind, to deal with cash flow difficulties ; or mill shop barter ; and casual labour to deal with expansions or contractions of demand.[25]

Smaller mills might hold down capital costs by other techniques : room and power sharing ; the use of older equipment or of cheaper machines. Bezaleel Taft Jr described these devices in Worcester County in the *McLane Report*, remarking that the business was "prosecuted with much of the looseness incident to our agricultural pursuits."[26] More detailed studies like that of Jonathan Prude,[27] but exploring the inner accountings within the early factories and setting the factories in their industrial as well as their social contexts, are needed in order to advance our understanding of the survival of rural businesses.

Other advantages of the rural cotton mills help to explain their survival. Their role was central to what Mokyr has called "growing up" in the industrial revolution or facilitating the transition from handicraft to factory production.[28] They offered low-skill labour higher wages than agricultural pursuits when prices of agricultural products were falling, and with none of the psychic costs of moving into the city. They formed part of a group of rural industries in a community and thus shared the externalities generated by other rural businesses. Contrary to one portrayal, they did not always dominate the village. They might occasionally create a mill village as at Webster but a rapid reading of Hayward's *Massachusetts Directory* of 1835 shows that few townships in Massachusetts were solely dependent on a cotton mill for employment.[29] Woolen mills were more numerous than cotton mills and there were besides many other rural industries — making paper, scythes, pegs, cabinet work, wagons, sleighs, palm leaf hats, axes, shoes, cheese, blacksmiths work and so forth. The reservoir of metal working and wood working skills and resources built up in the villages helped the rural manufacturer in several ways. Operatives with experience in the woollen mills might be attracted into the cotton factory and find use for their skills as spinning or weaving operatives or overseers ; mechanics might be employed to help maintain or build machinery. A realistic picture will come with closer studies. As a preliminary exercise I have done a rapid analysis of cotton mill villages in the Blackstone Valley in the mid 1830s, summarised in Table 19. This shows that even in a region with a strong clustering of cotton mills, they were not, in contrast to Lancashire, singularly responsible for population growth in their townships. In twenty miles of the Blackstone Valley (Massachusetts section) during much of the 1830s only two of nine cotton manufacturing villages exhibited the possibility of cotton mill employment keeping pace with population growth : Holden and Uxbridge. In Douglas the cotton mill population shrank at a faster rate than the village itself. In the village communities the cotton mill was clearly a marginal production unit.

The point is reinforced by the comparison of males in cotton mills with all adult males in the community in 1836-37 in Blackstone Valley cotton mill townships (Table 20). In only three communities could as much as a quarter of the adult male population possibly have worked in the village cotton mills. This would be a very high proportion of male employment if it were a reliable statistic. But because the source distinguishes only males from females and because the cotton factories employed boys, it is likely to be rather lower than it appears. At any rate it does suggest that cotton mills did not dominate the rural communities in central Massachusetts. Woollen mills may have done, but they were not in the same international markets as Lancashire's giants.

The cultural advantages of the rural cotton mill certainly constituted powerful reasons for its establishment and probably its longevity. As far back as 1788 Jefferson had warned Americans to confine themselves to agricultural pursuits and thereby complement the manufacturing, urban-based economy of European states. His agrarian ideal of small family firms would, he believed, allow the new nation to escape the twin perils embodied in the new industrial forms : the corruption of morals and the threat to law and order posed by urban mobs.[30] By the 1820s another threat was perceived in the emergence of "proud and wealthy capitalists" whose "extorted monopoly" reduced operatives to "machines, as much as the spindles they superintend" and served only "to increase the wealth of a few capitalists."[31] To protagonists of the factory like Tench Coxe and Alexander Hamilton, the New World environment would purify the factory system. Feudal rights, operative ill-health and immorality, mob rule, would all wither in the presence of a republican and democratic political structure, American ingenuity, a vigorous climate and, above all, abundance of land.[32] Rural cotton mills were therefore a manifestation of an American ideal. But, in the event, was the machine set in the garden any more subordinated to human values, any more widely distributive of wealth, than the machines set in the city ?

Contemporary arguments and the philosophical debate have been thoroughly surveyed ; the historical realities have scarcely been examined.[33] Little more can be contributed here to the issue than to point towards future lines of investigation. Firstly we need to discover how widely ownership of the rural mills was distributed. Prude, in his study of three Massachusetts towns, identified ninety-three men who invested in textile mills between 1810 and 1830 but, beyond Samuel Slater and half a dozen others, did not perform a systematic analysis of their social backgrounds, occupations and sizes of investment.[34] Comparisons with the shoe manufacturers studied by Faler would be instructive.[35] Such a survey for Massachusetts and even New England, of investment patterns, mill ownership, entrepreneurship and management, based on industrial records as well as state censuses, is essential if we are to advance our understanding of the cultural role of the rural mills. In other words, the studies of Wallace and Prude[36] need to be set in regional frameworks. Secondly, we need to learn a lot more about the proprietors most active in running these mills in order to explain more about their marketing strategies. Geographical location was not the sole determinant of a mill's product since most cotton goods were sold either in Massachusetts or else beyond the entrepôts of Boston and New York. These final destinations represented markets whose nature helped to determine the products made in the country mill. In this connection the letterbooks of the mills run by Boston proprietors (identified in Spalding's thesis)[37] need to be read alongside those of the records of the rural cotton mills under more localised ownership. Only then will entrepreneurship and markets be clarified as factors explaining the emergence and persistence of the phenomenon of the rural New England cotton mill.

Conclusion

This paper offers some material for further work on the role of rural-based cotton mills in a developing economy. Water power almost certainly was cheaper than steam power in the USA in this period. What other economic and social advantages did rural mills offer ? Early entry into an industry was an important one. They might also serve local markets, though in the 1830s rural mills in Massachusetts were competing in national and international markets. They appear to have had overhead cost advantages in a period of falling cotton prices. And they bridged what has been called industrial growing up, allowing merchant entrepreneurs the chance of imposing industrial technologies on a pre-industrial way of life with much less social upheaval than in an urbanised mill context.

NOTES

N.B. : Some sources are cited in the footnotes to Tables only.

1. For British population figures see B.R. Mitchell, and Phyllis Deane, *Abstract of British Historical Statistics*, Cambridge, Cambridge University Press, 1962, and for American populations figures, Jim Potter, "The growth of population in America, 1700-1860", in : David V. Glass and D.E.C. Eversley, eds, *Population in History : Essays in Historical Demography*, London, Edward Arnold, 1965, p. 631-688.
2. For Lancashire population figures see *Parliamentary Papers, House of Commons*, 1831 (348), XVIII, Comparative Account of the Population of Great Britain ; and *PP, HC*, 1843 (496), XXII, Census of 1841. For Massachusetts population figures see Horace C. Wadlin, *Census of the Commonwealth of Massachusetts, 1895*, Boston, Wright & Potter Printing Co, 7 vols, 1896-1900, I : *Population and Social Statistics.*
3. T.W. Freeman, H.B. Rodgers and R.H. Kinvig, *Lancashire, Cheshire and the Isle of Man*, London, Nelson, 1966, p. 95.
4. For locational aspects of the Lancashire cotton industry see Arthur J. Taylor, "Concentration and specialisation in the Lancashire cotton industry, 1825-1850", *Economic History Review*, 2nd ser., 1, 1949, p. 114-122 ; H.B. Rodgers, "The Lancashire cotton industry in 1840", *Transactions of the Institute of British Geographers*, 28, 1960, p. 135-153. His map on p. 139 showing relative horsepower per parish in 1835 is misleading because it shows Bolton (which had 1 343 hp) with more horsepower than Bury (which had 2 999 hp) ; Douglas A. Farnie, *The English Cotton Industry and the World Market, 1815-1896*, Oxford, Clarendon Press, 1979.

   It should be noted that Rodger's work compares the factory inspector's reports for 1835, 1838 and 1841 ; data which, as Jenkins has shown, cannot be compared because the 1838 returns counted mills more than once if they had more than one occupier while the 1835 returns counted mills only. The 1841 data relates to firms, not mills. See David T. Jenkins, "The validity of the factory returns, 1833-50", *Textile History*, 4, 1973, p. 26-46.
5. V.A.C. Gatrell, "Labour, power and the size of firms in Lancashire cotton industry in the second quarter of the nineteenth century", *Economic History Review*, 2nd ser., 30, 1977, p. 95-139.
6. Caroline Ware, *The Early New England Cotton Manufacture*, (first ed. 1931), repr. New York, Russell & Russell, 1966.
7. *Annual Statistics of Manufactures in Lowell and Vicinity, 1835 and 1882*, Lowell, Mass., Vox Populi Press, 1982.
8. Robert Brooke Zevin, "The growth of cotton textile production after 1815", in : Robert W. Fogel and Stanley L. Engermann, eds, *The Reinterpretation of American Economic History*, New York, Harper & Row, 1971.
9. G.N. von Tunzelmann, *Steam Power and British Industrialisation*, Oxford, Clarendon Press, 1978, p. 274. Higher speeds did not (as Tunzelmann argues, p. 267) lead to more labour per spindle because Americans quickly adopted stop motions to counter breakages, as shown in my book *Transatlantic Industrial Revolution : the Diffusion of Textile Technologies between Britain and America, 1790-1830s*, Cambridge, Mass., MIT Press, and Oxford, Blackwells, 1981, p. 198, 211.

   Higher operating speeds could well have saved labour relatively (in relation to capital — in requiring fewer workers per spindle ; or specified another way, fewer workers for the fixed cost of a spindle and the overhead costs of running it) as well as absolutely (in reducing labour per unit of output). For a much older locational study, which examines power sources in the early twentieth century, see J. Herbert Burgy, *The New England Cotton Textile Industry : a Study in Industrial Geography*, Baltimore, Waverly Press, 1932.

10. For this debate see my unpublished edition of James Montgomery's intended second edition of his *A Practical Detail of the Cotton Manufacture of the United States* (1840) and re-examination of the Justitia controversy (typescript of 461 p., in the Kress Library, Baker Library, Harvard University Graduate School of Business Administration, 1980). The following three paragraphs come from this source.
11. Peter Temin, "Steam and waterpower in the early nineteenth century", *Journal of Economic History*, 26, 1966, p. 187-205.
12. Jeremy Atack, "Fact in fiction ? The relative costs of steam and water power : a simulation approach", *Explorations in Economic History*, 16, 1979, p. 425. For the 1840s, Atack's 95 % confidence interval about the mean (of $83) ranges from $63 to $106 per hp for 100 hp prime movers driven by steam ; for water-driven prime movers his range is $67 to $147 per hp (about a mean of $99). See also Jeremy Atack, Fred Bateman and Thomas Weiss, "The regional diffusion and adoption of the steam engine in American manufacturing", *Journal of Economic History*, 40, 1980, p. 293. Here the range for New England in the 1840s is calculated at $56 to $163 per hp for steam power and $56 to $88 for waterpower.
13. United States Tenth Federal Census, vol. 16 : *Reports on the Water-Power of the United States*, Washington, Government Printing Office, 1885, part I, p. xxviii, xxxix.
14. Robert B. Gordon, "Cost and use of water power during industrialisation in New England and Great Britain : a geological interpretation", *Economic History Review*, 2nd ser., 36, 1983, p. 240-259.
15. Alfred Marshall, *Principles of Economics*, 9th ed., London, Macmillan, 1961, 2 vols, I, p. 315-316.
16. See *Davison's Textile Blue Book for 1949*, 84th ed., Ridgewood, NJ, Davison Publishing Co, 1949, p. 270-281. Ranked by population, the following Massachusetts towns contained cotton mill companies (of a wide variety of functions) in 1949 :

	Population under			
	5 000	*10 000*	*25 000*	*50 000*
N° of towns	24	33	47	55
N° of cotton mill companies in all these towns	26	41	65	80

In 1940 the Bureau of the Census defined urban areas as towns with a population of over 50 000.

17. United States Congress House, *Documents Relative to the Manufactures in the United States*, Executive Document, n° 308, 1st session, 22nd Congress, Washington, DC, 1833, 2 vols, repr. New York, Augustus M. Kelley, 1969 *(The McLane Report)*.
18. The first industrial census of Massachusetts is the indispensable source here : John P. Bigelow, *Statistical Tables Exhibiting the Condition and Products of Certain Branches of Industry in Massachusetts for the Year Ending April 1 1837*, Boston, Dutton & Wentworth, 1838.
19. See *McLane Report*, *op. cit.*, I, p. *(inter alia)*, 285, 341, 373, 379, 387, 403, 411, 523, 527, 537, 577, for clear evidence of national markets, and p. 497, 561 for references to the South American market which was opened up in the mid 1820s (see Ware, *op. cit.*, p. 189-197).
20. George Rogers Taylor, *The Transportation Revolution, 1815-1860*, New York, Holt, Rinehart & Winston, 1951, and New York, Harper & Row, 1968 ; John Hayward, *The Massachusetts Directory*, Boston, John Hayward, 1835.
21. Jeremy, *Transatlantic Industrial Revolution*, *op. cit.*, ch. 10.

22. *McLane Report*, *op. cit.*, I, p. 81 ; Paul F. McGouldrick, *New England Textiles in the Nineteenth Century : Profits and Investment*, Cambridge, Mass., Harvard University Press, 1968, p. 84.
23. Sir John Clapham, *An Economic History of Modern Britain : the Early Railway Age, 1820-1850*, Cambridge, Cambridge University Press, 1930, vol. I, p. 557.
24. See Thomas Dublin, *Women at Work : the Transformation of Work and Community in Lowell, Massachusetts, 1826-1860*, New York, Columbia University Press, 1979, especially ch. 3.
25. Two articles by John W. Lozier are especially illuminating on the tactics by which rural mill operators achieved operating profitability : "The forgotten industry : small and medium sized cotton mills south of Boston", *Working Papers from the Regional Economic History Research Center*, 2, n° 4, 1979, p. 101-124, and "Rural textile mill communities and the transition to industrialism in America, 1800-1840", *ibid.*, 4, n° 4, 1981, p. 78-96.
26. *McLane Report*, *op. cit.*, I, p. 80.
27. Jonathan Prude, *The Coming of Industrial Order : Town and Factory Life in Rural Massachusetts, 1810-1860*, Cambridge, Cambridge University Press, 1983.
28. Joel Mokyr, "Growing-up and the industrial revolution in Europe", *Explorations in Economic History*, 13, 1976, p. 371-396.
29. Hayward, *op. cit.* : this is an impressionistic assessment based on a rapid reading.
30. See Leo Marx, *The Machine in the Garden : Technology and the Pastoral Ideal in America*, Oxford, Oxford University Press, 1964, p. 116-141.
31. Thomas Cooper, *Two Tracts*, Charleston, 1823, p. 21-22, quoted in Michael B. Folsom and Steven D. Lubar, eds, *Documents in American Industrial History : the Philosophy of Manufactures. Debates over Industrialisation in the United States*, Cambridge, Mass., MIT Press, 1982, p. 256.
32. Marx, *op. cit.*, p. 152-165.
33. *Ibid.*, and Folsom and Lubar, *op. cit.*
34. Prude, *op. cit.*, p. 60-61.
35. Paul F. Faler, *Mechanics and Manufacturers in the Early Industrial Revolution : Lynn, Massachusetts, 1780-1860*, Albany, NY, State University of New York Press, 1981, ch. 4.
36. Prude, *op. cit.*, p. 60-61. Anthony F.C. Wallace, *Rockdale : the Growth of an American Village in the Early Industrial Revolution*, New York, Alfred Knopf, 1978.
37. Robert V. Spalding, *The Boston Mercantile Community and the Promotion of the Textile Industry, 1813-1860*, Yale, unpublished PhD, 1963.
38. McGouldrick, *New England Textiles*, *op. cit.*, 1968, p. 281, n. 14.
39. According to Duncan Bythell, *The Handloom Weavers*, Cambridge, Cambridge University Press, 1969, p. 115.
40. Jeremy, *Transatlantic Industrial Revolution*, *op. cit.*, p. 270.
41. *Ibid.*, p. 93.

APPENDIX

How Quickly Did Rural Cotton Mills in Massachusetts Adopt the Powerloom and Vertically Integrate Spinning and Weaving ?

Writing about the years 1825-33, Paul McGouldrick concluded that in southern New England half the industry's capital "was still represented by small mills which, from all the evidence, were not integrated spinning and cloth making units."(38) This note looks at census evidence and questions his assumption.

The 1820 Census of Manufactures shows that in the Massachusetts cotton industry the dominant organisational form was the spinning and weaving firm (47 out of 64 firms). However there is no indication how many of these firms had brought weaving out of the home and into the factory complex, either in handloom workshops or powerloom sheds. The only clue is that 13 out the 47 firms were running powerlooms, 321 in all (175 of them in the three mills of the Boston Manufacturing Co at Waltham).

The *McLane Report* of 1833 contains data which suggest that powerloom technology and vertical integration might well have been extensively adopted by 1832 in the Massachusetts cotton industry. McLane's questionnaire, unlike the 1820 one, did not ask for an inventory of manufacturing equipment. It did ask, however, for the value of investment in machinery (quite separately from buildings and land), and *inter alia* the weight of raw cotton annually consumed and the kinds and quantities of yarn and cloth produced. Knowing the ratio between the value of mechanised spinning and weaving plant (K) and the number of pounds of cotton processed per annum (throughput : T) at the vertically integrated and powerloom mills at Waltham and Lowell we have a yardstick by which to assess the country mills. A throughput much higher than the Waltham-Lowell range of T would indicate that K was being augmented by putting-out the weaving ; a much lower throughput would point to handlooms in the factory orbit.

Clearly we have to make a set of *ceteris paribus* assumptions : that firms had equal access to mechanised weaving technology ; that economic fluctuations affected all firms evenly ; that raw cotton had a uniform price throughout Massachusetts. Apart from the scale advantages of the giants at Waltham and Lowell, it does not seem too unreasonable to make these assumptions for most firms, and to accept the T/K range at Waltham and Lowell as the norm for any other firm using powerloom technology and vertically integrated manufacturing.

Throughput for the four firms at Lowell was 4.6 lb of cotton per annum for every $1 of machinery investment and at Waltham 6.8 lb. On a range of 4 to 7 lb throughput per $1 K only 8 of the 147 firms manufacturing cotton goods and reporting sufficient data (another 45 firms recorded insufficient data or made yarn only) were below the Waltham-Lowell range ; 25 were above. On a 4-8 lb T only 10 were above. Since some firms were reported in clusters by township we cannot be absolutely certain that these figures are unequivocal. However it does look as if by 1832 some 60-70 % of the whole of the cotton textile industry at Massachusetts was deploying capital equipment with a throughput level very similar to the range exhibited by the industry's leaders, and therefore implying the use of a similar powerloom technology and vertically integrated organisation.

There is the possibility that labour was still being used to augment capital in the small mills. The patchy nature of the reporting in the McLane returns means that labour costs for all the K-T firms is not known, though educated guesses could be made to complete the blanks. Two considerations suggest that labour was not augmenting capital however. Firstly extra labour costs would have significantly eroded unit profits in small mills, which needed to keep their margins above those of the Waltham-Lowell giants. Secondly, of the 19.5 million lb of cotton processed per annum in Massachusetts in 1832 10-12 million lb was going through the medium and small mills. Since 1 lb of cotton made about 3 yards of cloth and since a good handloom weaver could manage no more than 100 yards a week(39), it would take over 7 000 very good handloom weavers to handle the 12 million lb of throughput of the state. It is very doubtful whether anything like this number were active :

certainly 2 429 male weavers and 140 female weavers are known to have arrived in the USA from Britain in the years 1824-31[40] but there is no evidence that they gravitated to Massachusetts. Household weaving, the alternative to a body of handloom weavers working full time, was not an economic proposition for large scale cloth making either : putting-out in 1818 required the agent of a Massachusetts merchant to make itineraries of up to a hundred miles from his base to distribute yarn and collect cloth.[41]

In short, all the circumstantial evidence points towards the typicality of powerloom weaving and vertical integration in Massachusetts cotton spinning and weaving firms by 1832.

Table 1. *Lancashire : population density, 1811-1841, by parishes* (persons per square mile).

Cotton parishes	*1811*	*1821*	*1831*	*1841*	*Area (in sq.m.)**
Asthon-u-Lyne	1 311	1 787	2 312	3 187	14.53
Blackburn	546	748	839	1 006	71.28
Bolton	809	1 023	1 285	1 507	49.05
Bury	809	995	1 376	1 788	34.75
Cartmel	110	137	134	137	35.87
Chorley	1 066	1 505	1 910	2 704	4.86
Deane	534	626	761	868	30.22
Eccles	617	738	888	1 069	31.62
Garstang	149	178	167	184	41.53
Halton	130	172	139	116	5.98
Lancaster	169	188	216	234	103.28
Leigh	843	995	1 087	1 204	18.47
Leyland	389	462	495	500	28.05
Manchester	2 547	3 494	5 062	6 602	53.53
Middleton	579	711	800	861	17.98
Prescot	362	418	515	658	54.56
Preston	878	1 228	1 634	2 405	22.23
Prestwich-Oldham	1 250	1 588	2 044	2 376	33.06
Radcliffe	620	686	868	1 133	4.5
Rochdale	406	514	638	925	91.59
Ullverstone	129	156	171	193	45.47
Warrington	763	872	1 000	1 143	19.16
Whalley	373	499	579	662	168.75
Wigan	730	888	1 031	1 205	43.14
Winwick	371	421	466	471	38.53
Lancashire	469	596	752	944	1 766
Total populations					
Cotton parishes	618 906	792 901	1 000 233	1 240 063	
Lancashire	828 309	1 052 859	1 335 800	1 667 054	

*1 sq.m. = 640 acres.

Sources : PP, HC, 1831 (348), XVIII, p. 130-139, Comparative Account of the Population of Great Britain ; *ibid.*, 1843 (496), XXII, p. 140-150, Census of 1841.

Table 2. *County shares of Massachusetts' population, 1810-1840* (%).

County	*1810*	*1820*	*1830*	*1840*
Barnstable	4.7	4.59	4.67	4.41
Berkshire	7.58	6.8	6.18	5.66
Bristol	7.87	7.82	8.12	8.16
Dukes	0.7	0.63	0.58	0.54
Essex	15.23	14.27	13.57	12.88
Franklin	5.81	5.62	4.85	3.91
Hampden	5.17	5.35	5.18	5.06
Hampshire	5.2	5.06	4.96	4.19
Middlesex	11.18	11.75	12.77	14.45
Nantucket	1.44	1.39	1.18	1.22
Norfolk	6.62	6.97	6.88	7.2
Plymouth	7.45	7.29	7.05	6.42
Suffolk	7.28	8.4	10.18	12.98
Worcester	13.75	14.07	13.82	12.92
Massachusetts : N	472 040	523 287	610 408	737 700

Sources : Calculations of data in Horace G. Waldin, *Census of the Commonwealth of Massachusetts, 1895*, Boston, Wright & Potter Printing Co, 1896-1900, 7 vols, I : *Population and Statistics.*

Table 3. *Massachusetts : population density, 1810-1840, by counties* (persons per square mile).

County	*1810*	*1820*	*1830*	*1840*	*Area (in sq.m.)**
Barnstable	67	73	86	99	ca 330
Berkshire	42	41	44	49	860
Bristol	62	68	83	100	600
Dukes	27	27	29	33	120
Essex	200	207	230	264	360
Franklin	42	45	46	44	650
Hampden	42	48	54	64	585
Hampshire	46	50	57	58	532
Middlesex	66	77	97	133	800
Nantucket	136	145	144	180	50
Norfolk	78	91	105	133	ca 400
Plymouth	59	64	72	79	ca 600
Suffolk incl. Boston	31 434	40 174	56 835	87 564	109 375 (700 acres)
Worcester	43	49	56	64	1 500
Massachusetts	60	67	78	95	7 800
Total Mass. population	472 040	523 287	610 408	737 700	

*1 sq.m. = 640 acres

Sources : Horace G. Waldin, *op. cit.* ; and John Hayward, *The Massachusetts Directory*, Boston, John Haywara, 1835.

Table 4. *The relative size of Lancashire's cotton industry, 1838.*

Districts	*Mills*		*Horsepower*		*Workers*	
	N	*%*	*N*	*%*	*N*	*%*
Lancashire	1 125	69.7	33 467	72.6	152 150	75.9
Cheshire	154	9.5	4 590	9.9	15 253	7.6
Yorkshire	169	10.5	3 205 1/2	6.9	12 436	6.2
Derbyshire	85	5.3	3 037	6.6	10 462	5.2
All other districts	80	4.9	1.818	3.9	10 267	5.1
Total England & Wales	1 613	99.9	46 117 1/2	99.9	200 568	100

Sources : Derived from D.T. Jenkins, "The validity of the Factory Returns, 1833-1850", *Textile History*, 4, 1973, p. 40.

Table 5. *Lancashire cotton industry : distribution of horsepower and powerlooms, 1836.*

Cotton parishes	*% of steam and water hp (1)*	*% of steam 1835 hp (1)*	*% of power looms 1835-36 (2)*	*% of power loom mills 1835-36 (2)*
Ashton-u-lyne	7.46	7.96	6.57	3.79
Blackburn	4.06	4.16	6.96	5.52
Bolton	5.8	5.33	2.78	2.41
Bury	12.95	10.7	14.75	15.52
Cartmel	0.43	0.49		
Chorley	1.08	1.23	0.56	0.34
Deane	1.65	1.51	1.29	2.07
Eccles	0.97	0.99	0.68	0.69
Garstang	0.32	0.03		
Halton	0.43			
Lancaster	1.03	0.88	1.87	1.72
Leigh	1.13	1.29		
Leyland	0.63	0.44	0.31	0.34
Manchester	27.17	30.56	25.56*	22.76
Middleton	0.33	0.37	0.67*	0.34
Prescot	0.28	0.28		
Preston	3.99	4.55	3.85	2.76
Prestwich-Oldham	11.16	11.87	0.18*	0.69
Radcliffe	0.09	0.1	0.12	0.34
Rochdale	7.69	7.06	0.05	0.34
Ullverstone	0.24	0.11		
Warrington	0.67	0.77		
Whalley	6.73	4.63	8.21	12.09
Wigan	3.62	4.13	2.5	1.04
Winwick	0.47	0.53		
Macclesfield district in Lancs.			23.11	26.9
Total hp	23 153.5	20 302.5	61 176	290

* Only part of these parishes.

Sources : 1. *PP, HC*, 1836 (138), XLV, Factory Inspectors Returns, 1 May 1835, p. 18-21. 2. *PP, HC*, 1836 (24), XLV, Return of Powerlooms Made Sept. 1835 - Febr. 1836.

Table 6. *Percentages of steam engines per cotton parish by size of engine, 1835-36.*

Cotton parishes	*Under 10 hp*		*10-19 hp*		*20-29 hp*		*30-39 hp*		*40-49 hp*		*Over 50 hp*		*Total n° of steam engines*
	%	*N*	*%*	*N*	*%*	*N*	*%*	*N*	*%*	*N*	*%*	*N*	
Ashton-u-Lyne	5.9	3	9.8	18	13	25	6.3	8	3	2	7	7	63
Blackburn	3.9	2		7		6	6.3	8		2		5	30
Bolton	3.9	2	6	11	2.6	5	7.9	10	13.6	9	2	2	39
Bury	13.7	7	7.1	13	13.6	26	11.8	15	10.6	7	9	9	77
Cartmel										1		1	2
Chorley						5						2	7
Deane		1		7		4		1		2			15
Eccles				1		2		1		1		1	6
Garstang		1											1
Halton													
Lancaster		1		2		3		1		1			8
Leich				6		3		2		1			12
Leyland		1		2		1		1					5
Manchester	11.8	6	17.5	32	15.7	30	29.9	38	21.2	14	49.5	49	169
Middleton				1		1		1					3
Prescot				2				1					3
Preston		2		9		7		3		4		5	30
Prestwich-Oldham	5.9	3	15.8	29	15.7	30	11.8	15	13.6	9	8	8	94
Radcliffe						1							1
Rochdale	23.5	12	9.8	18	4.7	9	5.5	7	16.7	11	4	4	61
Ullverstone				2									2
Warrington						5		1					6
Whalley	11.8	6	9.8	18		14		9				1	48
Wigan		3		4		10		5		2		5	29
Winwick		1		1		4							6
Total	100	51	100	183	100	191	100	127	100	66	100	99	717

Sources : PP, HC, 1836 (138), XLV, Factory Inspectors' Returns, 1835-36.

Table 7. *Percentages of water wheels per cotton parish, by size of wheel, 1835-36.*

Cotton parishes	*Under 10 hp*		*10-19 hp*		*20-29 hp*		*30-39 hp*		*40-49 hp*		*Over 50 hp*		*Total n° of steam engines*
	%	*N*	*%*	*N*	*%*	*N*	*%*	*N*	*%*	*N*	*%*	*N*	
Ashton-u-Lyne		4		6									10
Blackburn		1		6		1							8
Bolton		2		2		4				1		1	10
Bury	26	31	17.6	12	30.7	8		2		2		3	58
Chorley													
Deane		3		3		1							7
Eccles		1		1									2
Garstang		1		2				1					4
Halton				1									1
Lancaster				1		1		1					3
Leigh													
Leyland		2		3		2							7

Cont.→

Table 7 (continued)

Cotton parishes	*Under 10 hp*		*10-19 hp*		*20-29 hp*		*30-39 hp*		*40-49 hp*		*Over 50 hp*		*Total n° of steam engines*
	%	*N*	*%*	*N*	*%*	*N*	*%*	*N*	*%*	*N*	*%*	*N*	
Manchester		3		3		1		1					8
Middleton													
Prescot		1											1
Preston													
Prestwich-Oldham		4						2				1	7
Radcliffe													
Rochdale	26.9	32		8		4							44
Ullverstone		1		2									3
Warrington													
Whalley	27.7	33	26.5	18		4	41.7	5					60
Wigan													
Winwick													
Total	100	119	100	68	100	26	100	12		3		5	233

Sources : PP, HC, 1836 (138), XLV, Factory Inspectors Returns, 1835-36.

Table 8. *Horsepower of all kinds (1) of cotton mills in Lancashire, 1841.*

Parishes	*Steam*		*Water*		*Total*	
	hp	*%*	*hp*	*%*	*hp*	*%*
Ashton	4 896	13.9	374	10.9	5 270	13.6
Blackburn	2 283[2]	6.5	201[2]	5.9	2 484	6.4
Bolton	2 515	7.1	474[2]	13.8	2 989	7.7
Bury	3 175	9	826	24.1	4 001	10.4
Eccles	357[2]	1	20[2]	0.6	377	1
Leigh	193[2]	0.5	40[2]	1.2	233	0.6
Manchester	7 794.5	22.1	55[2]	1.6	7 849.5	20.3
Preston	2 039[2]	5.8			2 039	5.3
Prestwich	2 437	6.9	424[2]	12.4	2 861	7.4
Oldham	3 570.5	10.1	23	0.7	3 593.5	9.3
Rochdale	2 513[2]	7.1	210[2]	6.1	2 723	7
Whalley	2 517	7.1	750	21.9	3 267	8.5
Wigan	904[2]	2.6	30[2]	0.9	934	2.4
Total	35 194[3]	99.7	3 427[4]	100.1	38 621	99.9

1. All kinds of mills : fine spinning ; coarse spinning ; integrated spinning and weaving ; powerloom ; yarn doubling ; small wares ; spinning and weaving cotton waste.
2. Each includes hp from one or more other parishes but never exceeding 100-200 hp.
3. This includes 5 323 hp in mills on short time and 2 959 hp in mills stopped.
4. This includes 345 hp in mills on short time and 354 hp in mills stopped.

Sources : PP, HC, 1842 (31), XXII, Reports of the Inspectors of Factories for the Half Year Ending 31 Dec. 1841.

Table 9. *Horsepower of cotton firms in Lancashire, 1841* (capacity only).

Parishes	*Fine spinning*		*Coarse spinning (under N° 60 yard)*		*Integrated spinning + weaving*		*Power loom*		*Yarn doubling*		*Small wares*		*Spinning + weawing waste cotton*		*Total*	
	Steam	*Water*	*S*	*W*	*S*	*W*	*S*	*W*	*S*	*W*	*S*	*W*	*S*	*W*	*S*	*W*
Ashton	346	29	985	142	1 419	30	46								2 796	201
Blackburn			190	64	1 458	60	93	13							1 741	137
Bolton	340	52[1]	589	131	509	121									1 438	304
Bury			573	256	2 057	397	59	83[5]							2 689	736
Eccles							137	20[6]							137	20
Leigh	105														105	
Manchester	938		1 770		3 129	2	233		44 1/2		181		44	45[8]	6 339.5	47
Preston	238[2]		287		1 258										1 783	
Prestwich			701	334[3]	1 673	60[4]					40	30[7]			2 414	424
Oldham	102		1 051	14	1 337	2	70		28 1/2				227	4	2 815.5	20
Rochdale			1 528	90	568	43									2 096	133
Whalley			296	286	1 676	393	109	27							2 081	706
Wigan			477												477	
Total	2 069	81	8 447	1 317	15 084	1 108	747	143	73		221	30	271	49		

1. Including Deane
2. ,, Leyland
3. And 14 other parishes
4. ,, 13
5. ,, 4
6. ,, 7
7. ,, 3
8. ,, 3

Sources : PP, HC, 1842 (31), XXII, Reports of the Inspectors of Factories for the Half Year Ending 31 Dec. 1841.

Table 10. *The structure of the Lancashire cotton industry, 1841.*

Firms	*Under 10 hp*		*10-49 hp*		*50-199 hp*		*Over 2000 hp*		*Total*	*average/ firm*
	i.e. under 5 000 sp.	*average/ firm*	*i.e. 5 000- 25 000 sp.*	*average/ firm*	*i.e. 25 000- 100 000 sp.*	*average/ firm*	*i.e. 100 000 sp. upwards*	*average/ firm*		
Fine spinners										
N° of firms	8		56		16				80	
Power (hp)	37	5	1 187	21	1 260	79			2 484	31
Workers	243	30	6 887	123	8 145	509			15 275	191
Coarse spinners										
N° of firms	121		273		73		3		470	
Power (hp)	609	5	6 520	24	5 156	71	700	233	12 985	28
Workers	2 837	23	36 024	132	10 620	145	2 079	693	51 560	110
Spinning + WG										
N° of firms	10		153		140		18		321	
Power (hp)	72	7	4 174	27	12 537	90	4 231	235	21 014	65
Workers	325	32	23 755	155	68 768	491	19 183	1 066	12 031	349

Cont.→

Table 10 (continued)

Firms	*Under 10 hp*		*10-49 hp*		*50-199 hp*		*Over 200 hp*		*Total*	*average/ firm*
	i.e. under 5 000 sp.	*average/ firm*	*i.e. 5 000- 25 000 sp.*	*average/ firm*	*i.e. 25 000- 100 000 sp.*	*average/ firm*	*i.e. 100 000 sp. upwards*	*average/ firm*		
Power weaving										
N° of firms	43		59		2				104	
Power (hp)	254	6	988	17	135	68			1 377	13
Workers	2 157	50	6 800	115	565	283			9 522	92
Total										
N° of firms	182		541		231		21		975	
Power (hp)	972	5	12 869	24	19 088	83	4 931	235	37 860	39
Workers	5 562	31	35 747	66	97 502	422	21 262	1 013	160 073	164

Sources : V.A.C. Gatrell, "Labour, power, and the size of firms in Lancashire cotton in the second quarter of the nineteenth century", *Economic History Review*, 2nd ser., 30, 1977, p. 130-139.

N.B. I have assumed 1hp drove 500 (mule) spindles but this would be 300 if there were many self actors and under 100 for throstle spindles. Mules dominated however.

Table 11. *United States cotton industry : spindleage, 1820-1840.*

State	*1820*[1]		*1831*[2]		*1840*[3]	
	N	*%*	*N*	*%*	*N*	*%*
Maine	3 946	1.15	6 500	0.52	29 736	1.3
Vermont	4 182	1.22	12 393	0.99	7 254	0.32
New Hampshire	18 621	5.44	113 776	9.13	195 173	8.54
Massachusetts	42 824	12.51	339 777	27.26	665 095	29.11
Rhode Island	70 308	20.54	235 753	18.91	518 817	22.71
Connecticut	36 189	10.57	115 528	9.29	181 319	7 94
New York	64 836	18.94	157 316	12.62	211 659	9.26
New Jersey	20 584	6.01	62 979	5.05	63 744	2.79
Pennsylvania	28 853	6.97	120 810	9.69	146 494	6.41
Delaware	14 204	4.15	24 806	1.99	24 492	1.07
Maryland	24 348	7.11	47 022	3.77	41 182	1.8
Elsewhere	18 371	5.36	9.844	0.79	199 666	8.74
Total	342 266	99.97	1 246 503	100.01	2 284 631	99.99

Sources : 1. David J. Jeremy, *Transatlantic Industrial Revolution : the Diffusion of Textile Technologies between Britain and America, 1790-1830s*, Cambridge, Mass., MIT Press, 1981, p. 277 : these do not include estimates.
2. Friends of Domestic Industry, New York Convention, *Report on the Production and Manufacture of Cotton*, Boston, J.T. & E. Buckingham, 1832, p. 16.
3. US Sixth Federal Census, *Compendium of the Enumeration of the Inhabitants and Statistics of the United States*, Washington, DC, Thomas Allen, 1841, p. 361.

Table 12. *Investment in major (over $1 million) industries in Massachusetts, 1836-1837.*

Industry	*Capital investment*	*Share %*
Cotton manufacturing	14 369 719	26.2
Fisheries	12 484 078	22.76
Woollen manufacturing	5 770 750	10.52
Sheep farming	2 842 778	5.18
Leather	2 033 423	3.71
Nails, brads, tacks	1 974 000	3.60
Cotton printing	1 539 000	2.81
Iron castings	1 516 025	2.76
Paper	1 167 700	2.13
Machine making	1 146 775	2.09
Oil (Whale)	1 133 500	2.07
Textile related Industries :		
Cordage and twine	285 375	0.52
Wool cards	148 340	0.27
Silk	137 000	0.25
Cotton batting, thread, warp, wicking	78 000	0.14
Total for all Industries (not listed)	54 851 643	100

Sources : Bigelow, *Statistical Tables (of Massachusetts) for the Year Ending April 1, 1837*, Boston, Dutton & Wentworth, 1838.

Table 13. *Massachusetts : capital investment in cotton manufacturing, 1800-1837* (% by county).

County	*Capital authorized 1800-20*[1]	*Capital reported 1820*[2]	*Capital in cotton machinery 1832*[3]	*Capital reported 1837*[4]
Barnstable		0.11	0.01	0.05
Berkshire	9.56	0.54	4	4.41
Bristol	18.86	31.67	18.75	11.29
Dukes				
Essex	2.65		1.88	2.35
Franklin			0.04	0.63
Hampden	8.72	9.85	12.85	11.81
Hampshire	1.81	1.19	1.88	1.5
Middlesex	6.72	29.7	32.24	48.01
Nantucket				
Norfolk	16.44		6.52	4.31
Plymouth	15.38		1.6	1.6
Suffolk	4.85			
Worcester	15	26.94	20.21	14.01
Total %	99.99	100	99.98	99.97
$	15 479 000	1 846 915	4 509 615	14 379 719

Sources : 1.US Congress, Senate, Report of the Secretary of State of Such Articles Manufactured in the United States as Would Be Liable to Duties if Imported from Foreign Countries ; as also... a Schedule of Factories Incorporated by State Laws, from 1800 to

1820, Inclusive, *Senate Report n°45,* 1st session, 18th Congress, Washington, DC, 1824.
2. US National Archives, RG 29, ms. *Returns of the 1820 Census of Manufactures,* USNA, Microfilm M 279, 27 rolls, Washington, DC, 1967, roll 2.
3. US Congress, House, *Documents Relative to the Manufactures of the United States,* Executive Document n° 308, 1st session, 22nd Congress, Washington, DC, 1833, 2 vols, I : *The McLane Report.*
4. Bigelow, *Statistical Tables, op. cit.*

Table 14. *Massachusetts : population (1840) compared to cotton mills, spindles, capital, in four leading cotton counties.*

Township	*Population 1840*	*N° of mills*	*Total spindles*	*Total capital investment %*
Middlesex county :				
Concord	1 784	1	640	NA
Hopkinton	2 245	3	3 428	50 000
Lowell	20 796	22	141 334	6 167 000
Newton	3 351	2	5 710	200 000
Shirley	957	3	2 568	30 000
Waltham	2 504	3	11 488	450 000
Watertown	1 810	1	700	12 000
Total		34	165 868	6 909 000
Average spindles + capitals			4 878	203 206
Worcester county :				
Ashburnham	1 652	1	512	8 000
Athol	1 591	1	1 024	15 000
Barre	2 751	1	2 550	60 000
Charlton	2 117	1	656	15 000
Douglas	1 617	2	4 000	56 000
Fitchburg	2 604	4	3 880	57 000
Grafton	2 943	5	14 054	192 000
Holden	1 874	4	5 800	83 000
Lancaster	2 019	3	256	7 000
Mendon	3 524	8	13 914	323 000
Milford	1.773	1	1 200	15 000
Millbury	2 171	1	1 848	20 000
New Brain	752	1	NA	6 000
Northborough	1 248	2	1 820	40 000
Northbridge	1 449	4	9 000	95 000
Oxford	1 742	4	6 226	107 000
Phillipston	919	1	1 512	20 000
Southbridge	2 031	3	6 844	80 000
Sturbridge	2 005	6	8 664	129 500
Sutton	2 370	4	7 356	108 000
Uxbridge	2 004	3	11 000	261 000
Warren	1 290	1	1 056	10 000
Webster	1 403	3	6 008	100 000
West Boylston	1 187	7	8 036	161 700
Winchendon	1 754	1	4 000	NA
Worcester	7 497	3	3 424	45 500
Total		74	124 720	2 015 100
Average spindles + capitals			1 685	27 231

Cont. →

Table 14 (continued)

Township	*Population 1840*	*N° of mills*	*Total spindles*	*Total capital investment %*
Hampden county :				
Brimfield	1 419	1	1 332	10 000
Chester	1 632	2	1 690	24 000
Holland	423	1	1 024	15 000
Ludlow	1 268	2	10 000	100 000
Morson	2 151	3	3 636	39 000
Palmer	2 139	2	11 020	60 000
Russell	955	1	150	500
Springfield	10 985	7	35 000	1 400 000
W. Springfield	3 626	1	2 700	50 000
Total		20	66 552	1 698 500
Average spindles + capitals			3 327	84 925
Bristol county :				
Attelborough	3 585	8	13 078	259
Dighton	1 378	3	3 564	50
Easton	2 074	4	1 824	31
Fairhaven	3 951	2	2 460	19 500
Fall River	6 738	10	31 000	700 000
Mansfield	1 382	6	3 412	48 550
Norton	1 545	4	1 993	57 228
Pawtucket	2 184	6	15 317	212 000
Rehoboth	2 169	1	1 440	12 000
Seekonk	1 996	3	5 709	81 000
Swansea	1 484	1	1 000	18 000
Taunton	7 645	8	20 638	84 500
Westport	2 820	1	3 072	50 000
Totals		57	104 507	1 622 778
Average spindles + capitals			1 833	28 470

Sources : Wadlin, *Census of the Commonwealth of Massachusetts, 1895* ; Bigelow, *op. cit.*

Table 15. *Waterpower costs at Lowell (Boot mills 1-4), 1841.*

Capital costs	$
Waterwheels, gearing	85 000
Foundation	20 000
Total	105 000
Interest at 6 %	6 300

Operating costs, per annum	$
Interest on capital	6 300
Waterpower for 29 248 spindles (at $12 per hp, pa)	5 340
Coal for heating (288 tons at $6)	1 728
Transport of coal, cotton, oil, starch, goods (at $1 to $1.50 a ton)	1 660
Total	15 028

Horsepower generated :
At Lowell one « mill » of 3 584 spindles was reckoned as equivalent to 54.54 horsepower. Therefore 29 248 spindles needed 445 horsepower.

Cost per horsepower :
This can be calculated by dividing operating costs per annum by the number of horsepower needed to keep the mill running :

15 028 : 445 = $33.77 per hp, pa

Sources : David J. Jeremy, unpublished edition of James Montgomery's second edition of his *A Practical Detail of the Cotton Manufacture of the United States*, (1840), (typescript in Kress Library, Harvard Business School), p. 215-216, "B" Letter n° 1, 26 March 1841.

Table 16. *Steam power costs at Newburyport (Bartlett mills 1-2), 1841.*

Capital costs	$
Two steam engines (169 hp combined)	25 000
Foundations for engines	3 000
Total	28 000
Interest at 6 %	1 680

Operating costs	$
Interest on capital	1 680
Coal	6 045
Wages for engineer and fireman	1 137
Total	8 862

Horsepower generated :
At average pressure the two engines combined reached 169 hp.

Cost per horsepower :
By formula previously used :

8 862 : 169 = $52 per hp, pa

Sources : Ibid., Justitia, n° 5, 12 March 1841 and Appendix n° 2, 11 May 1841.

Table 17. *Massachusetts : annual costs and receipts (at factory door), 1832.*

County or town	*Debits*						*Potential credits*	
	Fixed costs		*Variable costs*					
	Land 9%	*Machinery 21%*	*Labour*	*Other manuf'g costs*	*Stock*	*Cotton*	*Total costs (excluding stock and then including it)*	*Value of cotton goods (excluding stock and then including it)*
Berkshire (18 firms)								
Firm average	9 417	10017	5 922	658	2 711	6 780		
Annual rate	848	2 104	5 922	658	2 711	6 780	19 023	17 342
% of all costs	4.5	11	31	3.5	14.2	35.6	100	
Costs with stock shifted to receipts							16 312	20 053
							23 %	

Cont. →

Table 17 (continued)

County or town	Debits						Potential credits	
	Fixed costs		Variable costs					
	Land 9%	Machinery 21%	Labour	Other manuf'g costs	Stock	Cotton	Total costs (excluding stock and then including it)	Value of cotton goods (excluding stock and then including it)
Bristol (44 firms)								
Firm average	19 534	18 001	10 967	2 238	6 527	11 645		
Annual rate	1 758	3 780	10 967	2 238	6 527	11 645	36 915	33 945
% of all costs	4.8	10.2	29.7	6.1	17.7	31.5	100	
Costs with stock shifted to receipts							30 388	40 472
							31 %	
Worcester (47 firms)								
Firm average	22 963	19 233	12 209	1 446	3 799	13 147		
Annual rate	2 067	4 039	12 209	1 446	3 799	13 147	36 707	35 513
% of all costs	5.6	11	33.3	3.9	10.3	35.8	100	
Costs with stock shifted to receipts							32 908	39 312
							20 %	
Waltham (1 firm)								
Firm average	230 754	102 740	86 124	8 050	170 000	80 000		
Annual rate	20 768	21 575	86 124	8 050	170 000	80 000	386 517	218 500
% of all costs	5.4	5.6	22.3	2.1	44.0	20.7	100	
Costs with stock shifted to receipts							216 517	388 500
							79 %	
Lowell (4 firms)								
Firm average	335 064	305 939	144 390	24 773	231 174	159 952		
Annual rate	30 156	64 247	144 390	24 773	231 174	159 852	654 592	451 121
% of all costs	4.6	9.8	22	3.8	35.3	24.4		
Costs with stock shifted to receipts							423 418	682 295
							61 %	

Sources : Data from *The McLane Report.*

Table 18. *Sex ratios in Massachusetts cotton firms, 1832 and 1836-37.*

Township	1832			1836-37		
	Males per firm	N of firms	Males: females (M=1)	Males per mill	N of mills	Males: females (M=1)
Berkshire country :						
Adams	11	13	2.07	10	19	2.24
Cheshire	11	1	1.9	14	1	1.86
Stockbridge	46	1	1.8	38	1	2.16
Williamstown	8	1	1.9	11	1	2.73
County average M:F=1:1.9			1:1.9			1:2.25

Cont.→

Table 18 (continued)

Township	*1832*			*1836-37*		
	Males per firm	*N of firms*	*Males: females (M = 1)*	*Males per mill*	*N of mills*	*Males: females (M = 1)*
Bristol county :						
Attleborough	21	6	2	20	8	1.4
Dighton	22	3	2.1	10	3	2.6
Easton	3	5	4	3	4	4.1
Fairhaven	7	3	2.3	8	2	2.53
Pawtucket	24	5	2.25	21	6	1.94
Rehoboth	30	1	1.7	18	1	2.22
Seekonk	17	3	1.96	20	3	1.25
Taunton	22	5	3.86	15	8	3.8
Troy (fall river)	29	13	1.9	34	10	1.92
County average M:F			1:2.45			1:2.42
Worcester county :						
Athol	5	1	8	10	1	4.5
Barre	37	1	2.4	55	1	0.4
Douglas	91	1	1.2	33	2	1.15
Fitchburg	12	3	3.2	7	4	2.3
Grafton	20	4	1	27	5	1.69
Holden	10	2	2	17	4	0.97
Lancaster	5	2	3.8	1	3	1
Mendon	97	4	0.7	26	8	1
Milford	10	1	0.6	8	1	1.75
Millbury	11	3	1.75	20	1	1
Northborough	7	1	3.6	8	2	1.8
Northbridge	23	3	1.1	22	4	1.2
Oxford	16	4	1.5	16	4	1
Phillipston	8	1	2.5	10	1	3
Princeton	8	1	1.4			
Southbridge	25	3	1.6	28	3	1.1
Sturbridge	21	1	2.4	12	6	1.65
Sutton	36	2	1.75	13	4	1.1
Uxbridge	40	1	5.5	43	3	1.9
Webster	38	2	1.7	21	3	1.5
W. Boylston	20	3	2.5	13	7	1.9
Western	9	1	3.33			
Winchendon	16	1	2.5	25	1	5
Worcester	28	1	1.1	11	3	1.4
County average M:F			1:2.38			1:1.74
Middlesex county :						
Waltham	96		3.45	25	3	5.26
Lowell	144		3.45	39	22	6.59

Sources : McLane Report for 1832, figures 2 ; Bigelow, *Statistical Tables, op. cit.*

N.B. The 1832 figures for males include boys (though these are listed as a separate category in the original returns) and those for females include girls. The 1836-37 data distinguish between males and females only.

Table 19. *Massachusetts : cotton mill communities in the Blackstone Valley. Growth in cotton mill employment compared to population growth.*

Township (North-South sequence)	*Mills*		*Work forces in cotton mills*								*Population change pa 1830-37 (7 years)*
	1832	*1837*	*1832*				*1837*			*Change pa 1832-37 (5 years)*	
			Men	*Boys*	*Women*	*Total*	*Men*	*Women*	*Total*		
Holden	2	4	14	7	42	63	68	66	128	+65 (+13pa)	+10pa
Worcester	1	3	23	5	31	59	34	47	81	+22 (+4pa)	+421pa
Grafton	4	5	40	40	84	164	134	226	360	+196 (+39pa)	+146pa
Sutton	2	4	45	28	128	91	94	100	194	+103 (+21pa)	+39pa
Milbury	3	1	20	12	56	88	20	20	40	—48 (—10pa)	+77pa
North Bridge	3	4	43	27	75	145	90	107	197	+52 (+10pa)	+51pa
Uxbridge	1	3	30	10	220	260	130	250	380	+120 (+24pa)	+23pa
Mendon	4	8	203	185	287	675	207	205	412	—263 (—12pa)	+72pa
Douglas (on Mumford River, a branch of the Blackstone)	1	2	35	56	107	198	65	75	140	—58 (—12pa)	—4pa

Sources : Bigelow, *Statistical Tables, op. cit.*

Table 20. *Numbers of males in Blackstone Valley : cotton mills compared to adult males in communities.*

Township (North-South sequence)	*Population*			
	Total	*Adult men (assume 25% of total)*	*Males in cotton factories*	*Males in cotton mills as % of adult males in town*
Holden	1 789	447	68	15
Worcester	7 117	1 779	34	1.9
Grafton	2 910	727	134	18
Sutton	2 457	614	94	15
Milbury	2 153	538	20	3.7
Northbridge	1 409	352	90	26
Uxbridge	2 246	561	130	23
Mendon	3 657	914	207	23
Douglas (on the river Munford, a branch of the Blackstone)	1 713	428	65	15

Sources : McLane Report ; Bigelow, *Statistical Tables, op. cit.*

INDEX OF PEOPLE, FIRMS, AND PLACES